FRACTALS IN SOIL SCIENCE

Edited by

Philippe Baveye
Jean-Yves Parlange
Bobby A. Stewart

CRC Press
Boca Raton Boston London New York Washington, D.C.

Acquiring Editor: *Marsha Baker*
Project Editor: *Susan Fox*
Cover Design: *Dawn Boyd*
Prepress: *Kevin Luong and Walt Cerny*

Library of Congress Cataloging-in-Publication Data

Fractals in soil science / edited by P. Baveye, J.-Y Parlange, B.A. Stewart.
p. cm.
Includes bibliographical references and index.
ISBN 1-56670-105-8 (alk. paper)
1. Soil physics—Mathematics.. 2. Soil science—Mathematics.
3. Fractals. I. Baveye, P. (Philippe) II. Parlange, J.-Y. (Jean-Yves)
III. Stewart, B.A. (Bobby Alton), 1932- . IV. Series: Advances in soil science
(Boca Raton, Fla.)
S592.3.F725 1997
631.4'01'514742—dc21 97-35156
CIP

No claim to original U.S. Government works
International Standard Book Number 1-56670-105-8
Library of Congress Card Number 97-35156
Printed in the United States of America 1 2 3 4 5 6 7 8 9 0
Printed on acid-free paper

Preface

"Fractals everywhere!" This title of a successful book on fractals, published a few years ago, echoes probably very well the current views of many soil scientists. Every new issue of just about every soil science, or soil science related, periodical has a number of articles dealing in some fashion with the application of fractal geometry. This tidal wave shows no sign yet of abatement. Claims continue to be made that fractals and fractal geometry will allow the solution of unresolved questions, like the elusive and difficult quantification of soil structure. At the same time, however, critics accuse fractal enthusiasts of succumbing to yet another fad and forecast that, when the dust settles, fractals will prove of little value in describing the intricate geometry of soils and the complex processes that occur in them.

Tidal waves periodically sweep the shores of soil science research. Seeing that "universal" frameworks like the thermodynamics of irreversible processes, the theory of dissipative structures, catastrophe theory, percolation theory, chaos theory, the theory of self-organized criticality, to name only a few, were applied to systems ranging from biological organisms, to the history of human institutions (churches, governments), all the way to the description of the large-scale structure of the universe, some soil scientists were encouraged to try to apply them as well to soils and soil processes. This "jumping on the bandwagon" often is associated with an initial neglect of the intrinsic limitations of the theory being applied, a tendency to exaggerate and misrepresent its potential, followed by subsequent disappointment, if not even to a backlash and eventual condemnation by all but a minority of "die-hards". This pattern is perhaps clearest in soil science in the case of the use of the thermodynamics of irreversible processes.

Is the use of fractals in soil science another one of these fads? Certainly the claims made for it initially were formidable: nothing less than the encapsulation of the whole geometry of soils into a single parameter, the so-called fractal dimension. The starry-eyed predictions that, with this single dimension, one would be able to describe faithfully all the dynamical processes occurring in soils are now usually smiled at. Yet, along the way, fractal geometry has managed to reveal aspects of soils that were largely ignored before. As a result, the "fractal movement" is still going strong in soil science, some 16 years after its inception, unlike any other tidal wave in the past. It seems therefore appropriate to inquire about its current status and about the direction toward which it is heading.

The present book was designed to provide detailed and comprehensive information on the current status of the application of fractal geometry in soil science, and on prospects for its future use. Because applications of fractals in soil science are both relatively recent and in constant evolution, there occasionally are sharp differences among the viewpoints expressed in the various chapters. In order for the book to reflect accurately existing trends, nothing has been done to resolve these divergences of views nor to present a unified view. At the same time, with the inclusion of an introductory chapter, in which the theoretical foundations of fractal geometry are reviewed and illustrated, a special effort has been made to propose a self-contained coverage of the field, suitable as a textbook for an upper-level undergraduate or graduate course.

As a final word, we wish to thank the authors and reviewers for their dedication in producing the chapters contained in this book and Susan Fox (CRC Press LLC) for her thorough copyediting of the manuscript. Without their professionalism and commitment, this volume would not have been possible.

P. Baveye
J.-Y. Parlange
B.A. Stewart

Contributors

P. Baveye, Laboratory of Environmental Geophysics, Bradfield Hall, Cornell University, Ithaca, New York 14853-1901. [pcb2@cornell.edu]

F. Bartoli, Centre de Pédologie Biologique, UPR 6831 du CNRS, associé à l'Université Henri Poincaré, 54501 Vandoeuvre-les-Nancy, France.

C.W. Boast, Department of Natural Resources and Environmental Sciences, University of Illinois at Urbana-Champaign, 1102 South Goodwin Avenue, Urbana, Illinois 61801 [c-boast1@uiuc.edu]

G.M. Berntson, Department of Organismic and Evolutionary Biology, Biological Laboratories, Harvard University, 16 Divinity Avenue, Cambridge, Massachusetts 02138. [berntson@oeb.harvard.edu]

G.K. Bowman, Civil Engineering Department, Auburn University, Auburn, Alabama 36849.

H.W.G. Booltink, Department of Soil Science and Geology, Agricultural University, Wageningen, The Netherlands.

J.W. Crawford, Soil-Plant Dynamics Group, Cellular and Environmental Physiology Department, Scottish Crop Research Institute, Invergowrie, Dundee DD2 5DA, Scotland. [cepjc@scri.sari.ac.uk]

S. Crestana, Centro Nacional de pesquisa e Desenvolvimento de Instrumentação Agropecuária (CNPDIA, Empresa Brasileira de Pesquisa Agropecuária (EMBRAPA), Rua XV de Novembre 1452, Caixa Postal 741, Centro-CEP 13560-970, São Paulo, Brazil. [Crestana@npdia.embrapa.ansp.br].

M. Dubuit, Centre Interrégional de Resources Informatiques de Lorraine (CIRIL), Château du Montet, 54500 Vandoeuvre-les-Nancy, France.

P. Dutartre, Centre de Pédologie Biologique, UPR 6831 du CNRS, associé à l'Université Henri Poincaré, 54501 Vandoeuvre-les-Nancy, France.

C. Fuentes, Instituto Mexicano de Tecnologia del Agua (IMTA), Paseo cuauhnahuac No 8532, Col. Progreso, 62550 Jiutepec, Morelos, Mexico.

V. Gomendy, Centre de Pédologie Biologique, UPR 6831 du CNRS, associé à l'Université Henri Poincaré, 54501 Vandoeuvre -les-Nancy, France.

R. Hatano, Faculty of Agriculture, Hokkaido University, Kita 9, Niski 9, Kita-Ku, Sapporo 060, Japan. [Ryusuke.Hatano@A1.kines.kokudai.ac.jp]

R. Haverkamp, Laboratoire d'Etude des Transferts en Hydrologie et Environnement (LTHE)-CNRS URA 1512, INPG, B.P. 53, 38041 Grenoble Cédex, France. [Randel.Haverkamp@img.fr]

T.A. Hewett, Petroleum Engineering Department, Stanford University, Stanford, California 94305.

V.J. Homer, Lenoir City, Tennessee 37771.

C.-h. Huang, National Soil Erosion Research Laboratory, 1196 Soil Building, Purdue University, West Lafayette, Indiana 47907-1196.

J. Lynch, Department of Horticulture, 101 Tyson Building, Pennsylvania State University, University Park, Pennsylvania 16802. [jlynch@psupen.psu.edu]

F.J. Molz, Civil Engineering Department, Auburn University, Auburn, Alabama 36849. [fredmolz@eng.auburn.edu]

S. Niquet, Centre Interrégional de Resources Informatiques de Lorraine (CIRIL),

Château du Montet, 54500 Vandoeuvre-les-Nancy, France.

J.-Y. Parlange, Department of Agricultural and Biological Engineering, Riley-Robb Hall, Cornell University, Ithaca, New York 14853-1901. [jp58@cornell.edu]

E. Perrier, Laboratoire d'Informatique Appliquée, Centre ORSTOM Bondy, 70-74 Route d'Aulnay, 93143 Bondy Cédex, France. [Perrier@bondy.orstom.fr]

A.N. Posadas, Centro Nacional de Pesquisa e Desenvolvimento de Instrumentação Agropecuária (EMBRAPA), Rua XV de Novembre 1452, Caixa Postal 741, Centro -CEP 13560-970, São Paulo, Brazil.

M. Rieu, Laboratorie d'Hyrophysique des Sols, Centre ORSTOM Bondy, 70-74 Route d'Aulnay, 93143 Bondy Cédex, France.

S. Snapp, Rockefeller Foundation, PO 30721, Lilongwe 3, Malawi, SE Africa.

H. Van Damme, Centre de Recherche sur la Matière Divisée, CNRS and Université d'Orléans, 45071 Orléans Cédex 2, France. [hvd@cnrs-orleans.fr]

M. Vauclin, Laboratoire d'Etude des Transferts en Hydrologie et Environment (LTHE)-CNRS URA 1512, INPG, B.P. 53, 38041 Grenoble Cédex, France. [Michel.Vauclin@img.fr]

H. Vivier, Laboratoire des Sciences du Génie Chimique (LSGC), UPR 6811 du CNRS, ENSIC, BP 451, 54001 Nancy Cédex, France.

I. Young, Soil-Plant Dynamics Group, Cellular and Environmental Physiology Department, Scottish Crop Research Institute, Invergowrie, Dundee DD2 5DA, Scotland. [cepiy@scri.scot-agric-res-inst.ac.uk]

Contents

Fractal Geometry, Fragmentation Processes and the Physics of Scale-Invariance: An Introduction

P. Baveye and C.W. Boast

Contents

ISBN 1-56670-105-8

I Introduction

Since its formal introduction in 1975, the concept of fractal has captured the imagination, and has entered into the toolbox, of many scientists in a wide range of fields. Papers discussing fractals in various contexts, including geophysics and soil science, now appear almost daily. A rich and constantly growing panoply of textbooks describes the multiple facets of the theory of fractals.

In this flurry of publications, one of the most striking features is the extreme variety that exists in the mathematical background assumed of the readership. In some articles and textbooks, little or no knowledge of mathematics is needed; these are typically publications that emphasize the graphical aspects of fractals, the "hypnotically intricate visual patterns and images" (Jones, 1993) that one may base on fractals, or the use of fractals to describe natural systems. Contrastedly, another set of papers and books is grounded in the belief that much if not all of the "beauty of fractals" is to be found in their mathematics (Falconer, 1990). These publications usually require of the readers a solid background in geometry and set theory.

Rare are the articles or books that analyze the connection between the mathematical beings defined and manipulated by the geometricians, and the "natural" fractals identified in the real world. Reluctance to embark on this analysis appears to have caused much confusion in the literature; the term "fractal", when applied to natural systems, often means different things to different people, creating unnecessary difficulties in communication. A first objective of the present chapter, therefore, is to attempt to fill this gap and to make more explicit the connection between *theoretical* and *natural* fractals. A similar approach is followed in the last section of this chapter, which deals briefly with the increasingly important multifractal measures.

Another area in which researchers in geophysics and soil science have as yet seldom ventured deals with the reasons for the fractal behavior exhibited by natural objects. From a descriptive standpoint, it is of great interest to be able to look at an object or material as a fractal and to associate with it a particular fractal dimension, or a number of characteristic parameters if the object is best described as a multifractal. In some cases, this may result in significant advances in our understanding of the range of possible responses of the object or material to various imposed stimuli. However, one would ideally also like to know why this object or material exhibits a fractal behavior. Section VI of the present chapter briefly mentions various processes that, in a number of physical systems, have been shown to display a fractal behavior. One of these processes, diffusion-limited aggregation, is described in some detail for illustrative purposes.

This chapter is not meant to be a treatise on fractal geometry. Its objective is to introduce the aspects of fractal theory that have found application in soil science and are used in later chapters of this book. As much as possible, this chapter has been written in such a way as to be self-contained and accessible to readers with little more than an elementary background in calculus. Nevertheless, numerous pointers to other sources of information are provided in case the incursion into set and measure theory found here is not sufficiently basic. In particular, pages 3 to 16 of Falconer (1990) may be consulted in order to get up to speed; they introduce the mathematical concepts and the notation used throughout this chapter.

II A Gallery of Mathematical Monsters

II.A Cantor's Set and the Devil's Staircase

In a landmark letter, sent on June 20, 1877, to his friend Richard Dedekind, the German mathematician Georg Cantor started a revolution in the field of geometry (Mandelbrot, 1978). He mentioned that he had doubts about the meaning, and even the validity, of the concept of dimension. He also proved that a surface limited by a square, which like any regular surface has a topological dimension[1] equal to 2 (in Euclidian space), does not contain more points than any one of the square's sides (of topological dimension equal to 1)! Cantor's derivation showed that there is a one-to-one correspondence between points on the sides and points on the surface, so that a single number suffices to determine the position of any point in the square.

A few years later, Cantor (1884) gave the concept of dimension another serious jolt and created in the process the first of what, after Henri Poincaré, came to be called the mathematical "monsters". Known as the Cantor set, it is also commonly termed the "middle third", the "ternary" or the "triadic" Cantor set. It is constructed by a sequence of removal operations from a unit-length interval, termed the initiator. It is labelled I_o in Figure 1, where it corresponds to the interval [0,1] (*i.e.*, the set of numbers x such that $0 \leq x \leq 1$). If one removes from this initiator the segment (1/3, 2/3) (containing the real numbers x such that $1/3 < x < 2/3$), the set I_1 results. This set is sometimes termed the generator and consists of the two intervals [0,1/3] and [2/3,1]. Removing the middle thirds of these intervals, *i.e.*, applying the generator to each of them, yields I_2, which comprises 4 intervals of length 1/9. At the next stage (I_3), there are $2^3 = 8$ intervals of length $(1/9)/3 = 3^{-3}$. At the nth iteration of

[1] The topological dimension, D_T, of a set is always an integer and is 0 if the set is totally disconnected (*i.e.*, for isolated points), 1 if each point of the set has arbitrarily small neighborhoods with boundary of dimension 0 (*i.e.*, for lines), and so on for higher dimensions (Falconer, 1990, p. xx).

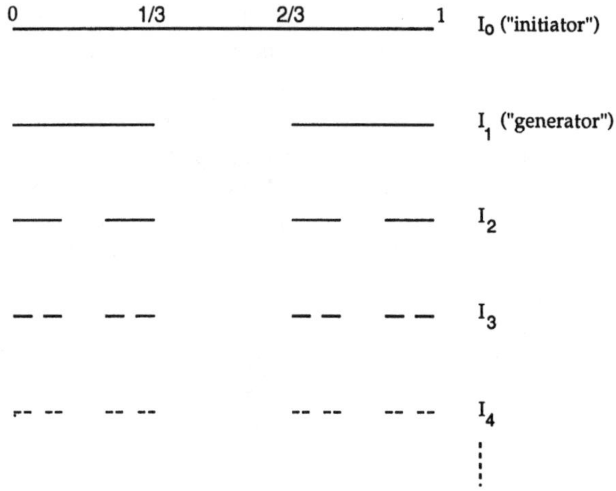

Figure 1: Initial steps in the construction of the Cantor set by repeated removal of the middle third of intervals.

this deletion procedure, the set I_n consists of 2^n intervals, each having a length 3^{-n}, and is included in all the preceding sets I_1, I_2, ...,I_{n-1} in the sequence.

The Cantor set, P, can be thought of as the limit of the sequence of sets I_n when n tends to infinity. Mathematically (Rudin, 1976), it is defined as $\bigcap_{i=1}^{\infty} I_n$, the intersection of all the sets I_n, with n going from 1 to infinity. It is obviously impossible to draw the set P itself, with its infinitesimal detail, so "pictures of the Cantor set P" are in fact only illustrations of one of the I_n's. It is apparent from Figure 1 that such representations are feasible only for relatively low values of n.

At first glance, it might appear that we have removed so much of the interval [0,1], during the construction of the Cantor set, that nothing remains. In a sense, this is true. The (Lebesgue) measure $m(I_n)$ of the set I_n, *i.e.*, practically, the total length of all the 2^n intervals included in I_n, is given by $(2/3)^n$. Since $P = \bigcap_{i=1}^{\infty} I_n$ and P is included in I_n for every n, the measure $m(P)$ is equal to the limit as $n \to \infty$ of $(2/3)^n$, which is 0. In other words, the Cantor set has zero length and its topological dimension is zero. Yet it is an uncountable set, containing infinitely many points in any neighborhood of each of its points. Furthermore, one can show that the points of the Cantor set can be put in one-to-one correspondence with those of the initiator [0,1]; mathematically, P and [0,1] have the same "cardinality". This raises some very serious questions about the use of traditional dimensions to characterize these two sets, and justifies the labels of "monstrous" and "pathological" that rapidly became associated with the Cantor set.

The Cantor set P has a number of striking features (Falconer, 1990):

- P contains exact copies of itself at many different scales. It is clear that the part of P in the interval [0,1/3], or that in the interval [2/3,1], are geometrically similar to P, except for the fact that they are scaled down by a factor 1/3. Again, the parts of P in each of the four intervals of I_2 are geometrically similar to P, but scaled by a factor 1/9, and so on. Some authors (*e.g.*, Falconer, 1990) call this feature of P "self-similarity". Strictly speaking however, as Feder (1988, p. 63) points out, P is not entirely self-similar, because of the finite length of its initiator. To obtain a self-similar set, one should enlarge P by an extrapolation procedure that covers the region [0,3] by two Cantor sets defined on the intervals [0,1] and [2,3]. Repetition of this procedure ad infinitum generates a self-similar set on the half-line $[0,\infty[$. Every time the expression "self-similar" is used in the remainder of this chapter, it will have to be interpreted within the context of Feder's (1988) comment.

- The set P has a "fine" structure; it contains detail at arbitrary small scales.
- Although P has an intricate and detailed structure, the actual definition of P is straightforward.
- P is obtained by a recursive procedure.
- The geometry of P is not easily described in classical terms; it is not the locus of the points that satisfy some simple geometric condition, nor is it the set of solutions of any simple equation.
- It is awkward to describe the local geometry of P – near each of its points are a large number of other points, separated by gaps of varying lengths.
- Although P is in some ways quite a large set (it is uncountably infinite), its size is not quantified by the usual measures such as length – by any reasonable definition, P has length zero.
- The actual set P cannot be represented graphically.

Many of these peculiar features of the Cantor set are shared by, or at least are similar to those of, the other mathematical monsters described later in this section. The above list will also be useful when we try to define as precisely as possible what is meant by the term "fractal".

Another mathematical "monster", sometimes called the Cantor singular function (Edgar, 1993), is closely related to the Cantor set. This singular function is constructed by integrating an appropriate distribution function defined on the Cantor set. One such distribution function is constructed by first considering the uniform distribution of mass on the interval [0,1], with total mass equal to 1 (in some arbitrary units). For example, one could visualize the initiator I_o of Figure 1, not as a line segment, but as a bar of some material with unit mass density ρ_o and length $l_o = 1$. The operation that resulted in I_1 in Figure 1 now consists of cutting the bar into two halves of equal mass (=1/2) and then hammering them so that the length of each part becomes $l_1 = 1/3$. The total mass is conserved in the operation, while the mass density in each part becomes $\rho_1 = 3/2$. Repeating this process, one finds that at the nth stage, the number of small bars is 2^n, each of length $l_n = (1/3)^n$ and of mass density $\rho_n = (3/2)^n$. Integrating the mass density along x, we obtain the total mass

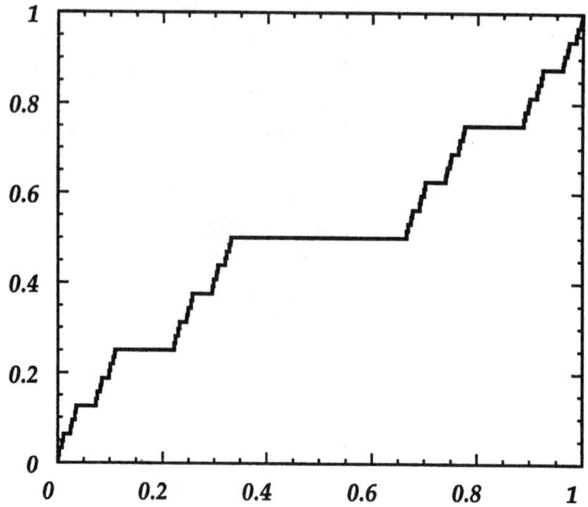

Figure 2: Intermediate stage ($n = 6$) in the construction of the Cantor singular function or "Devil's staircase".

contained between $x = 0$ and an arbitrary point x as:

$$M_n(x) = \int_0^x \rho_n(x)dx \tag{1}$$

where ρ_n equals $(3/2)^n$ on the 2^n intervals of length $l_n = (1/3)^n$ and is identically zero everywhere else. An example of such a function $M_n(x)$ is shown in Figure 2. The limit of Equation (1) as $n \to \infty$ is a singular function, discovered by Cantor (cf Edgar, 1993), and usually referred to as the "Devil's staircase". It is a continuous, non-constant function that is (almost) everywhere horizontal except on an uncountable set (the Cantor set), of Lebesgue measure equal to zero. At each point in this uncountable set, the derivative of the Cantor singular function is given by a Dirac delta distribution.

The Cantor singular function has many of the features of the Cantor set (listed above), except the first. That is, it does not contain *exact* copies of itself at different scales. Inspection of Figure 2 shows that the shape of the function $M_n(x)$ in the interval [0,1/3] is similar to that of the whole function (in the interval [0,1]), scaled down in the x direction (abscissa) by a factor of 1/3. However, it is also apparent that the scaling factor in the y direction (ordinate) in Figure 2 is not 1/3 but 1/2 (cf Figure 3). A set or function is said to be self-affine when its scaling factors are different in different directions. Self-similarity requires these scaling factors to be identical in all directions. Therefore, the Cantor singular function or "Devil's staircase" is not self-similar, but self-affine.

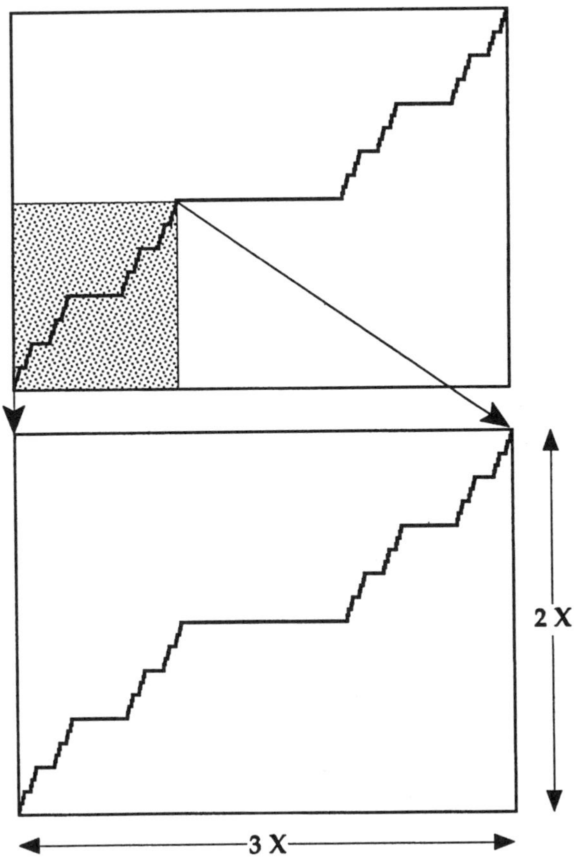

Figure 3: Schematic illustration of the self-affinity of the Cantor singular function or "Devil's staircase": the enlargement is identical to the original but the enlargement (scaling) factors are different in the x and y directions.

II.B The Peano-Hilbert Plane-Filling Curves

Following in Cantor's steps, another assault on the concept of dimension was made simultaneously by Peano (1890) and Hilbert (1891), in two short but influencial articles. Both describe polygons that appear at first glance to be perfectly innocent, but nevertheless happen to fill a square more and more completely, so that, in the limit, they pass through every single point in the square.

The construction of Peano's original curve begins with a single line segment, the initiator (Stage 0 in Figure 4). It is substituted by the generator (Stage 1), which touches (but does not cross) itself at two points labeled A and B in Figure 4. If each straight line segment in Stage 1 is replaced by a properly scaled-down

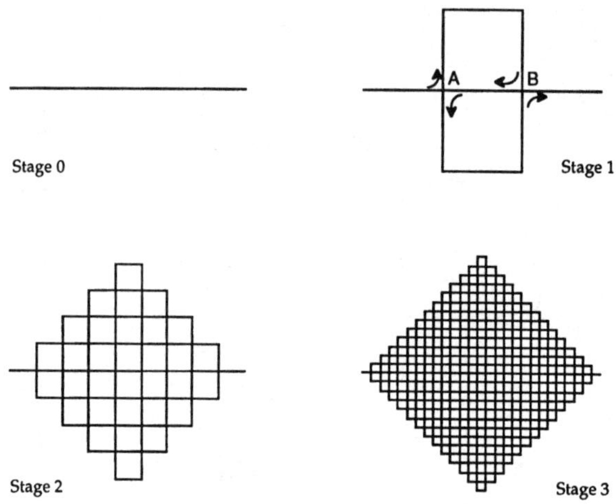

Figure 4: First stages of the construction of Peano's space-filling curve. At each step, each line segment is replaced by 9 line segments scaled down by a factor of 1/3.

generator, one obtains Stage 2. Independently of the direction (horizontal or vertical) of the line segments in Stage 1, the scaling factor is equal to 1/3. At the next stage, the scaling factor is equal to 1/9, and so on. If one assumes that the length of the original line segment constituting the initiator is 1, it is easy to calculate the length of the curves at each stage. In stage 1, there are 9 line segments of length 1/3 and the total length of the curve is 3. In stage 2, there are 9^2 line segments of length $1/3^2$, amounting to a total length equal to 9. Expressed as a general rule, in each step of the construction, the resulting curve increases in length by a factor of 3. In stage n, the length thus is 3^n.

If one pursues the above construction procedure to the limit as $n \to \infty$, the number of straight line segments and their total length tend to infinity. The result is generally termed Peano's curve and is of great interest mathematically because it is nowhere differentiable, *i.e.*, it does not admit a tangent at any of its points (which are all "corners"). As with the Cantor set (but for different reasons), it is impossible to visualize the structure of this curve. All one can see, at any scale, is a completely "filled out" square, which does not look in the least similar to the early steps of the construction (cf Figure 4). Nevertheless, the Peano curve, like P, contains exact copies of itself (*e.g.*, Peitgen *et al.*, 1992; Sagan, 1994). The same general characteristics are exhibited by Hilbert's (1891) plane-filling curve, except for the fact that it is self-avoiding, *i.e.*, it never intersects or touches itself.

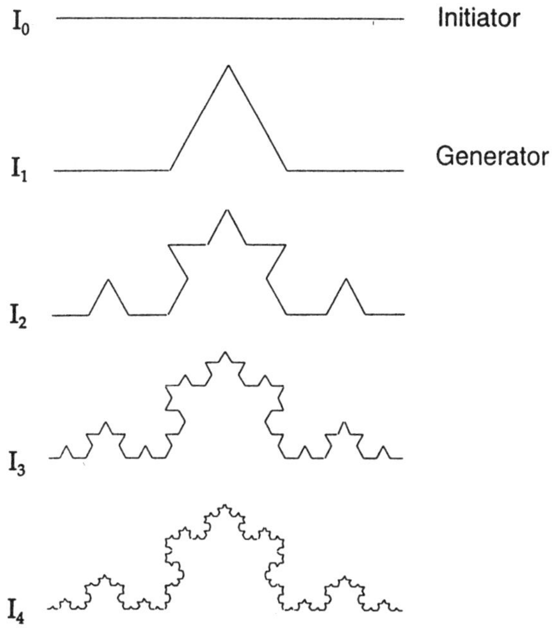

Figure 5: Construction of the triadic von Koch curve. At each step, the number of line segments increases by a factor of 4.

II.C The Triadic von Koch Curve

The Swedish mathematician von Koch introduced in 1904 what is now called the (triadic) von Koch curve (*e.g.*, Peitgen *et al.*, 1992, p. 89-93; Falconer, 1990). The construction of this curve, illustrated in Figure 5, starts with a line segment, I_o, of unit length. This initiator is replaced by the generator shown as the curve I_1, consisting of four segments of length 1/3. At the next stage, I_2 is constructed by replacing each line segment in I_1 by a properly scaled-down version of the generator. The scaling factor is equal to 1/3. The curve I_2 consists of $4^2 = 16$ segments each having a length of $1/3^2$=1/9, so that the total length of I_2 is equal to $(4/3)^2$=16/9. At the nth stage, there are 4^n segments of length $\delta_n = 3^{-n}$, with a total length of $L(I_n) = (4/3)^n$.

Parenthetically, this expression for the total length $L(I_n)$ may also be expressed in terms of the segment length δ_n. Taking the exponential of the natural logarithm of the total length $L(I_n) = (4/3)^n$ and substituting the value of n $(= -\ln \delta_n/\ln 3)$ obtained by solving the equality $\ln\delta_n = \ln 3^{-n}$ (*e.g.*, Feder, 1988, p.17), one obtains:

$$L(I_n) = \delta_n^{1-\ln 4/\ln 3} \tag{2}$$

Equation (2) corresponds to a *power-law* (or "Paretian") relationship between L and δ_n. Similarly, the number N_n of segments of length δ_n is also a power-law function of δ_n. Indeed, the equality $L(I_n) = N_n\delta_n$ implies that $N_n =$

I_0 I_1 I_2

Figure 6: First stages of the construction of the (triangular) Sierpinski gasket. I_o is the initiator. I_1 is the generator.

$\delta_n^{-\ln 4/\ln 3}$. We shall return to these important power-law dependencies in later sections of this chapter.

Even though we cannot represent it graphically, it is clear that if we carry on the iterative procedure illustrated in Figure 5 to the limit where $n \to \infty$, the resulting curve (the von Koch curve) has an infinite number of vanishingly small segments and has a total length tending to infinity.

If, in Figure 5, one takes the part of I_4 that corresponds to the interval [0,1/3] of the initiator, and scales this part up horizontally and vertically by 3, one obtains I_3. In the limit $n \to \infty$, however, this same scaling-up of any segment of I_n would reproduce the von Koch curve itself. Furthermore, like the Peano-Hilbert space-filling curves, the von Koch curve is nowhere differentiable. All of these curves have a topological dimension equal to 1; a simple stretching operation (or "rectification") transforms them into an infinite straight line.

At first sight, the von Koch curve appears less monstrous than the Cantor set or Peano's space-filling curve. Nevertheless, it has astonishing properties, which challenge the traditional concepts of dimension. Indeed, since the von Koch curve is infinite in length and contains exact replicates of itself, any scaled-down sub-image is also of infinite length (Jones, 1991). This leads to the conclusion that for any two points on the curve, no matter how close they are, the curve between them is of infinite length! In addition, if one joins together three initiators like I_o (in Figure 5) to form a triangle and one performs on each initiator the iteration procedure of Figure 5, the resulting geometrical construct, commonly known as the von Koch "island" or "snowflake", has infinite length, even though it fits within a finite area!

II.D The Sierpinski Gasket and Carpet

Cantor (1884) remarked that examples similar to the (Cantor) set P "can be easily constructed for higher dimensions". One such example is the Sierspinski gasket, also called "triangle" or "arrowhead". The iterative procedure leading to this gasket starts with an equilateral triangle, I_o, of side length r_o (Figure 6). In the next step, an inverted equilateral triangle, connecting the centers of the sides of I_o, is removed. This leaves three half-sized triangles (I_1 in

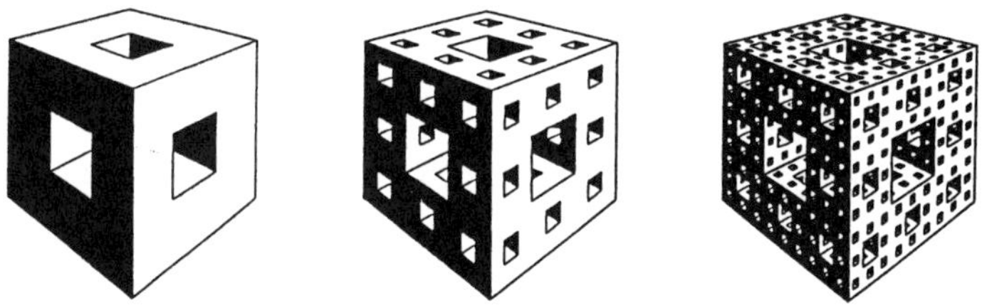

Figure 7: First three stages in the construction of the Menger sponge. (Harrison, 1995. Reprinted by permission of Oxford University Press.)

Figure 6). Repeating this procedure, one obtains 3^2 equilateral triangles in stage I_2, with $r = (1/2)^2 r_o$. In stage I_n, there are $(3)^n$ triangles of side length $r = r_o(2^{-n})$. The set resulting from an infinite iteration of this procedure, *i.e.*, the Sierpinski gasket, is Cantor-like in many respects, even though its initiator I_o is two dimensional. Like the Cantor set, the Sierpinski gasket is a self-similar, uncountable set with a topological dimension equal to zero. The two-dimensional Lebesgue measure of the Sierpinski gasket, *i.e.*, practically, its area in the plane, is equal to zero, so the traditional concept of area does not provide a very useful description of the spatial coverage of the Sierpinski gasket. Some other type of dimension is needed.

If instead of an equilateral triangle, one takes a square of side r_o as the initiator, divides it in 9 smaller squares of side length $r_o/3$, removes the central square, and applies the same procedure *ad infinitum* to the remaining squares, a structure known as the Sierpinski carpet is obtained (*e.g.*, Feder, 1988, p. 25). It has the same properties as the gasket. When the iterative process used for its construction is generalized to three dimensions, one obtains the so-called Sierpinski tetrahedron when the initiator is a tetrahedron and, when the initiator is a cube, a geometrical structure of particular interest in geophysics and soil science, the Menger sponge.

II.E The Menger Sponge

The construction of this mathematical monster is very similar to that of the Sierpinski carpet, described above, except that the initiator is a cube rather than a square. Each square face of the cube is treated in exactly the same way as the square initiator of the Sierpinski carpet. This time, extracting a square shape involves punching a hole directly through the cube at right angles to the face concerned. Thus, at the first stage (I_1), three holes are punched through. This leaves 20 subcubes at one-third scale, each of which is repeatedly subdivided to create, *ad infinitum*, the hollowed structure of the Menger sponge (Figure 7). Another way to describe the recursive construction process is to consider that, at the nth stage, each cube of size r_{n-1} is divided into 27 equal cubes of size $r_n = r_{n-1}/3$, and that the central small cube is removed along with the 6 cubes with which it shares faces.

This iterative construction may be envisaged from a physical standpoint. This produces mathematical relationships that will be useful later on. If the starting point is a cube of side length r_o made of some material of uniform mass density ρ_o, the first-order structure (obtained after one iteration) would have a porosity ϕ_1 (void volume divided by total volume) equal to $7/27$ and a mass density ρ_1 equal to $20\rho_o/27$. After the second iteration, the porosity ϕ_2 would increase to $329/729$ (≈ 0.45) and the density ρ_2 would decrease to $400\rho_o/729$ ($\approx 0.55\rho_o$). At the nth iteration, porosity and density would be given by

$$\phi_n = 1 - \left(\frac{20}{27}\right)^n = 1 - \left(\frac{r_n}{r_o}\right)^{3-\ln 20/\ln 3} \tag{3}$$

and

$$\frac{\rho_n}{\rho_o} = \left(\frac{20}{27}\right)^n = \left(\frac{r_n}{r_o}\right)^{3-\ln 20/\ln 3} \tag{4}$$

The second of these two equations again gives rise to a power-law relationship (as in the case of the von Koch curve).

Equations (3) and (4) will be mentioned again in a later section. For the time being, however, they can be used to illustrate some of the features of the Menger sponge, obtained when $n \to \infty$. Since the exponent in Equations (3) and (4) is strictly positive (indeed $3 - ln20/ln3 \approx 0.273$), ϕ_n and ρ_n/ρ_o will tend to 1 and 0, respectively, as $r_n \to 0$. Mathematically, the 3-dimensional Lebesgue measure of the Menger sponge, *i.e.*, practically, its volume, is zero. Like the Cantor set and the Sierpinski gasket, the Menger sponge is a self-similar, uncountable set of points with a topological dimension of zero. Therefore, totally against intuition, the Menger sponge is topologically equivalent to the Cantor set.

Anecdotically, there is another connection between the Menger sponge and the Cantor set. The intersections of the Menger sponge with medians or diagonals of the initial cube are triadic Cantor sets (Mandelbrot, 1982, p. 144).

II.F Bolzano-Weierstrass-Like Functions

Our rapid overview of mathematical "monsters" would not be complete without a brief reference to an interesting family of continuous, nowhere differentiable functions. They are occasionally referred to as "Weierstrass-like" but, for historical reasons, it seems more appropriate to call them "Bolzano-Weierstrass-like".

Indeed, Bernard Bolzano appears to have been the first (in a manuscript written around 1830 but published only in 1930) to provide an example of a continuous, nowhere differentiable function. A few years later, in 1872, Karl Weierstrass showed that the function

$$f(x) = \sum_{n=0}^{\infty} b^n \cos(a^n x\pi) \tag{5}$$

has these properties, provided the product ab exceeds a certain limit. Hardy

(1916) showed that the conditions to be satisfied by a and b are: $0 < b < 1$, $a > 1$ and $ab \geq 1$.

Over the years, many examples of continuous, nowhere differentiable functions have been published (cf Edgar, 1993, p. 7 and 341). One of them, the so-called Weierstrass-Mandelbrot function, assumes a particular significance in soil science because it constitutes the theoretical basis of the first article that used fractal geometry in connection with soils data (Burrough, 1981).

The Weierstrass-Mandelbrot function $w(t)$ is defined in complex form by (Berry and Lewis, 1980; Feder, 1988, p. 27):

$$w(t) = \sum_{n=-\infty}^{\infty} \frac{\left(1 - e^{ib^n t}\right) e^{i\varphi_n}}{b^{(2-D)n}} \tag{6}$$

where $1 < D < 2$, $b > 1$ and the φ_n's are arbitrary phase constants. Each choice of the phases φ_n defines a specific function $w(t)$. For a given $w(t)$, one may show that the variance of increments $V(t) = \left([w(t_o + t) - w(t_o)]^2 \right)$ is a power function, $= t^{4-2D}$, of time for $t \to 0$ (Burrough, 1981).

A simple function, the real part of $w(t)$ with the φ_n chosen as $\varphi_n = 0$, is the Weierstrass-Mandelbrot cosine function, $c(t)$:

$$c(t) = \sum_{n=-\infty}^{\infty} \frac{1 - \cos b^n t}{b^{(2-D)n}} \tag{7}$$

The shape of this function is reasonably smooth for low values of D (cf Figure 8). As D increases, however, $c(t)$ begins to fluctuate widely, as if larger and larger amounts of "noise" were added to an underlying trend.

Close inspection of Figure 8c reveals that $c(t)$ is self-affine, like Cantor's singular function (cf Feder, 1988, p. 29).

Both $w(t)$ and $c(t)$ are functions of a single variable, t. Multivariate Weierstrass-Mandelbrot functions have, however, been defined in the last decade (*e.g.*, Ausloos and Berman, 1985).

II.G Random Monsters and Fractional Brownian Motion

The various mathematical monsters described above have been singled out for their historical significance, not because they are somehow unique. For each of them, there is an infinity of variants. One could, for example, define the generator I_1 of a Cantor-like set to be the union of two intervals $[0,a]$ and $[b,1]$, for any a and b with $0 < a < b < 1$. It is easy to imagine extending this definition to include 3, 4, or 5 intervals, and so on. Similarly, instead of removing the middle-third segment in the initiator of the von Koch curve and replacing it by the other two sides of an equilateral triangle, one could remove in the middle of the initiator an interval of length a, such that $0 < a \leq 1/3$ and replace it, *e.g.*, by the other three sides of a square or by the other two sides of a suitable short isosceles triangle.

Obviously there is an infinity of such sets. In addition to sharing all of the general features of the Cantor set (listed above), they are also deterministic,

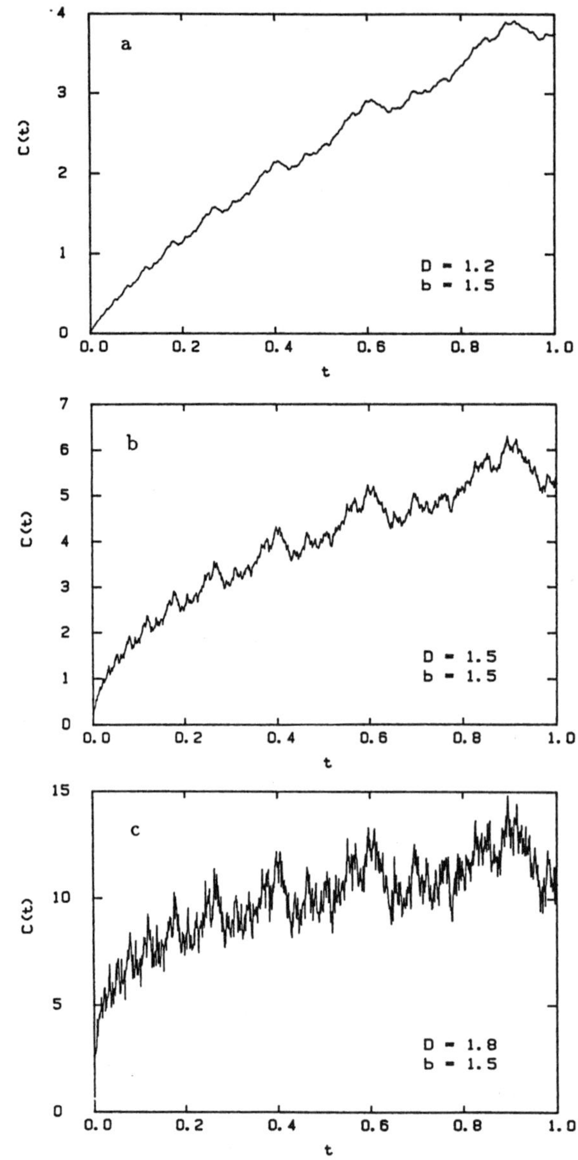

Figure 8: Graphical representations of the Weierstrass-Mandelbrot cosine function $c(t)$ with $b = 1.5$ and (a) $D = 1.2$, (b) $D = 1.5$, and (c) $D = 1.8$. (From Feder, 1988. Reprinted with permission.)

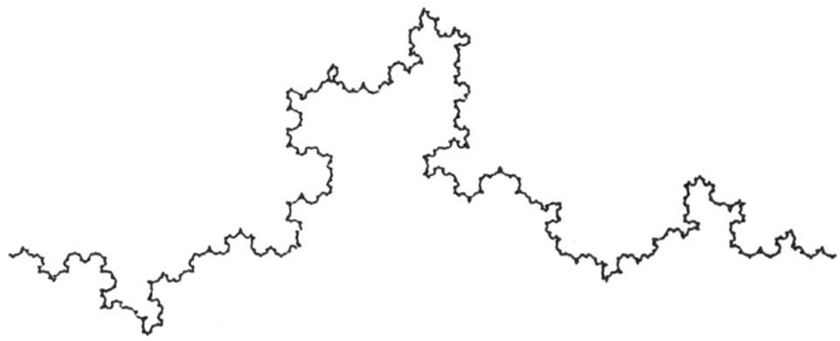

Figure 9: Intermediate stage (I_5) in the realization of a random version of the von Koch curve. (From Peitgen *et al.*, 1992, p. 460. Reprinted with permission.)

in the sense that their construction is based on the recursive application of an invariant and non-random generator.

It is straightforward to enlarge this family of monsters to include random analogues. For example, in the von Koch construction, each time the middle third of an interval is replaced by the other sides of an equilateral triangle, one might toss a coin to determine whether to position the new part above or below the removed segment. After a few steps, one gets an irregular-looking curve (see Figure 9) which nevertheless retains much of the "look" of the von Koch curve (Falconer, 1990; Peitgen *et al.*, 1992). It is no longer exactly self-similar, like its non-random counterpart, but it is said to be "statistically" self-similar. According to Feder (1988, p. 184), a set S is statistically self-similar when S is the union of N distinct (non-overlapping) subsets, each of which is scaled down by r from the original and is identical in all statistical respects to S. Put differently (Falconer, 1990, p. 225), enlargements of subsets of a statistically self-similar set have a statistical distribution identical to that of the whole set.

As with the deterministic monsters, it is possible to relax the requirement that the scaling factor r be identical in all directions. In this case, the random set or curve is said to be "statistically self-affine". Historically important examples of statistically self-affine sets are the Brownian- and fractional Brownian motions.

Brownian motion, also referred to as random walk or Wiener process, is named after the Scottish botanist Robert Brown, who in 1828 described the erratic motion of pollen in aqueous suspensions observed with a light microscope. A particle undergoing Brownian motion seems to wander around without any distinct pattern (Figure 10). Some regions of the plane are filled densely by the particle's trace. Increasing the resolution of the microscope and the time resolution produces a random walk that looks very much like that obtained at lower resolution.

Physically (cf Lavenda, 1985), the Brownian motion of a microscopic particle is due to its constant bombardment by the numerous smaller molecules of the medium (*e.g.*, water) in which it is suspended. A single molecule hardly ever has enough momentum for its effect on the suspended particle to become

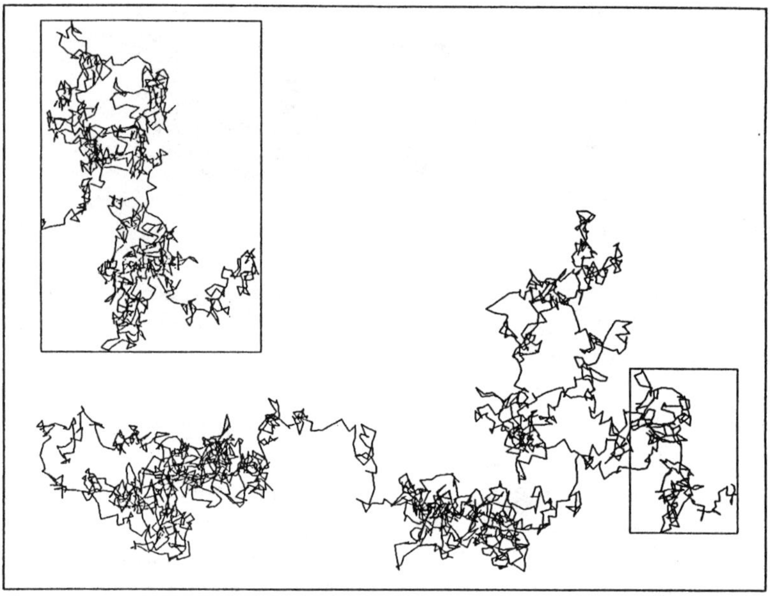

Figure 10: Trace of the Brownian motion of a particle in a plane. The boxed detail of the trace (magnified in the upper left portion of the figure) suggests an invariance of scale or self-similarity: the detail looks like the whole. (From Peitgen *et al.*, 1992. Reprinted with permission.)

visible under a microscope. Nevertheless, when many molecules collide with the particle from nearly the same direction at nearly the same time, they noticeably deflect it. These events are separated in time by what is commonly referred to as an "atomistic" or "collision" time τ. The displacement, ξ, of the particle is not fixed, since it depends on the number of molecules colliding with the microscopic particle, and on their direction. Furthermore, the sizes and directions of successive displacements are uncorrelated.

In 1923, Wiener proposed a rigorous mathematical model that exhibits a behavior similar to that observed in random motion. In this model, the displacement ξ is governed by a Gaussian (bell-shaped) probability distribution, with zero mean and unit variance. Therefore, if one plots the successive displacements of a particle (in one dimension for simplicity) over a certain period of time, the result (Figure 11a) corresponds to Gaussian, or "white", noise. The sum of successive steps ξ_1, ξ_2,..., ξ_n of the particle during n collision times is given by

$$X(t = n\tau) = \sum_{i=1}^{n} \xi_i \tag{8}$$

where X is the position of the particle at time t, relative to its initial position

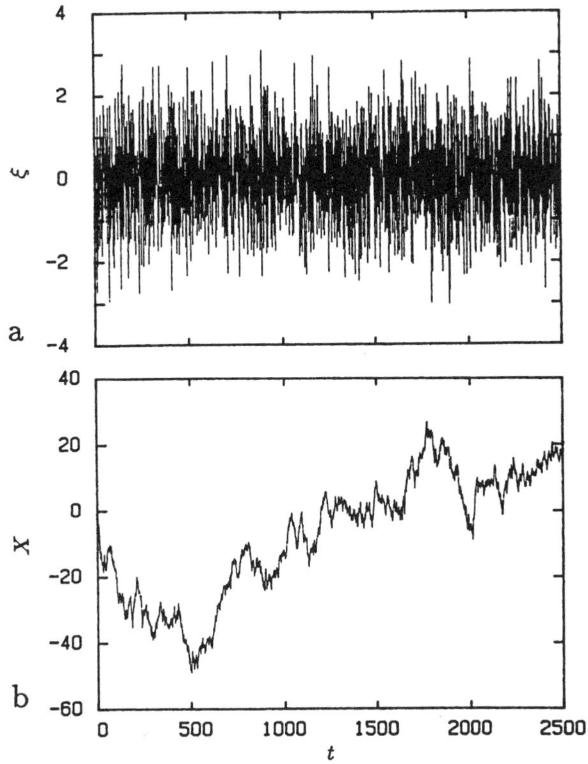

Figure 11: A sequence of independent Gaussian random variables with zero mean and unit variance (Gaussian noise) "observed" at every fourth time step, *i.e.*, at intervals of length 4τ. (a) Independent random steps of the "particle". (b) Position of the particle along the x axis. The time is in units of the atomistic time τ between steps. (From Feder, 1988. Reprinted with permission.)

at time $t = t_o$ (Figure 11b). The function $X(t)$ in Figure 11b is commonly referred to as the "graph" or the "record" of the Brownian motion and may be obtained by projection of the path of the Brownian motion onto one of the space axes.

In the limit of vanishingly small time steps ($\tau \to 0$), the position X in Equation (8) becomes a random function, usually denoted by $B(t)$. This function is such that for any two times t_1 and t_2, with $t_2 \geq t_1$,

$$|B(t_2) - B(t_1)| \propto (t_2 - t_1)^H \tag{9}$$

where $\propto$ denotes proportionality and H is the Hurst exponent. The variance of the increments is given by

$$var\,(B(t_2) - B(t_1)) \propto (t_2 - t_1)^{2H} \tag{10}$$

Early work concentrated on the ordinary "mathematical" Brownian motion, with $H = 1/2$, and showed that this motion has a number of remarkable features. First, it cannot be represented graphically as can motion for finite τ (Figure 11b). In addition, Wiener (1923) showed that $B(t)$ is continuous, as one expects, but nowhere differentiable. Thus, a particle undergoing mathematical Brownian motion does not have a well-defined velocity, and the curve $y = B(t)$ does not have a well-defined tangent anywhere. A related fact that illustrates the extreme irregularity of $B(t)$ is that in every time interval, no matter how small, a particle undergoing mathematical Brownian motion travels an infinite distance! Figure 10 suggests that Brownian motion is statistically self-similar, and it is indeed so in R^n for $n \geq 2$. However, graphs of Brownian motions, such as that illustrated in Figure 11b, are statistically self-affine rather than self-similar.

The exponent H in Equations (9) and (10) does not have to be set equal to $1/2$, but may instead vary arbitrarily. When $0 < H < 1$, this generalization leads to the *fractional Brownian motion* (fBm), to which is associated the so-called fractional Gaussian noise (fGn), in the same manner that Gaussian noise produces the ordinary Brownian motion (cf Figure 11).

A remarkable feature of fBm is that it generates infinitely long-run correlations (these correlations correspond, loosely, to the levels of dependence, in a probabilistic sense, of distinct spatial increments). The correlation function $C(t)$ associated with fBm is equal to $2^{2H-1} - 1$ (*e.g.*, Feder, 1988, p. 170), *i.e.*, it depends only on the value of H. When $H = 1/2$, fBm reduces to ordinary Brownian motion, in which successive displacements or increments are uncorrelated ($C(t) = 0$). For $0.5 < H < 1$, there is a positive correlation between successive increments and fBm exhibits a "persistent" behavior, characterized by clear trends and relatively little noise (*e.g.*, Feder, 1988, p. 181). For $0 < H < 0.5$, on the other hand, fBm displays "anti-persistence", the correlation is negative and as a result graphs of the fBm appear very noisy.

As in the case of the deterministic sets and functions considered in previous sections, the topological dimension of a random monster sheds very little light on its intricate geometry. For example, both the deterministic and random triadic von Koch curves have the same topological dimension, equal to one and identical to that of a regular line segment. Fortunately, mathematicians have introduced a number of dimensions that can characterize even the most pathological sets or functions. These dimensions are defined and analyzed in detail in the following section.

III Hausdorff Dimension and Alternatives

III.A Hausdorff Measure and Dimension

Of all the dimensions of sets, the one introduced by Hausdorff (1919) is undoubtedly the most useful for characterizing nowhere-differentiable sets. Familiarity

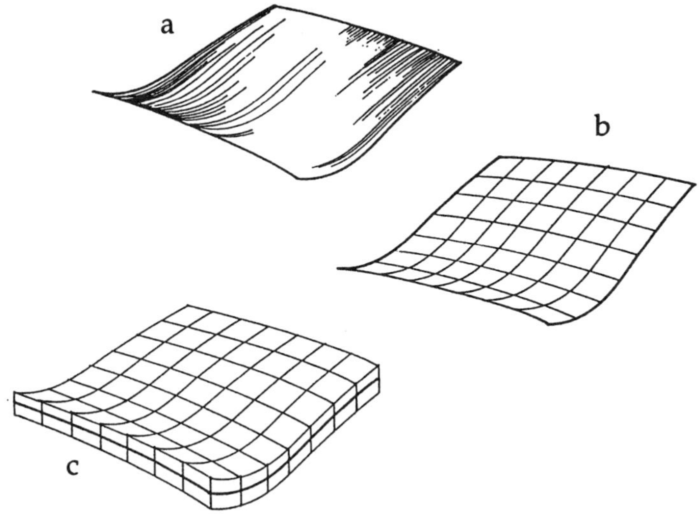

Figure 12: Schematic illustration of various approaches to measuring the "size" of a surface. (a) surface, (b) covering with squares, (c) covering with cubes. (Modified from Feder, 1988.)

with its definition, and with its limitations, is essential to understand the concept of fractals.

To understand the mathematical background of the Hausdorff dimension, it is useful to first consider as an illustration the process of measuring the "size" of a set of points defining a surface (Figure 12a) in 3-dimensional Euclidian space, R^3. The customary measure of this surface is its area. It can be approximated by the product $A \equiv N(\delta)\delta^2$, where $N(\delta)$ is the number of squares of side length δ needed to tile or cover the surface (Figure 12b). For "ordinary" surfaces, $N(\delta)$ tends to A_o/δ^2 in the limit of vanishing δ, so that

$$A \equiv N(\delta)\delta^2 \underset{\delta \to 0}{\longrightarrow} A_o\delta^0 = A_o \tag{11}$$

where A_o is the area of the surface.

We might also try to associate a volume with the surface by calculating the sum of the volumes of the cubes of side length δ needed to cover the surface (Figure 12c). Since $N(\delta)$ tends to A_o/δ^2 in the limit of vanishing δ,

$$V \equiv N(\delta)\delta^3 \underset{\delta \to 0}{\longrightarrow} A_o\delta^1 \tag{12}$$

which vanishes for $\delta \to 0$.

Formally (even though it is clearly not feasible!), we might also try to approximate the surface by the total length L of a finite number of line segments of length δ. It is easy to see, after replacement of $N(\delta)$ by A_o/δ^2, that

$$L \equiv N(\delta)\delta \underset{\delta \to 0}{\longrightarrow} A_o\delta^{-1} \tag{13}$$

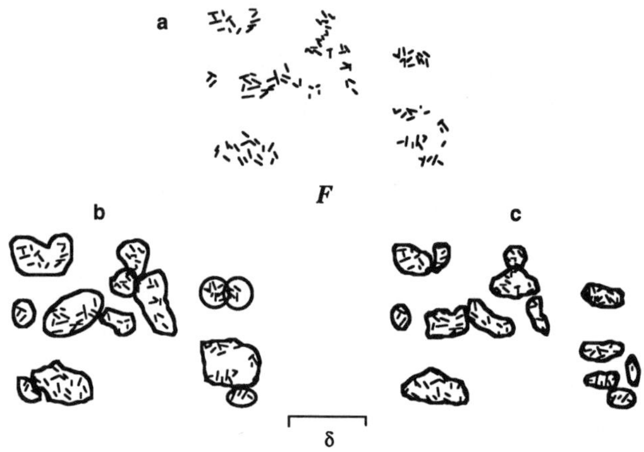

Figure 13: A set F and two possible δ-covers of F. (Modified from Falconer, 1990.)

which diverges (tends to infinity) for δ tending to 0.

The conclusion to be drawn from this (simplistic) illustration is of course that the only useful measure of a set of points defining an ordinary surface in R^3 is the area. Interestingly, the measure of this same set of points with a "yardstick" that has a lower dimension than the area (*i.e.*, a line segment) yields an infinite (divergent) measure, while a "yardstick" with a higher dimension than the area (*i.e.*, a volume) leads to a zero measure. Therefore, the practically useful yardstick corrresponds to a transition from divergent measures to zero measures. This same feature, as we shall see below, characterizes the Hausdorff dimension.

The definition of the Hausdorff dimension is based on the concept of Hausdorff measure, which itself makes use of the notion of δ-cover of a set.

If U denotes a non-empty subset of n-dimensional Euclidean space (*i.e.*, $U \subset R^n$), the *diameter* of U, $|U|$, is the greatest distance separating any pair of points in U. Mathematically, $|U| = \sup\{|x - y| : x, y \in U\}$, where sup is the conventional abbreviation of the supremum, the smallest upper bound of a set[2] and $|x - y|$ is the Euclidian distance between x and y. If F (shown in Figure 13a) is any subset of R^n, we can cover it with a (possibly large but countable) number of subsets U_i of R^n, such that each of the U_i has a diameter at most equal to a given δ ($0 < |U_i| \leq \delta$, for each i). This covering is done schematically in two different ways in Figures 13b and 13c. In each case, F is included in the union of all the U_i's, *i.e.*, $F \subset \bigcup_{i=1}^{\infty} U_i$. A set $\{U_i\}$ that has this property is said to be a δ-cover of F.

Among all the δ-covers of F, we might be interested in determining which

---

[2] The *supremum* of a set $S \subset R$ is the smallest $x \in R$ such that $x \geq s$ for every $s \in S$. Similarly, the *infimum* of a set $S \subset R$ is the largest $x \in R$ such that $x \leq s$ for every $s \in S$. If S is *compact*, $\sup S = \max S$ and $\inf S = \min S$.

one is the "smallest", in some sense. This δ-cover could be defined as the one with the smallest sum of diameters $\sum_{i=1}^{\infty}|U_i|$. It would be equally acceptable to minimize, *e.g.*, the sums $\sum_{i=1}^{\infty}|U_i|^2$ or $\sum_{i=1}^{\infty}|U_i|^3$. More generally, we could raise the diameters $|U_i|$ to any non-negative number s. For $\delta > 0$, we could then define (Falconer, 1990, p. 25)

$$\mathcal{H}_\delta^s(F) = \inf\{\sum_{i=1}^{\infty}|U_i|^s : \{U_i\} \text{ is a } \delta\text{-cover of } F\} \tag{14}$$

where inf represents the infimum, the largest lower bound of a set. In Equation (14), we look at all δ-covers of F and we seek to minimize the sum of the sth powers of the diameters.

As δ decreases, the class of permissible covers of F in Equation (14) is reduced. Therefore, the infimum cannot decrease as δ decreases. As $\delta \to 0$, the infimum approaches a limit which may be infinite or a real number ≥ 0. This limit is defined as

$$\mathcal{H}^s(F) \equiv \lim_{\delta \to 0} \mathcal{H}_\delta^s(F) \tag{15}$$

and is termed the *s-dimensional Hausdorff measure of* F. For integral values of s, the Hausdorff measure reduces to the traditional n-dimensional Lebesgue measure (*i.e.*, the usual n-dimensional volume), multiplied by a constant.

It is clear from Equation (14) that if $\delta < 1$, for any given set F, the terms $|U_i|^s$ tend to decrease when s increases. Therefore, $\mathcal{H}_\delta^s(F)$ is non-increasing with s for $\delta \leq 1$ and, by Equation (15), $\mathcal{H}^s(F)$ has the same property. In fact, it is possible to show that if $t > s$ and $\{U_i\}$ is a δ-cover of F

$$\sum_i |U_i|^t \leq \delta^{t-s} \sum_i |U_i|^s \tag{16}$$

Taking infima according to Equation (14), one finds that $\mathcal{H}_\delta^t(F) \leq \delta^{t-s}\mathcal{H}_\delta^s(F)$. Letting $\delta \to 0$, we see that if $\mathcal{H}^s(F) < \infty$, then $\mathcal{H}^t(F) = 0$ for $t > s$. Thus, a graph of $\mathcal{H}^s(F)$ against s (Figure 14) shows that there is a critical value of s at which $\mathcal{H}^s(F)$ "jumps" from ∞ to 0. This critical value of s (at which $\mathcal{H}^s(F)$ may in general be undefined) is D_H, the *Hausdorff dimension* of F.

The Hausdorff dimension of the triadic Cantor set is 0.631 and that of the von Koch curve equals 1.262. This latter number is entirely consistent with the von Koch curve being somehow "larger than 1 dimensional" (having infinite length within a bounded domain) and "smaller than 2 dimensional" (having zero area). In similar fashion, it is not surprising that graphs of Brownian motions (see section II.G) have a Hausdorff dimension equal to $1\frac{1}{2}$. Not all Hausdorff dimensions of mathematical "monsters" are fractional; that of the Cantor singular function (Devil's staircase), for example, is 1, whereas the Hausdorff dimensions of the Sierpinski tetrahedron, of the Peano plane-filling curve and of paths of the mathematical Brownian motion in R^n ($n \geq 2$) are all equal to 2.

To say the least, the above description of the theory leading to the definition of the Hausdorff dimension does not suggest a simple and intuitive way of evaluating this dimension in practical cases. Indeed, the calculation of Hausdorff

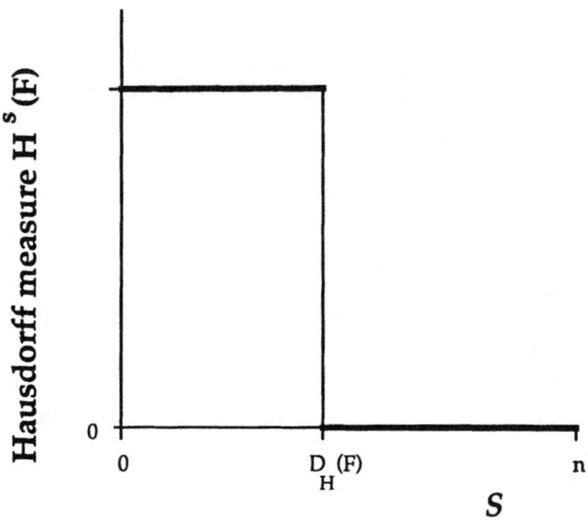

Figure 14: Graph of the Hausdorff measure against s for an arbitrary set F. The Hausdorff dimension is the value of s at which the Dedekind cut, or "jump" from ∞ to 0 occurs.

measures and dimensions is, in general, more than a little involved, even for simple sets (*e.g.*, Takayasu, 1990, p.156-157). Rigorous dimension calculations often involve pages of complicated manipulations. In some cases, for example the Weierstrass-Mandelbrot function, there remains some uncertainty on the correct outcome of these calculations (*e.g.*, Berry and Lewis, 1980; Feder, 1988, p. 27). Nevertheless, the Hausdorff dimension has the distinct advantage of being defined for any set (cf the handling of statistically self-similar sets in Hutchinson (1981) and Graf *et al.* (1988)). It can, therefore, provide a reference for comparison with values of dimensions computed via simpler, more straightforward methods (as described in the following subsections).

III.B Similarity Dimension

A heuristic method, which often results in a number equal to the Hausdorff dimension of sets of points, leads to the definition of another dimension. Following Falconer (1990, p. 31), let us take the Cantor set P (cf Figure 1) as an example. It can easily be split into a left part $P_L = P \cap [0, 1/3]$ and a right part $P_R = P \cap [2/3, 1]$. Both parts are geometrically similar to P but scaled down by a ratio 1/3, and P is the (disjoint) union of P_L and P_R. Because the Hausdorff measure is additive and scales according to $\mathcal{H}^s(\lambda F) = \lambda^s \mathcal{H}^s(F)$ where $F \subset R^n$ and $\lambda > 0$, one finds that

$$\mathcal{H}^s(P) = \mathcal{H}^s(P_L) + \mathcal{H}^s(P_R) = \frac{1}{3^s}\mathcal{H}^s(P) + \frac{1}{3^s}\mathcal{H}^s(P) \qquad (17)$$

Assuming that at the critical value $s = D_H(P)$, we have $0 < \mathcal{H}^s(P) < \infty$

(as stressed by Falconer (1990), this is a big assumption, but one that can be justified), we may then divide Equation (17) by $\mathcal{H}^s(P)$ to get $1 = 2(1/3)^s$ or $s = ln2/ln3 \approx 0.631$.

In this estimate of the Hausdorff dimension of the Cantor set, we end up with a ratio of two natural logarithms. In the denominator is the number of segments (3) in which the initiator is divided. In the numerator is the number of segments (2) that are retained in the generator. Alternatively, we could look at 2 as the number of copies of itself, scaled by a factor 1/3, that the Cantor set contains. One could extend this reasoning to other sets and consider that, in general, a set made up of m copies of itself scaled by a factor $1/r$ has a dimension ln m/ln r. The number obtained in this way is usually referred to as the *similarity dimension*, D_s, of the set.

It turns out that for virtually all the deterministic monsters introduced earlier, the similarity and Hausdorff dimensions coincide. For example, it is easy to see that the triadic von Koch curve has a similarity dimension $D_s = \ln 4/\ln 3 \approx 1.262$, identical to its Hausdorff dimension (see previous section). For the Sierpinski gasket, $D_s = \ln 3/\ln 2 \approx 1.585$, for the Sierpinski carpet, $D_s = \ln 8/\ln 3 \approx 1.89$ and for the Menger sponge, $D_s = \ln 20/\ln 3 \approx 2.73$; all three D_s values are identical to the corresponding Hausdorff dimensions. In the case of "non-monstrous" sets, the similarity dimension also behaves properly. For example, if we consider a straight line segment, divide it into four sub-segments of length scaled by a factor 1/4, keep all four segments and iterate this procedure indefinitely, the similarity dimension of the resulting set is $D_s = \ln 4/\ln 4 = 1$, which is the same as the set's Hausdorff dimension.

In subsection II.C, it was shown that after n iterations in the construction of the triadic von Koch curve, the number N_n of segments of length δ_n is a power-law function of δ_n: $N_n = \delta_n^{-\ln 4/\ln 3}$. With the introduction of the similarity dimension, and by virtue of the equality between Hausdorff- and similarity dimensions for the triadic von Koch curve, this power-law relationship may also be expressed as:

$$N_n = \delta_n^{-\ln 4/\ln 3} = \delta_n^{-D_s} = \delta_n^{-D_H} \tag{18}$$

from which it follows that $D_s = D_H$ may be obtained graphically from the slope of N_n versus δ_n in a log-log graph.

Even without having recourse to this graphical method, the similarity dimension is clearly very straightforward to compute. Unfortunately, it is meaningful only for a small class of strictly self-similar sets. It cannot be used to evaluate the dimension of self-affine, statistically self-similar or statistically self-affine sets. For these, other easily measurable dimensions are necessary, like the box-counting dimension.

III.C Box-Counting Dimension

In calculating the infimum $\mathcal{H}^s_\delta(F)$ in Equation (14), on the route to defining the Hausdorff dimension, the covering sets U_i were allowed to have various diameters, within the constraint $|U_i| \leq \delta$.

One possible approach to a dimension that would be equally general, yet simpler to compute than the Hausdorff dimension, would be to require that the diameters of all the U_i equal δ, or, equivalently, to replace $|U_i|$ by δ in Equation (14). Using this approach, one would obtain a number $N_\delta(F)$, associated with a given non-empty bounded subset F of R^n, such that

$$N_\delta(F)\delta^s = \inf\left\{\sum_i \delta^s : \{U_i\} \text{ is a (finite) } \delta\text{-cover of } F\right\} \tag{19}$$

In other words, $N_\delta(F)$ is the smallest number of sets of diameter at most δ that can cover F.

Based on the definition of Equation (19), it is convenient to define a new dimension, called the *box-counting dimension* and denoted here by D_{BC}, as

$$D_{BC}(F) \equiv \lim_{\delta\to 0} \frac{\ln N_\delta(F)}{-\ln \delta} \tag{20}$$

The motivation for this (apparently arbitrary) definition is that for many sets of points, there is a nearly linear relationship between $\ln N_\delta(F)$ and $\ln \delta$ for small δ, and, as $\delta \to 0$, the slope of this relationship approaches $D_{BC}(F)$. Nevertheless, $D_{BC}(F)$ as defined by Equation (20) has meaning only when the limit as $\delta \to 0$ exists (cf discussion in Falconer, 1990).

The dimension D_{BC} may be thought of as an indication of the efficiency with which a set may be covered by small sets of equal size, whereas the Hausdorff dimension D_H involves coverings by sets of small but perhaps widely varying size. The box-counting dimension is sometimes also referred to as the "Minkowski" (Falconer, 1990), "Minkowski-Bouligand" (Mandelbrot, 1982; Sapoval, 1991; Schroeder, 1991), "Bouligand-Minkowski" (Gouyet, 1992), or "capacity" dimension (Moon, 1992; Falconer, 1990) or the "Kolmogorov entropy" (Falconer, 1990).

Up to this stage, the geometrical shape of the U_i's has not been specified in any way, which suggests that the definition of Equation (19) can accommodate a variety of geometries. This variety is quite large indeed. It is possible to show mathematically (Falconer, 1990, p. 41) that $N_\delta(F)$ in Equation (20) can be any of the following options, illustrated in Figure 15:

i the smallest number of closed balls of radius δ that cover F. (A closed ball $B_r(\mathbf{x})$ is a set defined as $B_r(\mathbf{x}) = \{\mathbf{y} : |\mathbf{y} - \mathbf{x}| \leq r\}$, where $\mathbf{x}$ and r are called the center and the radius of the set.);

ii the smallest number of cubes of side δ that cover F;

iii the number of "δ-mesh" cubes of side δ that cover F (this number has been used for decades by geographers and cartographers, and justifies the "box-counting" qualifier);

iv the smallest number of sets of diameter at most δ that cover F;

v the largest number of disjoint balls of radius δ with centres in F. It is not readily obvious that this number is mathematically equivalent to the others, yet such is the case (*e.g.*, Falconer, 1990, p. 41).

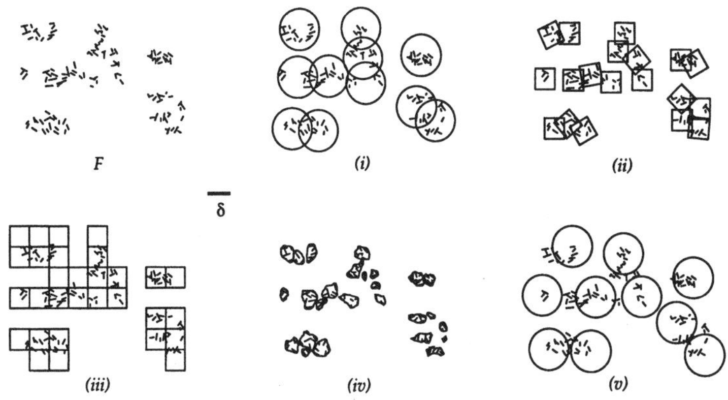

Figure 15: Five ways of finding the box dimension of F (see text for details on (i)-(v)). (Modified from Falconer, 1990.)

Mathematically, the definition of Equation (20) implies that $N_\delta(F)\delta^s \to \infty$ if $s < D_{BC}(F)$ and that $N_\delta(F)\delta^s \to 0$ if $s > D_{BC}(F)$. In other words, the product $N_\delta(F)\delta^s$ has a (Dedekind) cut similar to that experienced by the Hausdorff measure (cf Figure 14). The analogy does not extend further, however. The limit of $N_\delta(F)\delta^s$ as $\delta \to 0$ does not, unfortunately, correspond to a measure on subsets of R^n. This has a number of unpleasant consequences, one of them being that countable sets of points can have non-zero box-counting dimensions. For example, the countable (compact) set $\{0,1, 1/2, 1/3,...\}$ has a Hausdorff dimension equal to zero but a box-counting dimension equal to 0.5.

A number of alternative dimensions have been proposed to overcome these difficulties associated with the traditional box-counting dimension D_{BC}. They include the lower and upper modified box-counting dimensions (Falconer, 1990), and the packing or Tricot dimension (Tricot, 1995; Falconer, 1990; Le Méhauté, 1991). Unfortunately, these dimensions re-introduce all the difficulties of calculation associated with D_H, and in some cases are even more awkward to use!

One key advantage of the box-counting dimension D_{BC} over the similarity dimension D_s is that D_{BC} can be used to evaluate the dimension of self-affine sets. In these sets, however, D_{BC} is not uniquely defined but instead assumes two different values, a *local* or small-scale value and a *global* or large-scale value (*e.g.*, Feder, 1988, p. 187; Family and Vicsek, 1991, p. 8; Gouyet, 1992, p. 55). In the case of the fractional Brownian motion (cf section II.G above), the local D_{BC} value is equal to the Hausdorff dimension and is given by $2-H$, where H is the Hurst exponent, whereas the global D_{BC} value is equal to 1 (*e.g.*, Feder, 1988, p. 189).

III.D Divider Dimension

In the case of non-intersecting curves, it is possible to define another dimension, denoted here by D_D and generally referred to as the "divider" or "compass" dimension.

If C is a curve in R^n and $\delta > 0$, we define $M_\delta(C)$ as the maximum number of points x_o, x_1,..., x_m on the curve C, in that order, such that $|x_k - x_{k-1}| = \delta$ for k=1, 2,..., m. Thus, $(M_\delta(C) - 1)\delta$ may be thought of as the length of the curve measured using a pair of dividers with points set a distance δ apart, *i.e.*, with "yardsticks" of length δ. Under these conditions, the divider dimension D_D is defined as (Falconer, 1990):

$$D_D \equiv \lim_{\delta \to 0} \frac{\ln M_\delta(C)}{-\ln \delta} \tag{21}$$

assuming that this limit exists. It is possible to show that $D_D \geq D_{BC}$, when both these dimensions exist. In simple self-similar examples, such as the von Koch curve, D_D and D_{BC} are equal, but in other cases, the inequality has been shown to apply. For example, fractional Brownian motion (cf section II. G) has a local D_D value equal to $1/H$ where H is as before the Hurst exponent, which is always strictly larger than its local D_{BC} value, equal to $2 - H$ (*e.g.*, Feder, 1988, p. 189). Tricot (1995, p. 234) expresses reservations about the use of the divider method to evaluate the dimension of curves that are not exactly or statistically self-similar (*e.g.*, the west coast of Britain, which will be mentioned again in section IV.A); he indeed argues that the divider dimension does not have a precise metric meaning in these cases.

III.E Pointwise Dimension and Hölder Exponent

Another dimension has interesting properties. It is called the "*pointwise*" or "mass" dimension (*e.g.*, Moon, 1992, p. 334; Peitgen *et al.*, 1992, p. 736).

To define it, one considers a curve in R^n and samples it uniformly at a large number N_o of points, *i.e.*, one determines the spatial coordinates of N_o points uniformly distributed along the curve. Then, one centers a ball $B_r(\mathbf{x})$ of radius r at some location $\mathbf{x}$ on the curve, and one counts the number $N(r, \mathbf{x})$ of sampled points within this ball. The probability of finding a point in this ball is given by the limit

$$\mu(B_r(\mathbf{x})) = \lim_{N_o \to \infty} \frac{N(r, \mathbf{x})}{N_o} \tag{22}$$

This limit may also be viewed as the *measure* or "mass" of the curve that is contained in the ball $B_r(\mathbf{x})$.

By analogy with previous definitions of dimensions, one may define the pointwise dimension D_p at the location x as follows:

$$D_p(\mathbf{x}) = \lim_{r \to 0} \frac{\ln \mu(B_r(\mathbf{x}))}{\ln r} \tag{23}$$

In some cases, the limit as $r \to 0$ is not taken in Equation (23) and the ratio $\ln \mu(B_r(\mathbf{x}))/\ln r$ is termed in this case the *Hölder-* (Mandelbrot, 1989), *coarse Hölder-* (Evertsz and Mandelbrot, 1992) or *Lipschitz-Hölder exponent* (*e.g.*, Le Méhauté, 1990). It is traditionally denoted by α and it may be evaluated for any measure, defined or not by Equation (22). This exponent is useful to characterize "singular" measures, which have no local densities (*i.e.*, for which the limit in Equation (23) does not exist), and it plays an important role in the definition of multifractal measures (section VII below).

III.F Correlation Dimension

The definition of this dimension starts with a "sampling" of N_o points similar to that used in the previous section. One then calculates the distances between pairs of points, say $d_{ij} = |\mathbf{x}_i - \mathbf{x}_j|$, using either the conventional Euclidean measure of distance (square root of the sum of the squares of the vector components) or some equivalent measure such as the sum of absolute values of the components. The "correlation function" (Moon, 1992) or "correlation integral" (Korvin, 1992) associated with the N_o points is defined as

$$C(l) = \lim_{N_o \to \infty} \frac{(\text{number of pairs of points } (\mathbf{x}_i, \mathbf{x}_j) \text{ whose distance } d_{ij} < l)}{N_o^2} \tag{24}$$

This correlation function or integral may be used to define the *correlation dimension* D_c (Korvin, 1992; Moon, 1992; Peitgen *et al.*, 1992) as

$$D_c \equiv \lim_{l \to 0} \frac{\ln C(l)}{\ln l} \tag{25}$$

whenever this limit exists. An extensive study of this dimension is carried out by Grassberger and Proccacia (1983).

III.G An Arsenal of Dimensions

The previous subsections indicate that several concepts of dimension have been proposed over the years to characterize the geometrical properties of sets of points. Many more dimensions than those presented above exist. They include, for example, the Ljapunov dimension (Takayasu, 1990; Peitgen *et al.*, 1992), the spectral dimension (Le Méhauté, 1991), the information dimension D_I (Takayasu, 1990; Peitgen *et al.*, 1992; Korvin, 1992; Moon, 1992) and the Fourier dimension (Falconer, 1990).

Another dimension, occasionally used in soil science, might be called the "variogram" dimension. In section II.F, a relation was mentioned between the variance of increments, $V(t)$, and the parameter D appearing in the formulation of the Weierstrass-Mandelbrot function $w(t)$. In the case of spatial functions, it is more common to compute the semi-variance (half the variance of increments), also termed variogram. By analogy with the behavior of $V(t)$ at the origin, one may postulate for the variogram a power-law dependency on the spatial increment h (or "lag"), as $h \to 0$, and define on this basis a variogram dimension (*e.g.*, Burrough, 1981; Klinkenberg, 1994).

How many dimensions are there? Their number is in principle infinite, as was illustrated elegantly by Hentschel and Proccacia (1983), who defined a collection of dimensions D_q, for $q \geq 0$, now often referred to as the Rényi dimensions (*e.g.*, Peitgen *et al.*, 1992, p. 736). The first in this collection ($q = 0$) corresponds to the box-counting dimension D_{BC}. The second ($q = 1$) is equivalent to the information dimension D_I, and the third ($q = 2$) to the correlation dimension D_C. It can be proven mathematically that the Rényi dimensions D_q are decreasing with q, *i.e.*, $D_p \geq D_q$ if $p < q$. In particular, this implies that $D_C \leq D_I \leq D_{BC}$, where equality occurs only in special cases. In general, these dimensions are not equal.

IV Fractals

IV.A Physical Motivation

Section II of this chapter introduced a large array of "monstrous" mathematical beings, which exhibit pathological properties defying the traditional concept of dimension. In section III, we saw that various alternatives to the traditional topological dimension have been devised by mathematicians. In spite of their multiplicity, these dimensions have tended to make the mathematical monsters somewhat less terrifying.

Observations made by scientists over the years provided additional momentum in the same direction. The physical process whose study was perhaps most influential in stimulating interest in nowhere-differentiable functions is Brownian motion, described in section II.G above. Besides Brownian motion, several other physical processes contributed to foster interest in nowhere-differentiable functions. A few years after Wiener's work, Dedebant and Wehrlé (1935) studied a number of scale issues arising in meteorology and concluded that "many meteorological processes, observed at a very small scale, are like the continuous, nondifferentiable functions that seemed forever to belong to the realm of speculative mathematicians" (Dedebant and Wehrlé, 1935, p. 83).

More recently, Chepil (1950) observed a marked and systematic decrease of the densities of soil aggregates as the size of these aggregates increased (Figure 16). This behavior, which is not intuitive from a traditional geometric standpoint, is similar to that predicted by Equation (4) for intermediate steps in the iterative procedure leading to the Menger sponge. Indeed, if large aggregates are viewed as containing a greater variety of pore sizes than small aggregates - thus, corresponding to the complexity found in the later stages in the iterative construction of the Menger sponge - then in Equation (4) one could keep r_n and ρ_o constant, and observe that as r_o increases, the mass density ρ_n must decrease, since the exponent $3 - \ln 20/\ln 3$ is strictly positive (≈ 0.273). Therefore, at least some of the features of the Menger sponge may be useful to describe the

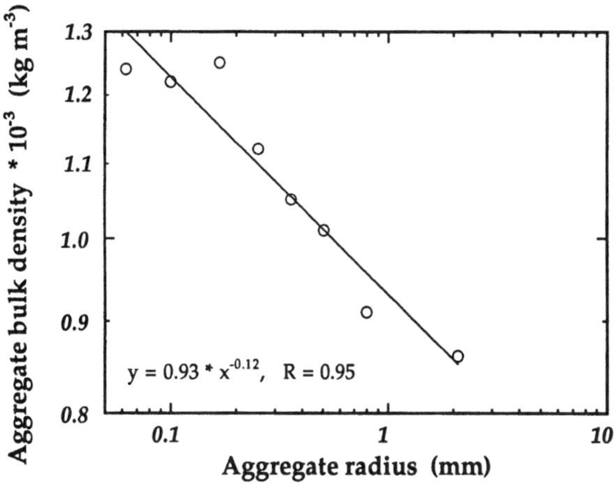

Figure 16: Bulk density of soil aggregates as a function of their radius. (Data from Table 4 in Chepil (1950), fine sandy loam.)

properties of soil aggregates.

Perhaps the most acute realization that the mathematical monsters of the previous section have properties very close to those routinely found in nature was made by geographers. Richardson's (1961) work is particularly significant in this respect. He measured a number of geographical curves (coastlines and country borders), using the divider or compass method, that led to the definition of D_D in section III.D above. For a circle, the total length tends to a limit as the segment length goes to zero (cf Figure 17). In all other cases, it increases as the side becomes shorter. This behavior is identical to that described above for the triadic von Koch curve (section I.C). In this latter case, the length of the nth iteration step I_n has a power-law or Paretian relationship with the "yardstick" length δ_n (cf Equation (2)). In other words, $L(I_n)$ and δ_n are related via a straight line in a log-log plot. This is precisely what is observed for the various geographical curves shown in Figure 17.

IV.B "Definition" of Fractals

As the examples of the previous section illustrate, many scientists observed over several decades that natural objects or processes often have features akin to those of the von Koch curve or Menger sponge. This greatly stimulated interest in these monstrous sets; in this case as in many others, interest in geometry was driven by its applications to nature (Falconer, 1990).

This movement led to the publication by Mandelbrot in 1975 of an essay in which he highlighted the similarities among the then-known continuous, nowhere-differentiable sets. He coined for these sets the term "fractal", to emphasize the fact that their Hausdorff dimensions are often fractional. In the

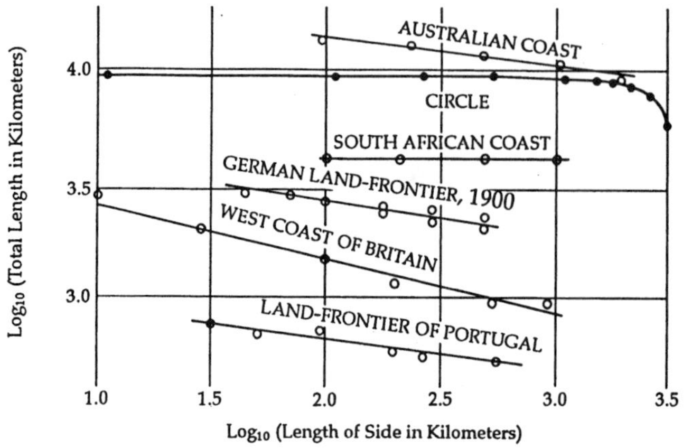

Figure 17: The length of coastlines as a function of yardstick length. (Data from Richardson, 1961.)

words of Dyson (1978), "fractal is a word invented by Mandelbrot to bring together under one heading a large class of objects that have [played] ... an historical role ... in the development of pure mathematics".

Mandelbrot believed initially that one would do better without a precise definition of fractals. His original essay (Mandelbrot, 1975) contains none. By 1977, however, he saw the need to produce at least a tentative definition. It is the now classical statement that "a fractal is a set for which the Hausdorff dimension strictly exceeds the topological dimension" (Mandelbrot, 1982; Falconer, 1990, p. xx; Feder, 1988, p. 11). For example, the Cantor set is a fractal, according to this viewpoint, since $D_H = 0.631 > D_T = 0$.

The above definition immediately proved unsatisfactory in that it excluded a number of sets with properties very similar to those of sets which satisfied the definition, and which therefore ought to be also regarded as fractals. Indeed, according to this definition, the Cantor singular function ($D_H = D_T = 1$) and the Peano plane-filling curve ($D_H = D_T = 2$) are not fractals. Various other definitions of fractals have been proposed (*e.g.*, Mandelbrot, 1982, p. 362), but they all seem to suffer from the same drawback.

Perhaps by modifying slightly the list of attributes of the Cantor set (cf section I.A), and by relaxing or deleting some of them, one could approach, as closely as possible, to a definition of the concept of fractal. This was done by Falconer (1990), in what appears to be the best approximation of this concept to date. According to this author, when one refers to a set F (of points) as a fractal, one typically has one or more of the following properties in mind:

- F has a fine structure, *i.e.*, detail on arbitrarily small scales
- F is too irregular to be described in traditional geometrical language, both locally and globally
- *Often* F has some form of self-similarity, perhaps approximate or statistical
- *Usually*, the "fractal dimension" of F (defined in some way) is greater than its topological dimension
- *In most cases* of interest, F is defined in a very simple way, *perhaps* recursively (in which case the various stages of the iterative construction are usually referred to as prefractals).

Falconer (1990, p. xx) has perhaps best captured the spirit with which the above definition of fractals has to be apprehended: "My personal feeling is that the definition of a "fractal" should be regarded in the same way as the biologist regards the definition of "life". There is no hard and fast definition, but just a list of properties characteristic of a living thing, such as the ability to reproduce or to move or to exist to some extent independently of the environment. Most living things have most of the characteristics on the list, though there are living objects that are exceptions to each of them."

While probably agreeing in principle with Falconer's (1990) perception, many authors (in the fractal geometry literature at least) tend to give preferential weight to the requirement that fractals exhibit some form of self-similarity (cf above, third point in the list of attributes of fractals).[3] Feder (1988, p. 11), for example, defines a fractal as a "shape made of parts similar to the whole in some way". In other words, a fractal appears the same regardless of the scale of observation; its "look" is scale-invariant[4].

Perhaps the most important aspect of the various definitions above is that they all consider fractals to be sets of points in R^n, *i.e.*, *geometric* constructs. This feature will assume particular significance in a later section (section V), where we shall discuss non-geometric fractals.

As with the concept of fractals itself, a certain level of vagueness characterizes the definition of the fractal dimension. The approach advocated by Mandelbrot in 1975, and reiterated in his 1982 book, is to use the expression "fractal dimension" as a generic term applicable to all the variants described in section III above, and to use in each specific case whichever definition is most appropriate. This suggestion is adopted by a number of authors (*e.g.*, Edgar, 1990). However, it could, potentially, lead to considerable confusion if it is followed inconsistently, particularly in cases where different dimensions assume different values (cf section III for examples). Therefore, many mathematicians consider it safer to refer to specific dimensions by name, instead of using the

[3] It should however be kept in mind, as stressed by Herzfeld (1993), that self-similarity cannot be the *sole* defining characteristic of fractals. A straight line segment is exactly self-similar, yet it hardly qualifies as a fractal!

[4] However, graphs like that of Figure 17 indicate that specific geometric features of fractals, such as perimeter length or surface area, depend strongly on observation scale. Instead of "scale-invariance" one might therefore consider, with Nottale (1993), that the expression "scale-covariance" captures better the essence of fractals. However, exploration of this notion of "scale-covariance" is beyond the scope of the present chapter.

generic term "fractal dimension" (*e.g.*, Falconer, 1990).

IV.C "Natural" versus Mathematical Fractals

Can fractals, defined as in subsection IV.B, serve as appropriate representations of natural objects or processes? The answer to this question is (surprisingly perhaps yet uncompromisingly) no. Strictly speaking, **there are no true (mathematical) fractals in nature** (*e.g.*, Falconer, 1990; Meakin, 1991, p. 319). Nevertheless, under specific conditions, physical objects, like soil aggregates or clouds, may have features that are very accurately described by fractals or prefractals. A similar situation pertains with other geometrical structures; there are no true straight lines or circles in nature, yet one would hardly claim that these concepts have absolutely no use in describing nature!

Two of the ways in which true fractals can fail to represent physical objects are illustrated by the attempt to use the Menger sponge as a model of soil aggregates. First, soil aggregates have a porosity strictly smaller than 1 and a mass density different from zero (otherwise there would be no aggregates to speak of!) but the Menger sponge has both porosity equal to unity and zero density (cf subsection II.E). Second, even if soil aggregates exhibit self-similarity over a range of observation scales (*e.g.*, in thin sections observed under the microscope at different magnifications), this self-similarity eventually disappears if the aggregates are viewed at sufficiently small observation scales (*e.g.*, in the extreme, at subatomic scales) or, at the other end of the spectrum, at observation scales commensurate with the size of the aggregate itself. By contrast, the Menger sponge has a fine structure at arbitrarily small scales.

For these two reasons, the Menger sponge cannot serve as a model of soil aggregates. In discussing Chepil's (1950) results, we have seen however that the prefractals associated with the Menger sponge (*i.e.*, the sets of points obtained at intermediate steps in the iterative procedure which leads *ad infinitum* to the Menger sponge) have properties that are closely related to those of real aggregates (cf Figure 16). These prefractals have a porosity < 1 and a mass density > 0. Analysis of other examples, such as coastlines, clouds or landscapes can lead to the same conclusion: fractal geometry is never an exact description of nature and often the geometry of a prefractal is a closer approximation to a physical object than is its associated fractal. Although physical systems are commonly referred to as "natural fractals" (and we shall uphold that usage in the following to conform to standard usage), the expression "natural prefractals" would probably be far more appropriate.

When describing the geometry of a given soil aggregate with a prefractal of the Menger sponge, one has to decide on the iteration step with which this prefractal is associated. To this iteration step corresponds a dimension r_n (the individual subcube size), which represents the lower value of the range of scales at which the prefractal exhibits self-similarity. This length, when related to natural fractals, is usually referred to as the *inner cutoff length* (*e.g.*, Meakin, 1991). Physically, in the case of soil aggregates, this length is associated with the size of individual particles. On the other hand, since a soil aggregate is necessarily of finite size, there must be an upper limit to the range of scales at which it may be observed. For a given system, this upper limit, referred to as the *upper-* or *outer cutoff length*, may be the actual size of the system itself. However, it may be more accurate to consider that it corresponds to the largest scale at which the system displays self-similarity.

The existence of an inner cutoff length has important consequences with respect to the evaluation of the dimensions of natural fractals. All the dimensions described in section III, except the similarity dimension, require a passage to a limit (*e.g.*, limit to vanishingly small ball radius, box side or divider length). For physical reasons, this taking of the limit is precluded for natural fractals. In consequence, the definitions of these dimensions have to be slightly modified when one applies them to natural fractals. For example, the box-counting dimension D_{BC} can no longer be defined as in Equation (20). At best, one would be able to determine the number of boxes $N_\delta(F)$ only in the range of δ values between the inner- and outer cutoff lengths. In practice, the range of δ values accessible via measurement may be much narrower. Under these conditions, Equation (20) may best be replaced by defining $D_{BC}(F)$ as the slope of the graph of $\ln N_\delta(F)$ vs $-\ln\delta$ over a sufficiently large range of δ or, equivalently, by finding a best-fit value of D_{BC} in the power-law relationship $N_\delta(F) \propto \delta^{-D_{BC}}$.

A direct consequence of this necessary change in the definition of the various dimensions is the fact that equalities or inequalities between dimensions that have been proven mathematically in the limit of vanishingly small ball radius, grid size or divider length, may no longer be valid when such limits are not taken. Furthermore, the finite length of the "yardsticks" may create serious practical difficulties. One of the key ones associated with the divider method (cf section III.D), for example, relates to the existence of a remainder. This remainder stems from the fact that, most often, a non-integer number of steps is required to cover a given curve. At present there is no general consensus on how to deal with this remainder (*e.g.*, Klinkenberg, 1994).

In addition to the above problems, the existence of an inner cutoff length raises the question of what is "fractal" in natural fractals. To understand this point, it is useful to take once again the example of the Menger sponge. As defined in section II.E, the Menger sponge is a set of points in R^3 with (Lebesgue) measure equal to zero. One may also consider that there is another set of points that is closely associated with this first one: the points that were removed during the iterative attrition process leading to the Menger sponge. In other words, to the "solid" structure of the Menger sponge is associated a volume of "voids". If one decides to interrupt at a certain level the iterative process leading to the Menger sponge, the "solid" structure (no longer of measure equal to zero!) and the voids now have an interface. This feature is common to many natural fractals. In general, if the interface is fractal and scales like the mass of solids (*i.e.*, follows a power law, or Paretian relationship, and has the same power-law, or Pareto-, exponent), the system is termed a *mass fractal* (*e.g.*, Pfeifer and Obert, 1989). If void- (or pore-) space and surface happen to scale alike, the system is called a *pore fractal*, whereas if only the surface is fractal, the system is called a *surface fractal*, or *boundary* fractal. In each case, one may use to characterize these fractals the dimensions introduced in section III, leading to, *e.g.*, "box-counting pore fractal dimensions" or "correlation surface fractal dimensions". This important distinction between various types of natural fractals is mentioned in several chapters in the present book.

V Is "Power Law" Equivalent to "Fractal"?

V.A Power Law or Paretian Distribution and Fractals

Repeatedly in the preceding sections, the analysis of the geometrical properties of mathematical and "natural" fractals has resulted in power-law or Paretian relationships between selected parameters. At no point in the text has this existence of a Paretian relationship been presented as a defining characteristic of fractals; it was instead obtained as a *consequence* of fractal geometry. A number of authors, however, consider that the essence of fractals is not the notion of an underlying geometry, as predicated above, but the evidence of Paretian behavior. From this viewpoint, a fractal is a set for which some statistical distribution function is a power law (*e.g.*, Turcotte, 1992). Some authors (*e.g.*, Crovelli and Barton, 1995) label the latter fractals as *probabilistic*, to distinguish them from the *geometric* fractals, defined in section IV.B.

To understand clearly the connections between these two types of fractals, it is worth reviewing briefly the development of the Pareto distribution and its relation to other statistical distributions in common use.

The Pareto distribution is named after the Italian-born Swiss professor of economics, Vilfredo Pareto (1848-1923). Originally, it dealt with the distribution of income over a population and may be stated as follows (Pareto, 1897; Mandelbrot, 1960, 1963; Arnold, 1983; Montroll and Schlesinger, 1983; Persky, 1992):

$$N = Ax^{-a} \tag{26}$$

where N is the number of persons having income $\geq x$. A and a are positive, real parameters. The relation of Equation (26) is now usually referred to as the "Pareto distribution of the first kind", to distinguish it from alternative forms (*e.g.*, Johnson and Kotz, 1970, p. 234). When applied to discrete data (*e.g.*, the length of words), it is also often referred to as the Zipf distribution (Arnold, 1983; Crovelli and Barton, 1995). In the following, we shall simply call it the Pareto distribution.

When N is plotted as a function of x, for given values of A and a, the distribution of Equation (26) is characterized by a very long right tail, or Paretian behavior. Over the years, this behavior has been observed in relation with many socio-economic and other naturally occurring quantities. Examples (*e.g.*, Zipf, 1949; Johnson and Kotz, 1970) are the distributions of city population sizes, insurance claims, occurrence of natural resources, stock price fluctuations, size of firms, and of error clusters in communication circuits, to list only a few.

Usage of the Pareto distribution to describe data exhibiting very long right tails has been criticized by many researchers on the grounds that the Pareto distribution is not the only distribution with a very long right tail and that, often, it does not convincingly outperform its competitors. Macaulay (1922), in particular, argues that "the approximate linearity of the tail of a frequency distribution charted on a double logarithmic scale signifies relatively little, because it is such a common characteristic of frequency distributions of many and various types". Indeed, in many cases, the exponential (Brown *et al.*, 1983; Korvin, 1992), Weibull (Turcotte, 1992; Korvin, 1992) and lognormal (Crovelli and Barton, 1995) distributions mimic the Pareto distribution over certain ranges, or even provide a statistically better fit to data than the Pareto distribution (*e.g.*, Turcotte, 1986; Montroll and Schlesinger, 1983).

In that context, whether or not one uses the Pareto distribution to fit experimental data often appears linked to one's belief in the universality of the Paretian behavior, and to the deviation from Paretian behavior that one is willing to tolerate. In some well-publicized cases of the use of the Pareto distribution in the literature (*e.g.*, Figure 4.7 in Crovelli and Barton, 1995, and discussion in Korvin, 1992, p. 205), there is enough deviation from linearity in log-log plots, and the deviation is systematic enough over the whole range of experimental data, that scepticism in the applicability of the Pareto distribution is warranted. Yet the authors of the original articles consider a unique Pareto exponent appropriate, implying that the systems under study exhibit power-law behavior. A related issue concerns the method one uses to fit the Pareto distribution to experimental data. There is a whole array of methods available, including weighted- or unweighted linear regression of log-transformed data, and nonlinear regression. Each of these methods may be applied to the full data set, or to a portion thereof, corresponding to points that are selected because they appear to fall on a straight line in a log-log plot. When the latter approach is adopted, one is faced with the problem of determining how big a range this linear segment should span to provide any confidence that the data indeed exhibit a power-law behavior. Brock (1971) argues that data points showing linearity should span at least two or three orders of magnitude in the abscissa before one tries to fit to them the Pareto distribution. A similar recommendation is made by other authors (*e.g.*, Korvin, 1992).

Figure 18 illustrates some of the practical difficulties frequently encountered when trying to determine if the Pareto distribution adequately describes experimental data, and when evaluating its parameters. Originally, Cargill *et al.* (1981) fitted a power law to their data via weighted linear regression, after elimination of the leftmost five data points (considered to be artifacts resulting from a less efficient extraction of copper from the ore). The weighting is proportional to the amount of ore. Cargill *et al.*'s (1981) regression line (dashed line in Figure 18) has a slope of -0.386, which, according to a formula derived by Turcotte (1992), corresponds to a fractal dimension D of 1.16. If an unweighted nonlinear regression is carried out on the same truncated data set, the slope of the resulting regression line (thin solid line in Figure 18) equals -0.406 (*i.e.*, $D = 1.22$), with an R value of 0.994. Finally, when the Pareto distribution

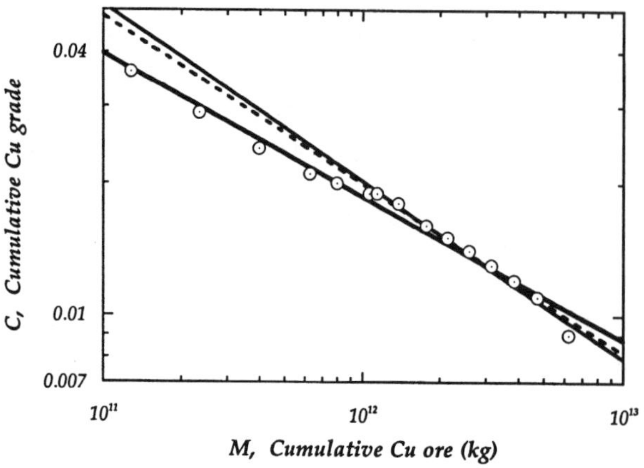

Figure 18: Dependence of cumulative copper grade C on cumulative copper ore M mined in the United States. (Data from Cargill *et al.*, 1981.) The meanings of the different lines are described in the text.

is fitted via an unweighted nonlinear regression to the full data set (thick solid line in Figure 18), the slope equals -0.327 (*i.e.*, $D = 0.98$). There is some waviness in the data, but the fit is quite close ($R = 0.995$), and is comparable to that found in the literature in situations where power-law behavior has been postulated. A consequence of this good fit is that the first 5 data points no longer appear artifactual or deviatory, as assumed by Cargill *et al.* (1981) and Turcotte (1992). However, the most interesting consequence is in terms of the fractal behavior of the cumulative copper grade C. If one uses the truncated data set to estimate D and one follows Brock's (1971) recommendation, C is *not* a probabilistic fractal, since its linear segment extends over less than one order of magnitude in M. On the other hand, if one considers the full data set, the linear segment spans very nearly two orders of magnitude in M and one may at least consider making the case that C is a probabilistic fractal.

A similar analysis could be performed on many data sets described in the literature (*e.g.*, in Turcotte, 1992). This clearly highlights the fact that the evaluation of fractal dimensions, whether of geometric- or probabilistic fractals, is not straightforward. It also shows that there is often some subjectivity involved in determining whether or not a given system has a Paretian behavior. These observations affect the practical use of both the geometric- *and* the probabilistic fractals. However, the subjectivity is particularly significant for probabilistic fractals, since evidence of Paretian behavior is their *sole* defining feature.

A further difference between geometric- and probabilistic fractals concerns the range of values that are acceptable for the Pareto exponent a in Equation (26). In the various illustrations of Paretian relationships associated with mathematical monsters and fractals in preceding sections of the present chapter (*e.g.*, Equations (2) and (4)), the value of the exponent was always constrained to be (inclusively) between the topological dimension of the set and the Euclidian dimension of the embedding space (R^n). No such restriction applies to the parameter a in Equation (26) when this equation is taken simply as the mathematical formulation of a particular statistical distribution. Clearly, when Equation (26) is applied to data related to the size of insurance claims or to stock price fluctuations, one cannot expect a to be restricted to a certain range (*e.g.*, $0 \leq a \leq 3$) for geometrical or physical reasons. Even when Equation (26) is used to describe some physical attribute of a real system, it is not guaranteed that either A or a, or even x for that matter, will have a clear physical meaning and that the values of a will be geometrically constrained. Some of the most convincing illustrations of this fact are found in applications of dimensional analysis to equations describing dynamical systems. Indeed, the so-called π-theorem, one of the cornerstones of dimensional analysis, states that if there is a relationship among $N+1$ variables involving N independent dimensional units, the relation can be expressed in terms of a single nondimensional parameter, giving a power-law dependence (Schmidt and Housen, 1995). This single nondimensional parameter, and consequently the coefficient and exponent of the power law, may not always have a clear physical interpretation, as is illustrated by Schmidt and Housen (1995) in the case of the cratering efficiency of conventional and nuclear explosives. In all these situations, and indeed for all probabilistic fractals, the range of values that the exponent a can assume is not restricted. Cases of physical systems where this exponent (after transformation whenever necessary) leads to "fractal dimensions" larger than 3 are routinely reported in the literature (*e.g.*, Brock, 1971; Turcotte, 1992; Turcotte and Huang, 1995; Van Damme, 1995). Such values are entirely acceptable for probabilistic fractals (*e.g.*, Brock, 1971; Matsushita, 1985), even though they would be meaningless for geometric fractals embedded in R^3. This distinction appears to have been frequently overlooked in the literature, as evinced by the lasting, yet groundless, controversy about supposedly "unphysical" fractal dimensions larger than 3 for natural fractals (*e.g.*, Avnir *et al.*, 1992).

Because of the constraints imposed on the exponent a in the case of geometric fractals, it is clear from the above analysis that whereas a geometric fractal is automatically also a probabilistic fractal, the reverse is not true in general. In soil science, this conclusion has particular significance with respect to the use of fragmentation models to account for the size distribution of soil particles (*e.g.*, Van Damme, 1995) or fragmented rocks and other geological materials (*e.g.*, Turcotte, 1992). These fragmentation models are described in the following subsection.

V.B Fragmentation Fractals

The earliest attempt to develop a fragmentation model that accounts for the Paretian distribution of the size of fractured or fragmented solids seems to have been Matsushita's (1985) fracture cascade model (cf Figure 19). The simplest derivation of this model, in two-dimensional space, starts with a square with a unit side length. At the first stage, the square is divided into four equal subsquares of side $1/2$. One of these subsquares, chosen randomly, is hatched

Figure 19: Fifth iteration in Matsushita's fracture cascade in two dimensions. The hatched squares are arranged regularly for convenience. (From Matsushita, 1985. Reprinted with permission.)

to indicate that it will not be fractured or fragmented further. At the second stage, the remaining 3 subsquares are divided each into 4 equal sub-subsquares of side $(1/2)^2$, and one of them (chosen randomly) is again hatched. The same procedure is in turn applied to the remaining 3^2 unhatched sub-subsquares, and so on.

As is easily seen, at the nth stage, there are 3^{n-1} newly hatched squares of side $r_n = 2^{-n}$. For finite $n \gg 1$, the cumulative number $N(r_n)$ of hatched squares of side length greater than $r_n = 2^{-n}$ may be expressed as follows (Matsushita, 1985):

$$N(r_n) = 1 + 3 + 3^2 + ... + 3^{n-1} \propto 3^n = r_n^{-D_{fragm}} \tag{27}$$

where $D_{fragm} \equiv \ln 3/\ln 2 \cong 1.585$.

Equation (27) amounts to a power-law relationship between $N(r_n)$ and r_n, so, as $n \to \infty$, it defines a probabilistic fractal. Following Kaye (1989), it is occasionally called a fragmentation fractal, and D_{fragm} is termed the fragmentation fractal dimension.

There is an interesting connection between Matsushita's (1985) fracture cascade model and some of the mathematical monsters (or geometric fractals) of section II. For example, if instead of dividing the initial square in Figure 19 into 4 equal subsquares and hatching one, we divide it into 9 equal subsquares and hatch one, it is easy to see that the collection of hatched squares that we are producing in this manner corresponds to the square holes that are punched during the attrition process leading to the Sierpinski carpet (cf section II.D). The exponent D_{fragm} of Equation (27), under these conditions, is equal to $\ln 8/\ln 3 \cong 1.89$, which is the value of the Hausdorff- and similarity dimensions of the Sierpinski carpet itself (cf section III.B).

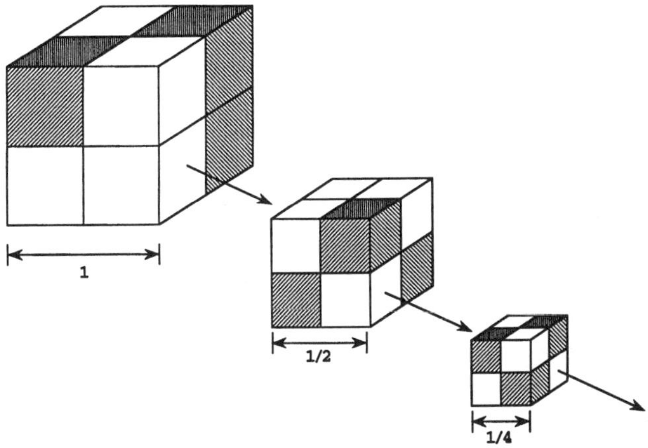

Figure 20: Schematic representation of a fracture cascade in 3 dimensions. Three cubes are hatched (*i.e.*, not fragmented further) at each fragmentation step.

The fracture cascade has been considered so far in two-dimensional space. To be applicable to the fragmentation of solid bodies, the model has to be extended to R^3. In this case (cf Figure 20), we could start with a cube with unit side length, divide it into 8 subcubes, hatch three, apply the same procedure to each of the remaining 5 subcubes, and so on. The resulting value of D_{fragm} would be $\ln 5/\ln 2 \cong 2.322$. More generally (*e.g.*, Korvin, 1992), one could consider that the initial cube is divided into b^3 equal-sized subcubes of side length $1/b$, and that i randomly chosen subcubes are hatched, where b and i are arbitrary integers satisfying $1 < b$ and $1 \leq i < b^3$. At the nth iteration of this procedure, the cumulative number of hatched cubes larger than $r_n = 1/b^n$ is:

$$N(r_n) = i\{1+(b^3-i)+(b^3-i)^2+.....+(b^3-i)^{n-1}\} \propto (b^3-i)^n = r_n^{-D_{fragm}} \tag{28}$$

where $D_{fragm} \equiv \ln(b^3 - i)/\ln b$.

The power-law relationship of Equation (28) may be expressed differently if one introduces a "probability of fragmentation" $p_c \equiv \left(b^3 - i\right)/b^3$, representing the fraction of the nth order cubes of size b^{-n} that are further fragmented. Under these conditions, D_{fragm} in Equation (28) becomes:

$$D_{fragm} = \frac{\ln\left(b^3 p_c\right)}{\ln b} \tag{29}$$

The inequality $1 \leq i < b^3$ (cf above) implies that $\frac{1}{b^3} \leq p_c < 1$, *i.e.*, that $0 \leq D_{fragm} < 3$. This restriction on the range of values taken by D_{fragm} is a definite sign that the model described above leaves much to be desired as a "physical" explanation of the general Pareto distribution, since values of a in Equation (26) as high as 4 or 5 are routinely measured in practice (*e.g.*, Turcotte,

1992; Van Damme, 1995). On the other hand, the inequality $D_{fragm} < 3$ has been considered by some to be an indication that geometric fractal concepts are appropriate to describe any fragmented material (Turcotte, 1992). This viewpoint deserves further analysis, since it assumes that every fragmented material was created by a process similar to that shown in Figures 19 and 20. If fragments or particulate matter are characterized using Equation (26), the exponent a characterizes only the size of the individual fragments. It provides **no** quantitative insight concerning the geometry of the arrangement of these same fragments. By piling up or aggregating the fragments in specific ways, it may be possible to create with them various pore-, mass- or surface fractals. There are probably infinitely many ways to do this. However, there are also many ways to assemble the fragments to create non-fractal structures. One such structure is in fact illustrated in Figure 20. Indeed, it is always possible to assemble the fragments resulting from a given fracture cascade in such a way as to reconstitute the original cube, which is definitely not fractal! Arguments that the dimension found when the size distribution of a fragmented material is characterized using Equation (26) should obey the inequality $a < 3$ (because $D_{fragm} < 3$) are, thus, without basis.

VI The Physics of Fractals

VI.A Going Beyond Fractal Dimensions

In Figure 16, we saw that the bulk density of soil aggregates decreased systematically as their radius increased. Using a formula provided by Turcotte (1992), one may associate with this behavior a particular value (= 2.88) of the "fractal dimension". If this value is shared by the aggregates of most soils, it may have some interest purely as an empirical parameter. It is only descriptive, so its usefulness is likely to be somewhat limited, particularly if it is not constant but varies from soil to soil. It would be much more satisfying from a practical standpoint if, armed with this fractal dimension, one could predict other features of soil aggregates, besides their bulk density, for example their chemical reactivity, their dielectric properties or their deformability as a function of aggregate radius. Later chapters in this book provide illustrations of the use of a fractal dimension to predict properties different than that used initially to evaluate the fractal dimension. This type of bridging will undoubtedly remain an important area of research in the future.

Another challenging, and very satisfying, way to proceed beyond the mere evaluation of fractal dimensions is to try to explain fractal properties in physical terms, *i.e.*, to find an answer to such questions as "Why do projections of clouds have perimeters of dimension 1.35 over a very wide range of scales?", "What are the geological processes that lead to a landscape of dimension 2.2?" (Falconer, 1990, p. 266) or "Why do the interstices in porous media tend to be fractal rather than smooth?" (Korvin, 1992, p. 306). To answer these questions, one needs to devise some sort of mechanism that explains natural phenomena, and to come up with a mathematically manageable model that, ideally, should be predictive as well as descriptive.

In the case of probabilistic fractals, we encountered in section V.B the embryo of a model dealing with fragmented materials. To acquire full predictive potential, however, this embryonic model would in addition need to involve physically-motivated fragmentation probabilities. In a different context, Brock (1971) accounts for the power-law distribution of atmospheric aerosols via a physical process of "condensational growth". Other phenomena or processes that lead to probabilistic fractals are chaotic dynamics (Crilly, 1991; Schroeder, 1991; Moon, 1992; Peitgen *et al.*, 1992; Turcotte and Huang, 1995) and self-organized criticality (Bak *et al.*, 1987; Bak and Chen, 1989, 1991; Hwa and Kardar, 1989). Reviewing them in detail is beyond the scope of this chapter.

Various processes and phenomena have also been advocated to explain the properties of geometric fractals and a number of mathematical models have been developed to simulate these processes. Those generally referred to as "cluster growth"- or "aggregation" models are of particular significance in soil science, because of their relevance to clay aggregation, fingering processes and invasion percolation (cf later chapters in present book). Following the introduction of the Diffusion-Limited Aggregation (DLA) model by Witten and Sander (1981), many aggregation models have been developed (*e.g.*, Meakin, 1991; Batty *et al.*, 1993), such as the dielectric breakdown model (DBM), the direct screening models (DSM), the potential field growth model (PFGM), and models involving (simple) cluster-cluster aggregation, diffusion-limited cluster-cluster aggregation, ballistic cluster-cluster aggregation, or reaction-limited cluster-cluster aggregation processes. In the following subsection, a short description of DLA is provided, primarily to illustrate some of the key features of these models and to give an idea of the type of fractal structure that they produce.

VI.B Illustration: Diffusion-Limited Aggregation (DLA)

The *diffusion-limited aggregation* model, in its most primitive formulation, is based on a lattice of small squares. A square is designated (solid square in Figure 21a) to represent the original seed or growth site of the cluster or aggregate. A large "maximum cluster radius" circle is drawn centered on this square, with a radius $r_{\max}$ set at some arbitrary value. An even larger concentric "launching" circle is drawn, with a radius larger than $r_{\max}$. Then a particle is released from a random square which intersects the launching circle, and is allowed to perform a random walk on the square lattice, with probability 1/4 each of moving left, right, up or down to a neighboring square. Eventually, this random walker reaches a square next to the original seed or next to a shaded square, in which case that square is also shaded (cf particle with trajectory t_1 in Figure 21a). Alternatively, it reaches the "killing circle" and its trajectory is terminated (t_2 in Figure 21a).

As this process is repeated a large number of times, a connected set of squares grows outward from the original one. Running the model for thousands

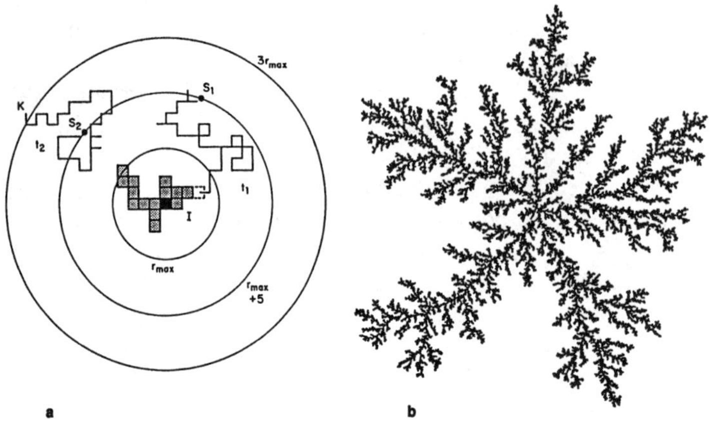

a b

Figure 21: (a) Schematic representation of a simulation of diffusion-limited aggregation on a square lattice. The innermost circle, with radius $r_{\max}$, corresponds to the maximum cluster size. The intermediate circle is the "launching" circle, and the outer one is termed the "killing circle". (From Meakin, 1991. Reprinted with permission.) (b) Computer realization of diffusion-limited aggregation. The cluster contains 50,000 particles and results from an off-lattice simulation. (From Feder, 1988. Reprinted with permission.)

of shaded squares gives a highly branched cluster (cf Figure 21b) that resembles a number of aggregation patterns found in nature or in laboratory experiments (*e.g.*, Meakin, 1991). Main branches radiate from the initial point and bifurcate as they grow, giving rise to subsidiary side branches, all tending to grow outwards. These structures appear geometrically self-similar. In general, however, they cannot be mapped exactly onto themselves after a change of length scales; only their asymptotic *statistical* characteristics are invariant under a change of scale. DLA clusters are therefore examples of statistically self-similar natural fractals.

Analysis of simulation results has shown that for clusters such as the one shown in Figure 21b, the growth of the cluster's radius of gyration, R_g, can be described quite well by the following power-law relation:

$$R_g = s^{\beta} \tag{30}$$

where s is the number of particles in the cluster. From this algebraic dependence of R_g on s, an effective fractal dimension or *cluster fractal dimension* D_β can be evaluated, defined by $D_\beta \equiv 1/\beta$ (*e.g.*, Meakin, 1991). For clusters in R^2, D_β has been found generally to be about 1.55 when the random walkers follow the lattice and about 1.71 when particles follow off-lattice random walk trajectories (Stanley et al., 1990). It is about 2.4 to 2.5 for three-dimensional DLA clusters.

Some interesting extensions of the simple DLA model above have been proposed (*e.g.*, Meakin, 1988; Peitgen *et al.*, 1992). First, one may introduce a sticking probability for the particles.[5] In addition, instead of following a single random walker at a time, many particles may be considered simultaneously. Moreover, the cluster may be allowed to move about, picking up particles that are close by. There is also another, seemingly unrelated DLA model. Instead of following individual particles, an equation is solved which reflects the fact that in actual diffusion situations there is a near infinity of particles moving about simultaneously. Thus in place of particles, a continuous density function is considered and it is assumed to be governed by a partial differential equation, the Laplace equation. It is possible to introduce a parameter in this equation to control the dimension of the cluster.

VII Multifractal Measures

The introduction to fractal geometry in the present chapter would not be complete without a short mention of an area that is conceptually challenging[6], yet is the object of considerable interest in the literature: multifractal measures.

As with fractals, many authors try to get by without having to provide a precise definition of multifractal measures. Consequently, the term "multifractal measure" often means different things to different people. At the same time, different terminologies are used to refer to the same concept; "multifractals" is often used in lieu of, or interchangeably with, the expression "multifractal measures". In the following, we shall use consistently the term "multifractal measures" to underline the fact that, unlike geometric (mono)fractals, which are self-similar sets of points, multifractal measures are self-similar *measures* defined on specific sets of points.[7]

VII.A The Binomial Fractal Measure

The self-similarity of measures is best illustrated with the binomial measure (*e.g.*, Mandelbrot, 1989; Moon, 1992; Evertsz and Mandelbrot, 1992) which is generated recursively via a process known as a *multiplicative cascade*. This process starts (at the stage $k = 0$) with a uniformly distributed unit of mass on the unit interval $I = [0, 1]$ (cf Figure 22a). At the next stage ($k = 1$), a fraction m_o of the mass is distributed uniformly on the left half ($I_0 = [0, 1/2]$) of the unit interval, and the remaining fraction $m_1 = 1 - m_0$ is distributed uniformly on the right half ($I_1 = [1/2, 1]$). At this stage, the left half thus carries the measure $\mu(I_0) = m_0$ and the right half carries the measure $\mu(I_1) = m_1$. Of course, the total mass is conserved, so that $\mu(I_0) + \mu(I_1) = \mu(I) = 1$. For that

[5] This is the probability that a particle which encounters the cluster actually sticks to it, and this probability has in general no influence on the dimension D_β of the cluster. This feature reveals a relative insensitivity of the cluster growth process to microscopic details, a property referred to as "universality".

[6] Korvin (1992) humorously comments that multifractal measures are not for the squeamish!

[7] This definition of multifractal measures follows the treatment of Mandelbrot (1988), Falconer (1990) and Evertsz and Mandelbrot (1992). Other definitions will be briefly mentioned at the end of this section.

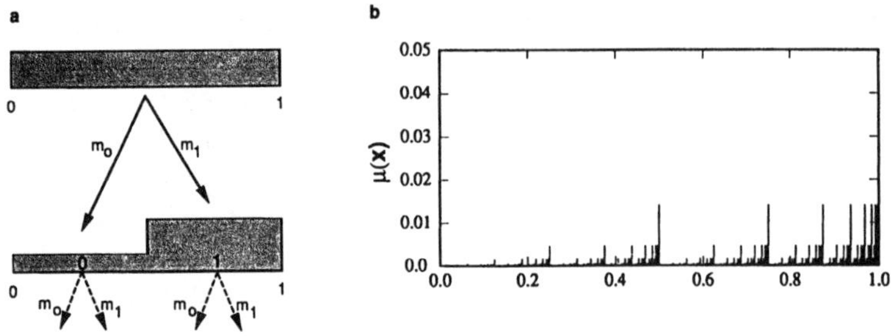

Figure 22: (a) First two stages ($k = 0$ and $k = 1$) of the multiplicative cascade leading to the binomial measure, with $m_o = 0.25$ and $m_1 = 0.75$. (b) Resulting measure $\mu(x)$ for $k = 11$, as a function of the segment position $x = i \cdot 2^{-11}$. (Modified from Feder, 1988.)

reason, the μ's appear like probabilities and could be considered as probability measures (*e.g.*, Evertsz and Mandelbrot, 1992).

At the next stage ($k = 2$) of the multiplicative cascade, the subintervals I_0 and I_1 receive the same treatment as the original unit interval. For example, I_0 is split into two intervals $I_{00} = [0, 1/4]$ and $I_{01} = [1/4, 1/2]$ of size 2^{-k}, and the mass is also fragmented as it was at stage $k = 1$. It is easy to see that, *e.g.*, $\mu_{01} \equiv \mu(I_{01}) = m_0 m_1$, with similar relations for the other measures μ_{00}, μ_{10} and μ_{11}. In other words, the measure associated with a given interval is obtained by multiplication of suitable combinations of the measures m_0 and m_1, hence the adjective "multiplicative" that characterizes the cascade process.

At the kth stage, the measure of a given interval, say $I_{\beta_1 \beta_2 ... \beta_k}$, is given by

$$\mu_{\beta_1 \beta_2 ... \beta_k} = m_{\beta_1} m_{\beta_2} ... m_{\beta_k} = m_0^{n_0} m_1^{n_1} \tag{31}$$

where n_o is the number of β_i's equal to 0 and $n_1 \equiv k - n_0$. For example, the "peak" just to the left of $x = 0.5$ (in fact, in the interval $[0.5 - 2^{-11}, 0.5]$) in Figure 22b corresponds to a measure $\mu(x)$ equal to $(0.25) \cdot (0.75)^{10} \cong 0.0141$, three times smaller than that at $x = 1$ where $\mu(x) = (0.75)^{11} \cong 0.0422$ (the latter peak is hidden by the right ordinate axis). Even though it is not possible to count them all on Figure 22b, there are 11 peaks with a measure equal to that of the peak near $x = 0.5$. In general, if we write $\xi \equiv n_0/k$, the number of peaks with measure $\mu = \left(m_0^{\xi} m_1^{(1-\xi)}\right)^k$ is given by the number of ways one can distribute $n_0 = \xi k$ zeros among k positions in the addresses $\beta_1 \beta_2 ... \beta_k$ of the intervals $I_{\beta_1 \beta_2 ... \beta_k}$. This number of peaks is given by the binomial coefficient

$$N_k(\xi) = \binom{k}{\xi k} \equiv \frac{k!}{(\xi k)!((1-\xi)k)!} \tag{32}$$

which explains why the term *binomial* is applied to the binomial measure (*e.g.*, Evertsz and Mandelbrot, 1992, p. 929).

The limit of this process, as $k \to \infty$, called the binomial measure, is exactly self-similar. The mass of an arbitrary interval $I_{\beta_1 \beta_2 ... \beta_k}$ is $\mu_{\beta_1 \beta_2 ... \beta_k}$ times smaller than that, equal to 1, of the entire unit interval I. But the distribution of mass within these intervals $I_{\beta_1\beta_2\ldots\beta_k}$ and I is distributed in exactly the same way. Indeed, by spatially rescaling the subinterval $I_{\beta_1\beta_2\ldots\beta_k}$ by a factor 2^k and renormalizing its mass by a factor $(\mu_{\beta_1\beta_2\ldots\beta_k})^{-1}$, one recovers the mass distribution in the interval I. It is in this sense that the multiplicatively-generated binomial measure is said to be (exactly) self-similar.

Self-similar measures like the binomial measure are "singular", *i.e.*, their local density, defined as $\lim_{\epsilon\to 0}\mu[x, x+\epsilon]/\epsilon$, is (almost) everywhere undefined. To illustrate this, let us consider the value of the binomial measure in the neighborhood of the point 0. Because of Equation (31), $\mu[0, 2^{-k}] = m_0^k = \left(2^{-k}\right)^{v_o}$ with $v_0 = -\log_2 m_0$. That is, the measure in the neighborhood of 0 scales as $\mu[0,\epsilon] \propto \epsilon^{\alpha}$, where $\alpha = v_o$ is the coarse Hölder exponent when ϵ is of finite length (cf section III.E). The (non-local) density μ/ϵ, consequently, scales like $\epsilon^{\alpha-1}$, and if $\alpha \neq 1$ (*i.e.*, if $m_0 \neq \frac{1}{2}$), the local density (the limit of the μ/ϵ as $\epsilon \to 0$) is degenerate, equal to either 0 (if $m_0 < \frac{1}{2}$) or ∞ (if $m_0 > \frac{1}{2}$). This is true not only in the neighborhood of 0; the local density of the binomial measure is either 0 or ∞ almost everywhere in the interval [0,1].

VII.B Parameterization of Multifractal Measures

For multifractal measures to be useful in practice, we need to find ways to characterize and parameterize their geometrical properties. This is achieved in particular by the "$f(\alpha)$ curve", whose theoretical foundation is now described. Evertsz and Mandelbrot (1992) and Lavallée *et al.* (1993) describe various methods for its evaluation. By simple arithmetic (*e.g.*, Evertsz and Mandelbrot, 1992, p. 932), one may show that, at stage k of the construction of the binomial measure the coarse Hölder exponent associated with a given interval $I_{\beta_1\beta_2\ldots\beta_k}$ is:

$$\alpha = \xi v_0 + (1-\xi)v_1 \tag{33}$$

where $v_1 = -\log_2 m_1$ and, as above, $v_0 = -\log_2 m_0$, $\xi \equiv n_0/k$, and n_0 is the number of β_i values which equal zero. If, as in Figure 22, $m_0 < m_1$, then $v_1 < v_0$ and, by virtue of Equation (33), $v_1 \leq \alpha \leq v_0$. The extreme values of α are usually denoted by $\alpha_{\min}$ and $\alpha_{\max}$, so that $\alpha_{\min} \equiv v_1 \leq \alpha \leq v_0 \equiv \alpha_{\max}$. The limits $\alpha_{\min}$ and $\alpha_{\max}$ are independent of the size 2^{-k} of the subintervals at stage k, *i.e.*, are independent of the scale at which the measure is probed. The fact that the value of α is bound within a fixed range makes it an ideal index to associate with the intervals within the support of the binomial measure, i.e., within the geometrical domain in which the measure assumes non-zero values.[8]

From that viewpoint, it is interesting to inquire about the values that the coarse Hölder exponent assumes over the support of the measure and about how these values are distributed. For a given k, this distribution is given by the

[8] Of course, given the one-to-one relationship between α and ξ in Equation (33) and the fact that the values of ξ are bound (within [0,1]) like those of α, one could conceivably also use ξ as an index (*e.g.*, Feder, 1988). However, ξ is strongly tied to the concept of multiplicative cascade, which is a given in the case of the binomial measure but does not necessarily exist for other measures (Evertsz and Mandelbrot, 1992). By contrast, α may be evaluated for *any* measure.

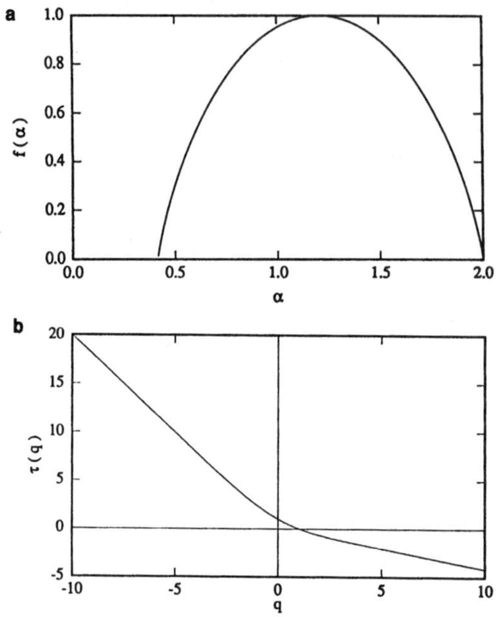

Figure 23: (a) $f(\alpha)$ curve for the binomial measure with $m_o = 0.25$ and $m_1 = 0.75$. (b) Sequence of mass exponents $\tau(q)$ as a function of moment order q, for the binomial measure with $m_o = 0.25$ and $m_1 = 0.75$. (Both figures modified from Feder, 1988.)

binomial coefficient $N_k(\xi)$ of Equation (32). Because of Equation (33), each ξ corresponds to a unique α. Therefore the number of intervals I_k of size 2^{-k} with coarse Hölder exponent α is given by (Evertsz and Mandelbrot, 1992):

$$N_k(\alpha) = \begin{pmatrix} k \\ \xi(\alpha)\, k \end{pmatrix} \propto \left(2^{-k}\right)^{-f(\alpha)} \tag{34}$$

where $f(\alpha)$ equals $\log_2[\xi^{-\xi}(1-\xi)^{-(1-\xi)}]$ with, based on Equation (33), $\xi = (\alpha - \alpha_{\min})/(\alpha_{\max} - \alpha_{\min})$.

The proportionality relationship in Equation (34) is an approximation based, in part, on the application of Stirling's formula to the factorials in the binomial coefficient. Although the coarse Hölder exponent α has meaning only for finite k, as k becomes very large $N_k(\alpha)d\alpha$ is approximately "the number of intervals with a coarse Hölder exponent between α and $\alpha+d\alpha$". (Evertsz and Mandelbrot, 1992, p. 935) Thus the function $f(\alpha)$ provides the information needed to find the frequency distribution of α for k large. The graph of the $f(\alpha)$ curve for the binomial multifractal measure is shown in Figure 23a. Several of the features of this particular curve (*e.g.*, symmetry around its maximum, local quadratic behavior near its maximum) are not generally typical of self-similar measures. Nevertheless, the $f(\alpha)$ curve of Figure 23a shares two characteristics with other $f(\alpha)$ curves: first, $f(\alpha) \leq \alpha$ for all α and, secondly, the maximum value of $f(\alpha)$ corresponds to the box-counting dimension D_{BC} of the support of the measure ($D_{BC} = 1$ for the binomial measure because its support is a line segment).

The $f(\alpha)$ curve is not the only way to characterize the properties of multifractal measures. Another, widely used, approach to describe their features involves sequences of the *mass exponents* $\tau(q)$. To define these exponents, it is convenient to consider a set S covered with δ-mesh "cubes", as in Figure 15 (iii). Within each one of these δ-mesh cubes, one may define a measure μ which is similar to that of Equation (22) and which represents the probability of finding an element (a point) of set S within the δ-mesh cube. With this probability, one may construct, for any real number q, the measure

$$M_d(q,\delta) = \sum_{i=1}^{N(\delta)} \mu_i^q \delta^d \equiv N(q,\delta)\delta^d \underset{\delta\to 0}{\longrightarrow} \begin{cases} 0, \text{ for } & d > \tau(q), \\ \infty, \text{ for } & d < \tau(q) \end{cases} \tag{35}$$

where the summation limit $N(\delta)$ is the total number of δ-mesh cubes that cover the set S, q plays the role of a moment order, and $N(q,\delta)$ is the weighted number of δ-mesh cubes in the cover. In close analogy with the Dedekind cut in Figure 14, the measure $M_d(q,\delta)$ as $\delta \to 0$ approaches zero when d is large and approaches infinity when d is small . The value of d at the division point is called the mass exponent and is denoted $\tau(q)$ to indicate its dependency on the moment order q.

The measure $M_d(q,\delta)$ is characterized by a whole sequence of exponents $\tau(q)$ that controls how the moments of the probabilities $\{\mu_i\}$ scale with δ. It follows from Equation (35) that the weighted number of δ-mesh cubes $N(q,\delta)$ has the form

$$N(q,\delta) = \sum_{i=1}^{N} \mu_i^q \propto \delta^{-\tau(q)} \tag{36}$$

and the mass exponent $\tau(q)$ is given by

$$\tau(q) = -\lim_{\delta\to 0} \frac{\ln N(q,\delta)}{\ln \delta} \tag{37}$$

This relationship between $\tau(q)$ and q is illustrated in Figure 23b for the binomial measure described previously.

If the probabilities μ_i are normalized (*i.e.*, $\sum_{i=1}^{N(\delta)} \mu_i = 1$), then by Equations (36) and (37), $\tau(1) = 0$ (cf Figure 23b). Another interesting special case is $q = 0$ where all the $\mu_i^q = 1$ in Equation (36) and therefore, $N(q = 0, \delta) = N(\delta)$ is simply the number of δ-mesh cubes needed to cover the set, so $\tau(0)$ equals the box-counting dimension of the set. This suggests that the value $q = 0$ corresponds to the value of α for which $f(\alpha)$ is maximum, and that there are therefore "bridges" between the two ways to characterize multifractal measures. In fact, one may show that $\tau(q)$ and $f(\alpha)$ are intimately related via the relationships

$$\alpha(q) = -\frac{d\tau(q)}{dq} \tag{38}$$

and

$$f(\alpha(q)) = q\alpha(q) + \tau(q) \tag{39}$$

This pair of equations in effect constitutes a "Legendre" transformation from the independent variables τ and q to the independent variables f and α and may be very valuable in practical situations, when one of these two sets of independent variables turns out not to be particularly easy to evaluate directly (cf Folorunso *et al.*, 1994).

Another interesting relationship exists between the sequence of exponents $\tau(q)$ and the Rényi dimensions D_q mentioned in section III.G. One can indeed show (*e.g.*, Feder, 1988, p. 87) that $\tau(q) = (1 - q)D_q$.

VII.C Alternative Definitions of Multifractal Measures

At this juncture, one may wonder why, *e.g.*, the binomial measure is termed multifractal. Indeed, in our coverage of the theory in the preceding two subsections, we have provided a rationale neither for the "multi-" nor for the "-fractal" parts of this qualifier.

Following Frisch and Parisi (1985), who coined the term "multifractals", and a number of other authors (*e.g.*, Halsey *et al.*, 1986; Aharony, 1990), we might look at the Paretian relationship of Equation (34) as a manifestation of fractal behavior. If the length $\delta \equiv 2^{-k}$ of the intervals I_k is shrunk to zero, then indeed the exponent $f(\alpha)$ in Equation (34) is defined by an expression that is formally similar to Equation (20), defining the box-counting dimension D_{BC}. On that basis, one could consider that the geometrical support S of a multifractal measure is the union $\bigcup_\alpha S^\alpha$ of fractal sets S^α, each with a particular Hölder exponent equal to α and a "fractal" dimension equal to $f(\alpha)$. In much of the literature on multifractal measures, this statement is reversed[9], and multifractal measures, or "multifractals" are defined as measures whose geometric support is the union of one or more fractal sets.[10]

From a mathematical standpoint, this alternative definition of multifractal measures, due to Frisch and Parisi (1985), is frought with difficulties. Indeed, a closer look at $f(\alpha)$ shows that it cannot be looked at as a box-counting dimension, in particular because the intervals counted by Equation (34) when $\delta \equiv 2^{-k}$ is small need not be contained in those counted when δ is much larger

[9] This reversal is very similar to the one discussed in subsection V.A in the case of fractals.

[10] Yet another definition of multifractals, occasionally found in the literature, is even farther removed from the concept introduced in subsection VII.A. It stems from the observation that the log-log plots of specific parameters exhibit several distinct linear segments. When there are two such segments, the system is sometimes called "bifractal" (*e.g.*, Korvin, 1992). When there are more than two linear segments, or when the curve in the log-log plot is nonlinear, the system has been called "multifractal", but the use of the word in that context has nothing to do with the multifractal measures described here.

(since the Hölder exponent associated with an interval δ_1 is not necessarily equal to that of a larger interval which contains δ_1!) (*e.g.*, Falconer, 1990, p.255). In the case of the binomial measure, one finds (e.g., Evertsz and Mandelbrot, 1992, p. 935) that the (true) box-counting dimension of each of the sets S^α is 1, whereas $f(\alpha)$ varies between 0 and 1 (cf Figure 23a). Interestingly, it turns out that for a *special* class of multifractal measures, including the binomial measure, the value of $f(\alpha)$, for each α, is equal to the Hausdorff dimension of S^α (Evertsz and Mandelbrot, 1992). Therefore, some of the sets S^α may be geometric fractals, but this does not appear to be a general characteristic of multifractal measures.

VII.D Beyond the Binomial Measure

In the multiplicative cascade used to generate the binomial measure, the number of intervals was increased by two at each step. One could easily lift that restriction and consider instead a fixed number b of intervals, larger than 2. The resulting, more general, measures are usually termed *multinomial* measures (*e.g.*, Evertsz and Mandelbrot, 1992, p. 935). Another way to expand the notion of multiplicative cascade is to use random multipliers. At each step in the cascade, the multipliers are obtained by using a random number generator, subject of course to the constraint that the total measure be conserved. This process leads to random multifractal measures, which are statistically self-similar.

The various cases considered so far have all involved multifractal measures whose supports are not fractal. It is easy to imagine measures of practical interest in soil science that might belong to this category. Evertsz and Mandelbrot (1992, p. 922) provide the example of the "quantity of groundwater" down to some prescribed depth in a given (in general non-fractal) geographical area, such as an agricultural field, a watershed, state, country, continent or island. As one subdivides the support more and more finely, following the multiplicative pattern of Figure 22a, it may turn out that the distribution of the quantity of groundwater be self-similar at different stages of the cascade, *i.e.*, at different scales of observation. In this case, the quantity of groundwater is a multifractal measure, according to the definition given above.

There are also situations where the support of a multifractal measure is itself a fractal. It is very easy to conceive a multiplicative cascade that would generate such a multifractal measure. Indeed, if in Figure 22a, one distributes the fractions m_0 and m_1 of the original mass, not over the two intervals $[0, 1/2]$ and $[1/2, 1]$, but over the intervals $[0, 1/3]$ and $[2/3, 1]$, respectively, it is clear that as $k \to \infty$, the support of the measure is the Cantor set. Clusters generated by a DLA model can also serve as a basis for constructing multifractal measures with fractal support. In this case, one may define on the DLA cluster a measure associated with the growth probability of the cluster. It is evaluated as follows. One starts with a cluster obtained by running the DLA model a certain number of times (*e.g.*, 10^4 or 10^5). The model is then run several thousand more times, but without growing the cluster further; instead, each random walker that reaches the cluster is recorded as a "hit" on the element it reaches and is removed. For each site of the cluster, the sum of the number of hits may be divided by the total number of random walkers considered. In this process, a measure is defined that quantifies the probability for a given site to be contacted by a random walker. This measure assumes its highest values at the periphery of the cluster (*e.g.*, Stanley and Meakin, 1988) and in many cases, is multifractal.

This existence of multifractal measures with a fractal support raises an interesting question that has practical consequences. One may look at a DLA cluster, for example, as a fractal, or as the support of a multifractal measure. Even though these two viewpoints are not exclusive, one of them might lead to a much deeper insight into the properties of the cluster. Which one is it? In other words, is it more fruitful to concentrate on characterizing the fractality of given physical objects, or to identify multifractal measures defined on these objects and investigate their features in detail? Obviously, the answer to this question will vary from situation to situation. In the context of soil science, it is not clear at this stage what the answer may be.

Acknowledgments

Sincere gratitude is expressed by the authors to the several generations of soil physics students at Cornell University and at the University of Illinois, and to the DEA students at the Université Henri Poincaré (Nancy, France, in 1994-1995) who challenged them to provide a rigorous yet physically-motivated account of fractal geometry. The present chapter owes much to their raised eyebrows, (occasionally) puzzled looks, and (always) loaded questions. It has also benefited much from the comments and suggestions made by Caroline Benaron (University of Cambridge) and M. Cedric Williams (Cornell University).

References

Aharony, A. 1990. Multifractals in physics: successes, dangers and challenges. *Physica A* 168:479-489.

Arnold, B.C. 1983. *Pareto Distributions.* International Cooperative Publishing House, Fairland, Maryland.

Ausloos, M. and D.H. Berman. 1985. A multivariate Weierstrass-Mandelbrot function. *Proceedings of the Royal Society of London* A400:331-350.

Avnir, D., D. Farin and P. Pfeifer. 1992. A discussion of some aspects of surface fractality and of its determination. *New J. Chem.* 16:439-449.

Bak, P. and K. Chen. 1989. The physics of fractals. *Physica D.* 38:5-12.

Bak, P. and K. Chen. 1991. Self-organized criticality. *Scientific American* 264(1):46-53.

Bak, P., C. Tang and K. Wiesenfeld. 1987. Self-organized criticality: an explanation of 1/f noise. *Physical Review Letters* 59(4):381-384.

Batty, M., A.S. Fotheringham and P. Longley. 1993. Fractal geometry and urban morphology. p. 228-246. In N.S.-N. Lam and L. De Cola (eds.) *Fractals in Geography.* PTR Prentice Hall, Englewood Cliffs, New Jersey.

Bedford, T. 1989. On Weierstrass-like functions and random recurrent sets. *Math. Proc. Camb. Phil. Soc.* 106:325-342.

Berry, M.V. and Z.V. Lewis. 1980. On the Weierstrass-Mandelbrot fractal function. *Proc. Royal Soc. A* 370:459-484.

Bolzano, B. 1930. *Funktionenlehre*, Prague.

Brock, J.R. 1971. On size distributions of atmospheric aerosols. *Atmospheric Environment* 5:833-841.

Brown, W.K., R.R. Karpp and D.E. Grady. 1983. Fragmentation of the universe. *Astrophys. Space Sci.* 94:401-412.

Burrough, P.A. 1981. Fractal dimensions of landscapes and other environmental data. *Nature* 294 (5838):240-242.

Brown, R. 1828. On the existence of active molecules in organic and inorganic bodies. *Phil. Mag.* 4:162-173.

Chepil, W.S. 1950. Methods of estimating apparent density of discrete soil grains and aggregates. *Soil Sci.* 70:351-362.

Crilly, T. 1991. The roots of chaos-A brief guide. p. 193-209. In A.J. Crilly, R.A. Earnshaw and H. Jones (eds.) *Fractals and Chaos.* Springer-Verlag, New York.

Crovelli, R.A. and C.C. Barton. 1995. Fractals and the Pareto distribution applied to petroleum accumulation-size distributions. p. 59-72. In C.C. Barton and P.R. La Pointe (eds.) *Fractals in Petroleum Geology and Earth Processes.* Plenum Press, New York.

Dedebant, G. and P. Wehrlé. 1935. *Le Rôle de l'Echelle en Météorologie.* Gauthier-Villars, Paris.

Dyson, F.J. 1978. Characterizing irregularity. *Science* 200:677-678.

Edgar, G.A. (Ed.). 1993. *Classics on Fractals.* Addison-Wesley Publishing Company, Reading, Massachusetts.

Edgar, G.A. 1990. *Measure, Topology and Fractal Geometry.* Springer-Verlag, New York.

Evertsz, C.J.G. and B.B. Mandelbrot. 1992. Multifractal measures. p. 921-953. Appendix B. In H.O. Peitgen, H. Jürgens and D. Saupe, *Chaos and Fractals. New Frontiers of Science.* Springer-Verlag, New York.

Falconer, K.J. 1988. The Hausdorff dimension of self-affine fractals. *Math. Proc. Camb. Phil. Soc.* 103:339-350.

Falconer, K.J. 1990. *Fractal Geometry. Mathematical Foundations and Applications.* John Wiley & Sons, Ltd., Chichester.

Family, F. and T. Vicsek. 1991. *Dynamics of Fractal Surfaces.* World Scientific, Singapore.

Feder, J. 1988. *Fractals.* Plenum Press, New York.

Folorunso, O.A., C.E. Puente, D.E. Rolston and J.E. Pinzón. 1994. Statistical and fractal evaluation of the spatial characteristics of soil surface strength. *Soil Sci. Soc. Amer. J.* 58(2):284-294.

Frisch, U. and G. Parisi. 1985. Fully developed turbulence and intermittency. p. 84. In M. Ghil, R. Benzi and G. Parisi (eds.) *Turbulence and Predictability of Geophysical Flows and Climate Dynamics*. North Holland, New York.

Gouyet, J.-F. 1992. *Physique et Structures Fractales*. Masson, Paris.

Graf, S., R.D. Mauldin and S. Williams. 1988. *The Exact Hausdorff Dimension in Random Recursive Constructions*. Memoirs of the American Mathematical Society 71, Providence, Rhode Island.

Grassberger, P. and I. Proccacia. 1983. Characterization of strange attractors. *Phys. Rev. Lett.* 50:346-349.

Halsey, T.C., M.H. Jensen, L.P. Kadanoff, I. Procaccia and B.I Shraiman. 1986. Fractal measures and their singularities: the characterization of strange sets. *Phys. Rev.* A 33:1141.

Hardy, G.H. 1916. Weierstrass's non-differentiable function. *Trans. Amer. Math. Soc.* 17:301-325.

Harrison, A. 1995. *Fractals in Chemistry*. Oxford University Press, Oxford, U.K.

Hausdorff, F. 1919. Dimension und usseres Mass. *Math. Annalen* 79:157-179.

Hentschel, H.G.E. and I. Procaccia. 1983. The infinite number of generalized dimensions of fractals and strange attractors. *Physica D* 8:435-444.

Herzfeld, U.C. 1993. Fractals in geosciences - Challenges and concerns. p. 217-230. In J.C. Davis and U.C. Herzfeld (eds.) *Computers in Geology - 25 Years of Progress*, Oxford University Press, New York.

Hutchinson, J.E. 1981. Fractals and self-similarity. *Indiana Univ. Math. J.* 30:713-747.

Hwa, T. and M. Kardar. 1989. Fractals and self-organized criticality in dissipative dynamics. *Physica D* 38:198-202.

Johnson, N.L. and S. Kotz. 1970. *Distributions in Statistics. Continuous Univariate Distributions-1*. John Wiley and Sons, New York.

Jones, H. 1991. Fractals before Mandelbrot. A selective history. p. 7-33. In A.J. Crilly, R.A. Earnshaw and H. Jones (eds.) *Fractals and Chaos*, Springer-Verlag, New York.

Kaye, B.H. 1989. *A Random Walk through Fractal Dimensions*. VCH Verlagsgesellschaft, Weinheim, Germany.

Klinkenberg, B. 1994. A review of methods used to determine the fractal dimension of linear features. *Mathematical Geology* 26(1):23-46.

Korvin, G. 1992. *Fractal Models in the Earth Sciences*. Elsevier, Amsterdam.

Lavallée, D., S. Lovejoy, D. Shertzer and P. Ladoy. 1993. Nonlinear variability of landscape topography: multifractal analysis and simulation. p. 158-192. In N.S.-N. Lam and L. De Cola (eds.) *Fractals in Geography*. PTR Prentice Hall, Englewood Cliffs, New Jersey.

Lavenda, B.H. 1985. Brownian motion. *Scientific American* 252(2):70-85.

Le Méhauté, A. 1991. *Fractal Geometries. Theory and Applications*. CRC Press Inc., Boca Raton, Florida.

Macaulay, F. 1922. Pareto's laws and the general problem of mathematically describing the frequency distribution of income. Chapter XXIII. In: *Income in the United States, its Amount and Distribution, 1909-1919.* National Bureau of Economic Research, New York.

Mandelbrot, B.B. 1960. The Pareto-Levy law and the distribution of income. *International Economic Review* 1(2):79-106.

Mandelbrot, B.B. 1963. New methods in statistical economics. *The journal of Political Economy* 71(5):421-440.

Mandelbrot, B.B. 1967. How long is the coast of Britain? Statistical self-similarity and fractal dimension. *Science* 155:636-638.

Mandelbrot, B.B. 1975. *Les Objets Fractals: Forme, Hasard et Dimension.* Flammarion, Paris.

Mandelbrot, B.B. 1978. Les objets fractals. *La Recherche* 9(85):5-13.

Mandelbrot, B.B. 1982. *The Fractal Geometry of Nature.* W.H. Freeman and Company, New York.

Mandelbrot, B.B. 1989. Multifractals measures, especially for the geophysicist. *PAGEOPH* 131:5-42.

Mandelbrot, B.B. 1990. New "anomalous" multiplicative multifractals: left sided f(α) and the modelling of DLA. *Physica A* 168:95-111.

Matsushita, M. 1985. Fractal viewpoint of fracture and accretion. *J. Phys. Soc. Japan* 54(3):857-860.

Meakin, P. 1991. Fractal aggregates in geophysics. *Rev. Geophys.* 29(3):317-354.

Montroll, E.W. and M.E. Schlesinger. 1983. Maximum entropy formalism, fractals, scaling phenomena, and 1/f noise: a tale of tails. *Journal of Statistical Physics* 32(2):209-230.

Moon, F.C. 1992. *Chaotic and Fractal Dynamics.* John Wiley and Sons, New York.

Nottale, L. 1993. *Fractal Space-Time and Microphysics.* World Scientific, Singapore.

Pareto, V. 1897. *Cours d'Economie Politique.* Volume 2. F. Rouge, Lausanne, Switzerland.

Peitgen, H.O., H. Jürgens and D. Saupe. 1992. *Chaos and Fractals. New Frontiers of Science.* Springer-Verlag, New York.

Persky, J. 1992. Pareto's law. *Journal of Economic Perspectives* 6(2):181-192.

Pfeifer, P. and M. Obert. 1989. Fractals: basic concepts and terminology. p. 11-43. In D. Avnir (ed.) *The Fractal Approach to Heterogeneous Chemistry.* John Wiley and Sons Ltd., New York.

Richardson, L.F. 1961. The problem of contiguity: an appendix of statistics of deadly quarrels. *General Systems Yearbook* 6:139-187.

Rogers, C.A. 1970. *Hausdorff Measures.* Cambridge University Press, Cambridge, U.K.

Rudin, W. 1976. *Principles of Mathematical Physics* (3rd ed.). McGraw-Hill Book Company, New York.

Sagan, H. 1994. *Space-Filling Curves.* Springer-Verlag, New York.

Sapoval, B. 1991. Fractal electrodes, fractal membranes, and fractal catalysts. p. 207-226. In A. Bunde and S. Havlin (eds.) *Fractals and Disordered Systems.* Springer-Verlag, Berlin.

Schmidt, R. and K. Housen. 1995. Problem solving with dimensional analysis. *The Industrial Physicist* 1(1):21-24.

Schroeder, M. 1991. *Fractals, Chaos, Power Laws. Minutes from an Infinite Paradise.* W.H. Freeman and Company, New York.

Smirnov, B.M. 1990. The properties of fractal clusters. *Phys. Rep.* 188(1):1-78.

Stanley, H.E., A. Bunde, S. Havlin, J. Lee, E. Roman and S. Schwarzer. 1990. Dynamic mechanisms of disorderly growth: recent approaches to understanding diffusion limited aggregation. *Physica A* 168:23-48.

Stanley, H.E. and P. Meakin. 1988. Multifractal phenomena in physics and chemistry. *Nature* 335:405-409.

Steinhaus, H. 1954. Length, shape, and area. *Colloquium Math.* 3:1-13.

Takayasu, H. 1990. *Fractals in the Physical Sciences.* John Wiley and Sons, Chichester.

Tricot, C. 1995. *Curves and Fractal Dimension.* Springer-Verlag, New York.

Turcotte, D.L. 1986. Fractals and fragmentation. *J. Geophys. Res.* 91(B2):1921-1926.

Turcotte, D.L. 1989. Fractals in geology and geophysics. *PAGEOPH* 131:171-196.

Turcotte, D.L. 1992. *Fractals and Chaos in Geology and Geophysics.* Cambridge University Press, Cambridge, U.K.

Turcotte, D.L. and J. Huang. 1995. Fractal distributions in geology, scale invariance, and deterministic chaos. p. 1-40. In C.C. Barton and P.R. La Pointe (eds.) *Fractals in the Earth Sciences*, Plenum Press, New York.

Van Damme, H. 1995. Scale invariance and hydric behaviour of soils and clays. *C. R. Acad. Sci. Paris, série II a* 320:665-681.

Wiener, N. 1923. Differential-space. *J. Math. and Phys.* 2:131-174.

Witten, T.A. and L.M. Sander. 1981. Diffusion-limited aggregation, a kinetic critical phenomenon. *Phys. Rev. Lett.* 47:1400-1403.

Zipf, G. 1949. *Human Behavior and the Principle of Least Effort.* Addison-Wesley, Reading, Massachusetts.

Structural Hierarchy and Molecular Accessibility in Clayey Aggregates

H. Van Damme

Contents

I Introduction

Soil is a complex bio-mineral reactor within which gas, liquid and solute reside and interact among themselves and the solid surfaces. More than any other porous medium, it has a strong textural-based response to ions or molecules (this is the basis for the action of soil conditioners). It is also subject to a number of mechanical stresses leading either to fragmentation (frost, drying, roots, human action) or to compaction (animals, human machinery). In the long term, soil evolution depends on solid-solution reactions involving dissolution and/or precipitation. The location of such reactions and the kinetics are, to a large extent, the result of an interplay between texture (defined as the distribution of voids and solid) and diffusion or hydrodynamics (wetting, imbibition). In other words, the dynamic complexity of the medium is related to its textural complexity. The key question is: at what rate can a given reactant penetrate the soil, reach a given microscopic place, and react? To begin answering this question, it is necessary to know what portion of the solid interface is accessible to a given reactant, whatever the time required to reach this interface. This basic question of accessibility *versus* non-accessibility may be addressed in simple fractal dimension terms. Then the next step is to assign a rate distribution to this accessible interface, taking into account the rate of the transport process and the rate of the chemical reaction itself. This second kind of problem typically

ISBN 1-56670-105-8

requires a multifractal measure.

In this chapter, it will be shown how the concept of fractal dimension may help to quantify the relationship between microtexture and accessibility to an inert gas. This will be illustrated mainly on the most active mineral components of soils (as far as hydromechanical and adsorptive properties are concerned): smectite clays. On a more qualitative level, the relationship between accessibility and reaction rate will also be addressed.

II Microtexture and Accessibility: Trivial Models

Although water, water vapor or oxygen gas are more relevant to soil chemistry, the simplest accessibility measurement that one could perform would involve an inert gas such as nitrogen for instance. By performing the experiment at low temperature, one is then led to a straightforward physical adsorption study such as those commonly performed to measure surface area or micro- and meso-porosity. The surface area and the pore volume calculated from the adsorption isotherm are a direct assessment of the accessible area and void volume, respectively.

Rather than measuring the surface area and porosity on samples (lumps, aggregates) of a given size, one may look to the way the results depend on the lump size. This is typically a *scaling* approach. Rather than measuring the average properties or characterizing the texture of a material at a given scale, one looks to their evolution as the observation scale is changed. The anticipated benefit is that the rules that allow one to go from one scale to another may reveal features that are more "universal" than those measured at a single scale. However, as far as surface area of soil or clay lumps are concerned, performing a scaling study like this may seem useless at first sight. Indeed, like many other natural particulate materials, a macroscopic piece of solid clay or soil is in fact a (weakly) cohesive packing of submicron particles held together by capillary or short range surface forces. The common feeling is that the surface area and the porosity per unit mass (or volume) of such materials does *not* depend on the lump size, because they are determined by the size and the arrangement of much smaller structural units. In other words, soil and clay lumps of various sizes are usually considered as "ideally porous" particulate solids, which will be defined more precisely hereafter. Before showing that this is wrong in most cases, let us discuss a few simple (non-fractal) situations.

Figure 1 illustrates the structure of an "ideally porous" particulate solid. It is a consolidated sand composed of submillimetric quartz grains. Although there is no long-range order in the system, it is quite clear that to have a statistically satisfying description of the texture, one does not need to describe the size and the position of *all* the quartz particles and *all* the pores of the system. Knowing the average particle size and the average local texture in a representative elementary volume is enough to understand the texture and the properties of the whole system. In other words, the system is characterized by a well-defined characteristic length scale, L_c, such that the whole system

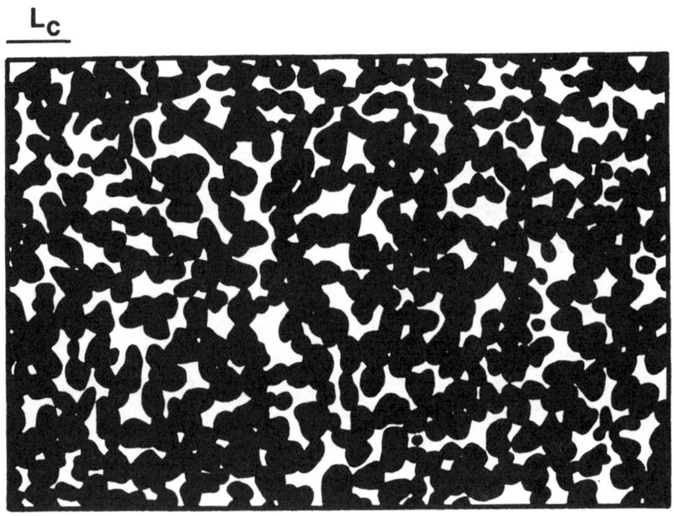

Figure 1: Polished section of a consolidated sand, as example of "ideally porous" particulate solid.

can be statistically reproduced by simple translation of cells of size L_c. In the example, L_c is of the order of one average particle diameter plus one average pore diameter, *i.e.*, much smaller than a macroscopic piece of material. Furthermore, in this example the pore space is fully connected (not necessarily on the figure, which is a 2-D cut, but through the third dimension). Thus, whatever the size, *R*, of a piece of material, properties such as its surface area per unit mass, $S_m(R)$, or its pore volume per unit apparent volume, $P_v(R)$, are expected to remain constant as long as R is large compared to L_c. This is the definition of an "ideally porous" particulate solid. The surface area per unit mass, S_m, of a lump of rock, sediment or soil like this, containing N subunits, is very close to the surface area per unit mass of each subunit, s_m:

$$S_m = \frac{N \cdot \text{area(subunit)}}{N \cdot \text{mass(subunit)}} = s_m \tag{1}$$

Thus, S_m is independent of N and of R. In other words, even though N is scaling as R^3, S_m is scaling as R^0.

A limiting case that is opposite to the ideal porous solid is the "ideally compact" particulate solid, in which the subunits are so tightly packed that no molecule can enter the interstitial space, as sketched in Figure 2. No "internal" surface area is accessible. Only the envelope of the packing contributes to the adsorption of molecules. S_m is scaling as $N^2/N^3 = N^{-1}$ or R^{-1}, as for any non-porous granular material. The totally compact model is clearly unrealistic for particulate materials and should only be considered as a limiting case. With undeformable particles, it could only be reached by packing regular polyhedra with translational order (Figure 2a; like in crystal lattices), but this

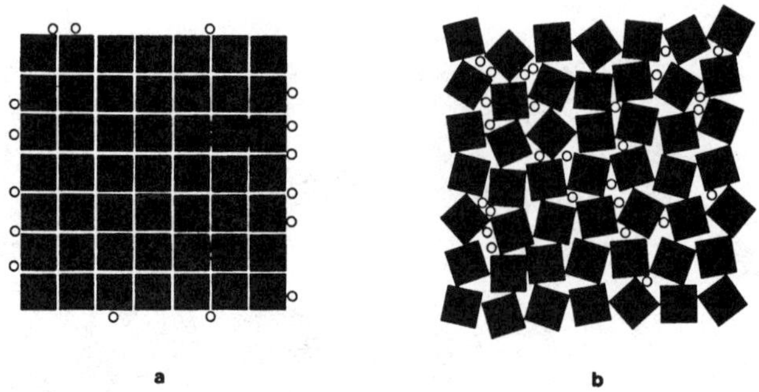

Figure 2: (a) Example of "ideally compact" particulate solid made of rigid and ordered particles, converted to (b) an "ideally porous" solid by the introduction of disorder.

situation is destroyed by disorder (Figure 2b). However, with highly deformable particles pressed together by sufficiently strong forces, one might approach total compactness, as illustrated in Figure 3 with platelets.

With clays, which are composed of highly anisotropic particles (fibers, needles, rigid platelets or flexible sheets), one may imagine situations between the limiting cases discussed above. For instance, by randomly packing fibers (palygorskite, sepiolite) or rigid platelets (kaolinite, illite), one builds "ideally porous" architectures, as sketched in Figure 4, but with smectite clays (montmorillonite, saponite, beidellite, hectorite), which are made of thin (≅1 nm), flexible silicate sheets held together by short-range hydration forces, one expects lump textures intermediate between "ideally porous" and "totally compact". In spite of this, surface area data obtained on smectite clays or vertisols are still interpreted commonly in terms of the first, ideally porous, model of the book-house type (Murray *et al.*, 1985; Murray and Quirk, 1990). It is assumed that the sheets form ordered stacks, also termed "quasi-crystals" (Aylmore and Quirk, 1967), in which the sheets are in close contact or separated from each other by a small number of water molecule layers. The one-dimensional orientational and translational order is assumed to be strong within the stacks, along the axis of the stacks, whereas there is symmetry breaking when leaving a given stack and passing from one stack to another. The stacks are assumed to be totally compact for molecules that are not able to swell the interlayer space (like nitrogen or oxygen, but not water). The lumps (or aggregates) are assumed to be "ideally porous" and fragile assemblies of quasi-crystals, as illustrated in Figure 5a.

Assuming that the thickness, e, of the stacks is much smaller than their width and, consequently, that the area of the lateral surfaces is negligible compared to the area of the basal surfaces, it is easy to show for each individual sheet, and hence for the entire stack (Tessier, 1990), that S_m is inversely related to e:

$$S_m = \frac{2}{e\,q} \tag{2}$$

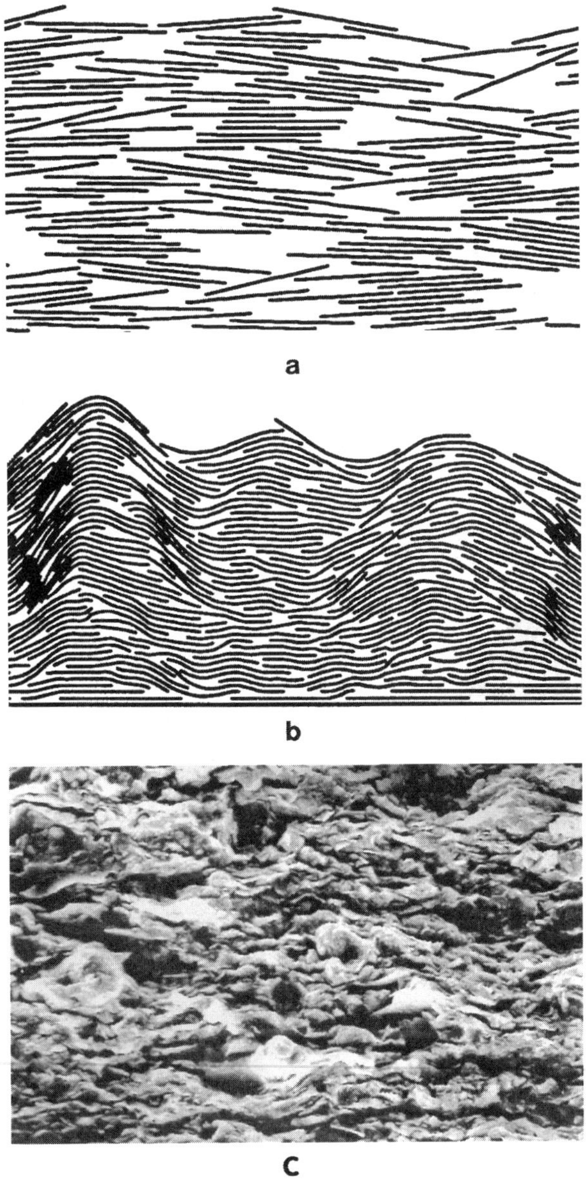

Figure 3: A disordered and ideally porous collection of rigid platelet (a) may be converted to an ideally compact structure in the case of very deformable particles (b), as illustrated in (c) for a real shale rock.

where q is the sheet density, known from crystal structure data. Knowing S_m from adsorption measurements, it is then very tempting to calculate the average

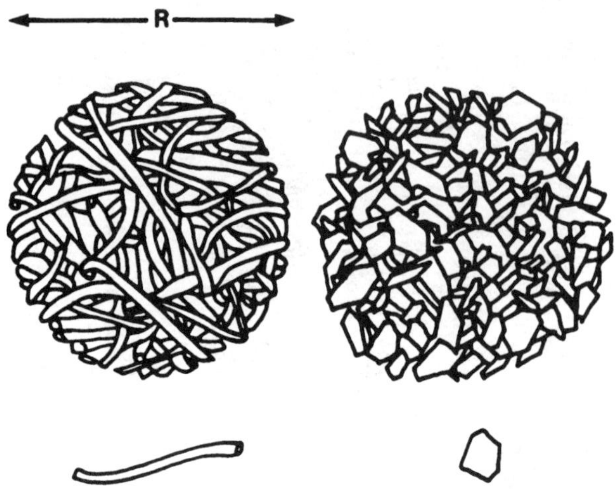

Figure 4: Three-dimensional, ideally porous particles made of fibers or rigid platelets which are characterized by scaling exponents for surface area, porosity and mass distribution equal to 3.

number of smectite sheets in each "quasi-crystalline" stack and to model the structure of the material as a disordered assembly of quasi-independent quasi-crystals of thickness *e*. As we shall see, this way of modeling the structure of smectite clays is, in many instances, very deceptive.

III Microtexture and Accessibility: a Scaling Approach

What might be wrong with using the ideally porous packing model for interpreting surface area measurements on clays? Very little in the case of more or less rigid fibrous clays and in the case of rigid kaolinite or illite platelets, except perhaps the assumption that the lateral surface area is negligible compared to the basal surface area. The orientational disorder between kaolinite or illite platelets may also be smaller than what is sketched in Figure 4, as shown by electron microscopy studies (Tessier, 1990) (a dense packing of platelets tend to form orientationally ordered domains), but this does not seriously modify molecular accessibility since even a small angular difference between two micrometer-sized rigid platelets opens a wedge-shaped pore totally accessible to small molecules. On the other hand, very much may be wrong with the ideally porous packing model in the case of smectites. The main source of error comes from neglecting the deformability, which, as shown in Figure 3, may further enhance the connectedness of the solid and limit the void space. Thus, there may be some close contact regions between the quasi-crystals. This may lead to a drastic reduction of molecular accessibility as first suggested by Stul and Van

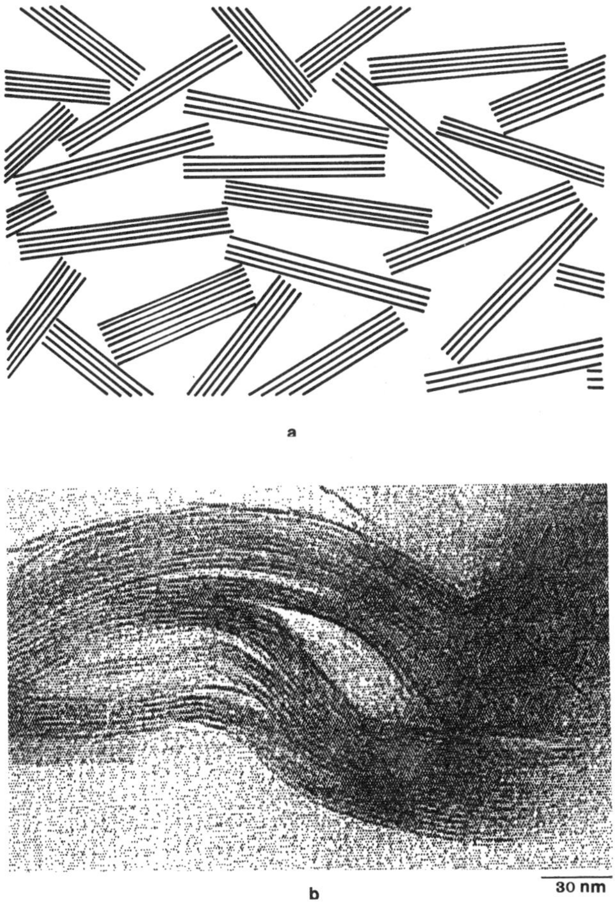

Figure 5: (a) Sketch of the texture of smectite clays according to the quasi-crystal model. The elementary clay sheets are assumed to form ordered and compact stacks (quasi-crystals) that are connected to each other by very localized edge-to-face contacts. (b) Illustration of the general face-to-face connectivity and layer entanglement in a real Ca-montmorillonite (courtesy D. Tessier). The same elementary clay sheet may be involved in different stackings at different places and is connected to the others over large areas.

Leemput (1982). As a result, the textural modeling of surface area data looses its simplicity. In fact, it is quite obvious, from a look at electron micrographs of smectites (Tessier, 1990) that the assumption of quasi-independence of the structural subunits (sheet stacks) is a very poor assumption. Even the simple identification of these subunits is not obvious. The same sheet may be involved in one stack at some place and in another stack of different thickness at another place (Figure 5b). Multiple face-to-face connectivity of the sheets and their flexible stacks is the rule even in muds and, *a fortiori*, in dry media.

Rather than assuming *a priori* a given type of model, one can try to gain information on the type of model to be used from experimental data on the relationship between accessible area and particle size. As an operational approach, one may for instance look for simple power-law relationships between the size of macroscopic particles (which are in fact aggregates or lumps), R, and their mass, $M(R)$, their surface area, $S(R)$, and their pore volume, $P(R)$. This is a generalization of the analysis extensively applied by Avnir and coworkers to the surface area of powders (Avnir *et al.*, 1984, 1985). In practice, the measurements are made on an aggregate bed containing N_g aggregates belonging to the same granulometric fraction. If the aggregates have a spheroidal shape, N_g is related to the apparent volume of the bed, V_a, and to the size of the aggregates by $N_g(R) \propto V_a/R^3$ (where the sign "$\propto$" means "proportional to"). Thus, if $M(R)$, $S(R)$ and $P(R)$ are scaling as R^{D_m}, R^{D_s} and R^{D_p}, respectively, for single grains, the following relations can be derived for the apparent density, $\mathbf{M_v}(R)$, for the specific surface area, $\mathbf{S_m}(R)$, and for the internal porosity, $\mathbf{P}(R)$, of a bed of $N(R)$ grains (Van Damme *et al.*, 1988; Ben Ohoud *et al.*, 1988):

$$\mathbf{M_v}(R) = \frac{N(R)M(R)}{V_a} \propto R^{(D_m-3)} \tag{3}$$

$$\mathbf{S_m}(R) = \frac{N(R)S(R)}{N(R)W(R)} \propto R^{(D_s-D_m)} \tag{4}$$

$$\mathbf{P}(R) \simeq \frac{N(R)P(R)}{N(R)R^3} \propto R^{(D_p-3)} \tag{5}$$

In this framework, the ideally porous model is characterized by $D_s = D_m = D_p = 3$, whereas the ideally compact model is characterized by $D_s = 2$ and $D_m = 3$.

IV Experiments

The experimental study reviewed in this section was performed by Ben Ohoud and Van Damme (1990). One kaolinite, one sepiolite, one palygorskite and twenty monoionic montmorillonites were prepared by performing an ion exchange (for 19 materials) or an adsorption (for humic acid) treatment on the same parent material (a Na-exchanged bentonite). The samples were prepared by allowing several liters of slurries containing about 10% dry matter to dry by slow evaporation at 353 K. The resulting lumps were first fragmented with a hammer, and with a ball mill afterwards. For each material, about twenty granulometric fractions were isolated by sieving, with mesh sizes ranging from 20 to 4500 μm. The shape of the particles was examined by scanning electron microscopy (SEM) and found to be roughly spheroidal in all cases, as shown in Figure 6.

$\mathbf{M_v}(R)$ was calculated by measuring the volume occupied by a given mass of powder in a simple measuring glass tube, after gentle vibration for a fixed

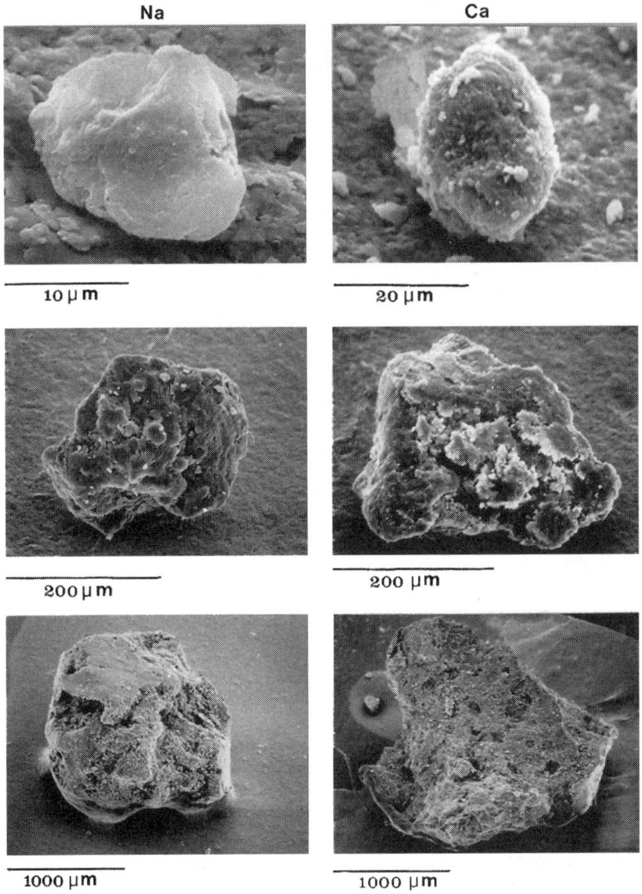

Figure 6: Scanning electron micrographs of Na- (left column) and Ca- (right column) montmorillonite particles in three granulometric fractions.

time. The accessible surface area was measured by nitrogen adsorption at 77 K, using the classical Brunauer-Emmett-Teller (BET) model of multilayer physical adsorption to calculate monolayer coverage (Gregg and Sing, 1982). The open porosity of the grains was also determined by adsorption, by measuring the total volume of nitrogen adsorbed at a relative vapor pressure of 0.99. At this relative pressure, all the pores of radius smaller than 10 nm are expected to be filled with liquid nitrogen by capillary condensation (Gregg and Sing, 1982).

Within experimental error, the slope of $\log \mathbf{M_v}(R)$ *versus* $\log R$ plots was found to be zero in all cases, so that $D_m = 3$ for all the clays that were studied. In other words, the mass distribution and the *total* (open + closed) pore volume distribution within the grains are homogeneous. Thus, solid clay samples are neither mass- nor *total* pore volume fractals. As one might expect, they behave,

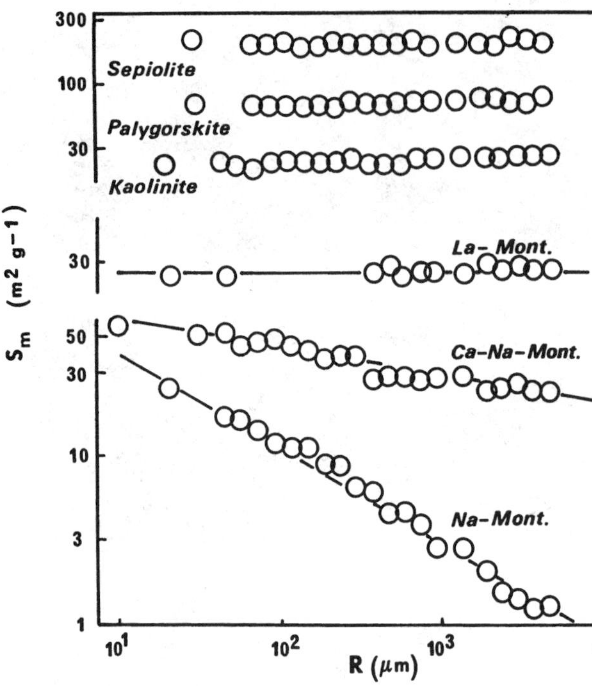

Figure 7: Particle-size dependence of the surface area of various clays measured by nitrogen adsorption. Note that all the montmorillonite samples were prepared from the same <2 μm fraction of parent material.

as far as *total* mass and porosity distributions are concerned, as ideal porous solids.

Much more interesting is the behavior of the accessible surface area and open porosity of the grains. For fibrous clays and kaolinite, $\mathbf{S}_m$ and $\mathbf{P}$ are independent of the grain size (Figure 7). Thus, for these clays the ideally porous model is applicable. However, for most smectite samples, *neither* $\mathbf{S}_\mathrm{m}$ nor $\mathbf{P}$ are independent of the grain size. Only a few of the twenty samples follow the ideally porous model. All the other samples are characterized by linear $\log \mathbf{S}_\mathrm{m}$ *versus* $\log R$ plots, over almost three decades, but, in the overwhelming majority of cases, the slopes lead to non-trivial D_s exponents for the accessible surface between 2.4 and 3. No samples are found to follow the ideally *compact* model (which would require $D_s = 2$).

Equally remarkable is the fact that the same power laws are obtained for the open porosity, with scaling exponents that are in good agreement with the surface fractal exponents (Figure 8). Thus, most solid smectite clays are accessible surface- and open pore volume fractals.

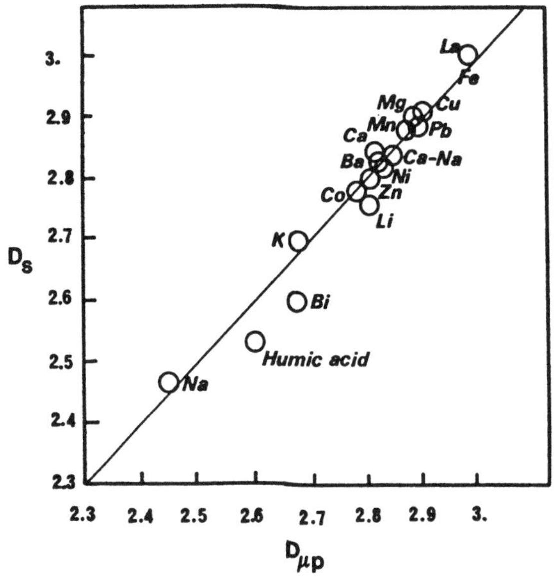

Figure 8: Correlation between the surface fractal exponents of ion-exchanged montmorillonite particles and the fractal exponents describing the scaling of porosity (pore size less than 10 nanometers) in the same samples.

V Scaling, Local Order and Morphomechanics

What are the structural and textural features that are at the origin of these non-trivial scaling properties of most smectite clays? One knows, often from simulation studies, that many physical processes are able to produce fractal structures. Diffusion-limited aggregation is among the best known (Vicsek, 1992), but it does not correspond to what is observed with clays. Indeed, it leads either to mass fractals or to compact surface fractals, but not to surface and pore volume fractals. Cracking by capillary stresses is a better candidate. It has been shown by Skjeltorp and Meakin (1988), in two-dimensional experiments and simulations with spherical particles, to lead to a fractal hierarchy of cracks, within certain limits, and the cracking patterns of smectite-rich soils (vertisols) have been observed for a long time by soil scientists! More recently, Rieu and Sposito (1991a,b) showed evidence that an incompletely fragmented fractal model of crack network captures the essential hydric properties of many soils.

In any case, a close look at the SEM micrographs of the smectite grains shows that the structural features responsible for the scaling properties of smectite clay aggregates cannot be large voids. Clearly, they ought to be much

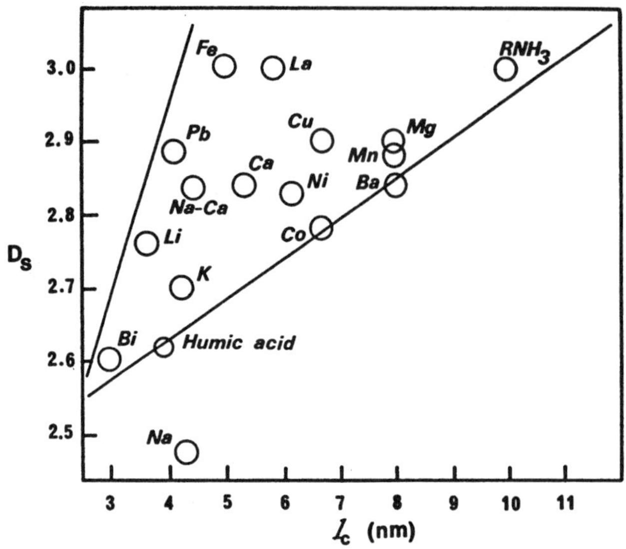

Figure 9: Correlation between the surface fractal exponents of various ion-exchanged montmorillonites, determined by nitrogen adsorption, and their coherence length (thickness of the ordered stacks of elementary clay sheets), determined from X-ray diffraction.

smaller than the aggregate size. A plausible hypothesis is that they are stress-induced micro- or nano-cracks or, more generally, defect zones or zones of weak connectivity separating more compact nano- or micro-(stacked) domains. In this logic, one expects a direct correlation between the formation of cracks and the deformability of the structure. Such a correlation was indeed found between the scaling exponents and the nanometric stacking order of the elementary lamellae in the solid (Ben Ohoud and Van Damme, 1990). This is shown in Figure 9, where the scaling exponents derived from the adsorption data have been plotted *versus* the coherence length, l_c, for X-ray diffraction, calculated from the line broadening of the 001 reflection lines using the Debye-Sherrer formula (Guinier, 1964). The 001 reflection is due to diffraction by parallel individual clay sheets (1 nm thick) in stacks. Its sharpness is directly related to the so-called coherence length, *i.e.*, the distance perpendicular to the sheets over which the stacking is strictly periodic. The narrower the diffraction line, the larger the coherence length l_c, which is a direct estimate of the thickness (hence the stiffness) of the ordered layered stacks in the solid. Although the data are scattered, the general trend is clear: D_s and D_p increase as l_c increases. Thus, as l_c increases, each aggregate looks increasingly like a disordered coherent packing of quasi-independent building blocks. Mechanical coherence is maintained thanks to the incomplete connectivity of the crack network between the building blocks.

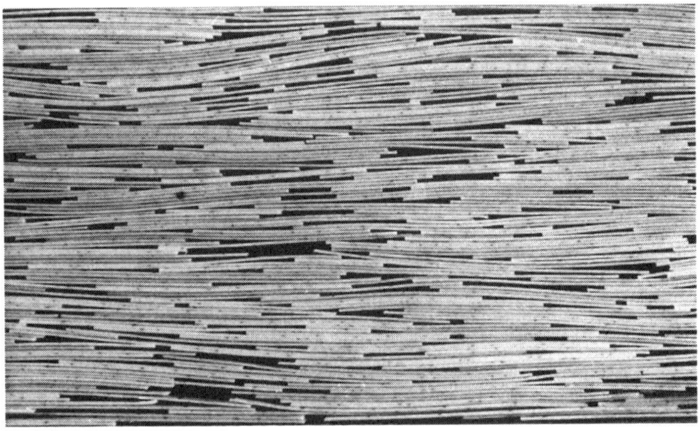

Figure 10: Simulation with pieces of paper of the type of sheet organization which may be present in "entangled" smectite clays. Local order is weak, but the overall compactness of the packing is large.

All this can be rationalized in morphomechanical terms, since the thickness of stacks is expected to be related to their stiffness. This will be explicitly shown later. The two reference microstructural situations that we have to consider are the following. The first, which is valid only with a few cations (La^{3+}-, Fe^{3+}-, Cs^{+}-bentonite), is the first classical model introduced in section II (Figure 5a): an open book-house-type structure of thick and rigid quasi-crystals, with (trivial) scaling exponents equal to 3. The other, which would be closer to reality in many more cases (particularly in natural smectites and, probably, vertisols) is a much more compact architecture obtained by packing thin and flexible units, but with a non-zero crack-like open porosity along the interfacial defect zones between more ordered domains. In the former case, the orientational correlation between elementary layers is at its maximum within the quasi-crystals, but drops suddenly beyond $l_c = L_c$. In the latter case, the orientational correlation is weaker but the characteristic length is larger, up, eventually, to the size of the macroscopic lump: $l_c << L_c$, and D_s and $D_p < 3$. An idea of what the latter structure might be is given by Figure 10, where a compacted two-dimensional packing of thin and flexible pieces of paper is shown.

An open question is the relation between the properties of the exchangeable cations and their capacity to generate stacks of smectite lamellae. Initially, one might think that highly charged and/or small cations, with a large polarizing power, would have a strong local ordering tendency and would be able to form thick stacks. As shown in Figure 11, the trend is satisfying since highly charged (trivalent) ions do indeed yield large (trivial) scaling exponents ($= 3$), divalent ions yield intermediate values (between 2.9 and 2.8), whereas most monovalent ions yield generally even lower values. However, a closer look shows that the correlation between the polarizing power and l_c, D_s or D_p is not simple. Large but less hydrated monovalent ions like Cs^{+} for instance lead also to high D_s

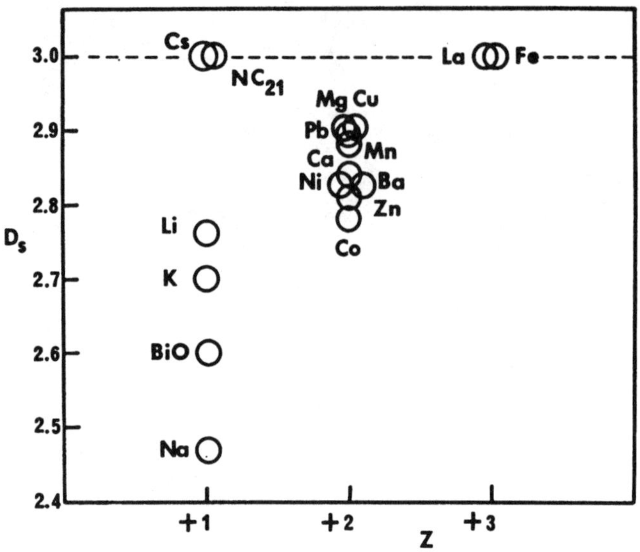

Figure 11: Correlation between the surface fractal dimension of ion-exchanged montmorillonites and the cationic charge. "NC_{21}" is a sample exchanged with a long chain quaternary ammonium ion.

and D_p values. The fine structure of the hydration shells in the interlayer space most probably has to be taken into account.

VI From Solids to Muds

It would be very surprising not to find a close relationship between the textural features of solid smectite lumps and the properties of the muds that they form after redispersion in water. Generally speaking, one expects the following trends:

- smectites that give rise to fragile book-house-type arrangements of thick and well ordered stacks in the solid state with $D_s = D_p = D_m = 3$ ("stacked" smectites) are also expected to form thick quasi-crystals in muds, with a small water-tactoid interface area. Those that form compact packings of flexible units ("entangled" smectites) with non-trivial scaling exponents are expected to come from and to form well-dispersed, delaminated muds, with a large solid-liquid interface area.
- since the total excluded volume of one thick quasi-crystal containing n elementary lamellae is much smaller than the total excluded volume of n individual lamellae, one expects muds of "stacked" smectites to have lower plastic viscosity and gel strength (yield stress) than the "entangled" smectites.

- assuming that thick rigid units form less coherent and more porous sediments than thin flexible units, "stacked" clays should form more permeable filtration cakes than "entangled" clays. Upon drying, "stacked" filtration cakes are also expected to crack extensively under the action of the capillary stress, since the units have only very limited contact areas. "Entangled" clays should withstand stresses much better.

All these trends are confirmed experimentally, in spite of a relatively large scatter in the data. Na- and Li-bentonites are known to have a much higher gel strength, viscosity and a much lower water loss rate than Ca- or Al-materials (van Olphen, 1963). On the other hand, the cracking pattern develops extensively as D_s and D_p approach 3 (Figure 12). Hence, the accessible surface and pore volume fractal exponents measured on the solid samples prove to be good parameters to trace the microtextural evolution of the drying muds.

VII Chemical Reactivity and Diffusion-Limited Accessibility

In the previous sections, we examined the accessibility of the medium under conditions where kinetics did not matter. Adsorption was measured at equilibrium, whatever the time required to reach that equilibrium. When transport to the interface becomes the rate-limiting step, the active part of the interface may be considerably smaller than the total accessible interfacial area. Only those regions that are the most readily accessible contribute significantly to the overall activity. This problem of diffusion towards and eventually across a fractal interface is not only encountered in soil science but in related fields as well, such as catalysis, membrane technology and biology.

The consequence of the non-equivalence of all surface sites in rate processes is that one can define a reaction dimension, D_r, from the particle-size dependence of the reaction rate (Farin and Avnir, 1987):

$$a_m \propto R^{(D_r-3)} \tag{6}$$

where a_m is the activity per unit mass. For instance, the oxidation of sulphur by *Thiobacillus albertis* was found to scale with the sulphur particle radius as $R^{2.16}$ (Farin and Avnir, 1987). The dissolution of alkali feldspar in HCl goes as $R^{2.95}$ (Farin and Avnir, 1987), whereas the ion exchange of Co^{3+} ions in zeolite X is scaling as $R^{2.65}$ (Ignatzek *et al.*, 1987). Oxidation, dissolution and ion exchange are all reactions that occur also in soils. It is usually found that D_r is smaller than D_s (Farin and Avnir, 1989), which can be intuitively understood by realizing that the active sites are only a subset of the total number of surface sites. However, the relationship between D_r and D_s is not straightforward.

The field that controls diffusion in a non-fractal medium (a solution) *towards* a fractal interface (eventually a soil lump) is a Laplacian field. If we consider a stationary situation between an infinite source of reactants far from the interface and a sink (the interface) where the reactants disappear (are instantaneously transformed), the boundary conditions are $C = C_o$ far from the interface and

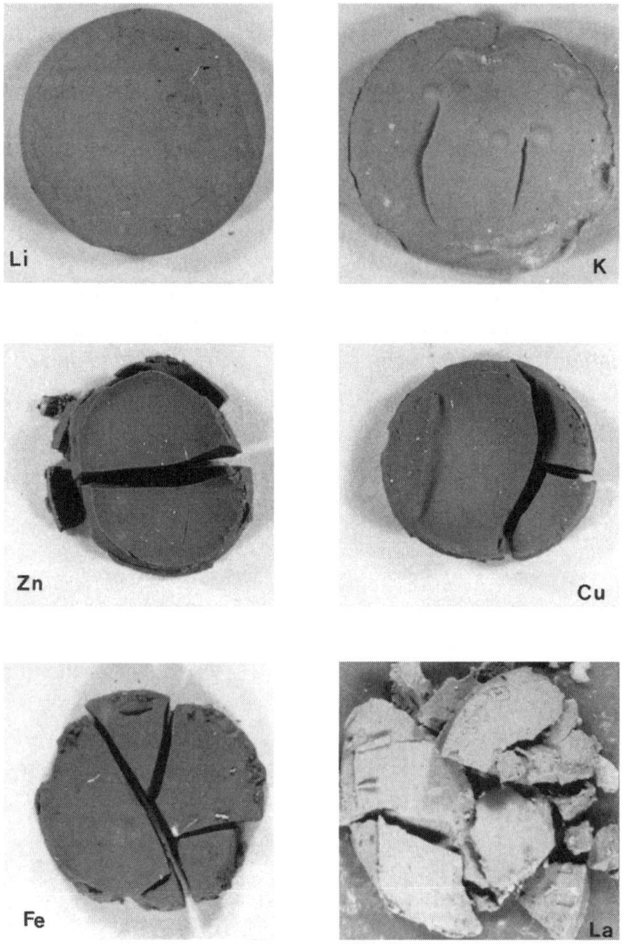

Figure 12: Cracking patterns obtained by drying the filtration cakes of a few ion-exchanged montmorillonite muds. Note the extensive cracking of the "stacked" clays and the absence of cracking in the "entangled" clays.

$C = 0$ on the interface. This is equivalent to calculating the electrical potential map between a circular electrode at potential $V = V_o$ and a central counter-electrode at potential $V = 0$. In the diffusion problem, the "potential" is the probability of presence or, equivalently, the concentration of diffusing species. The law that governs the diffusion process is Fick's second law $\nabla^2 C = dC/dt$, which, in the stationary state, reduces to Laplace's equation $\nabla^2 C = 0$. On the other hand, the flux towards each point of the interface is controlled by the gradient of concentration, ∇C, in front of that point. Thus, the flux will be highest in the zones where the isopotential or isoprobability curves are closest to each

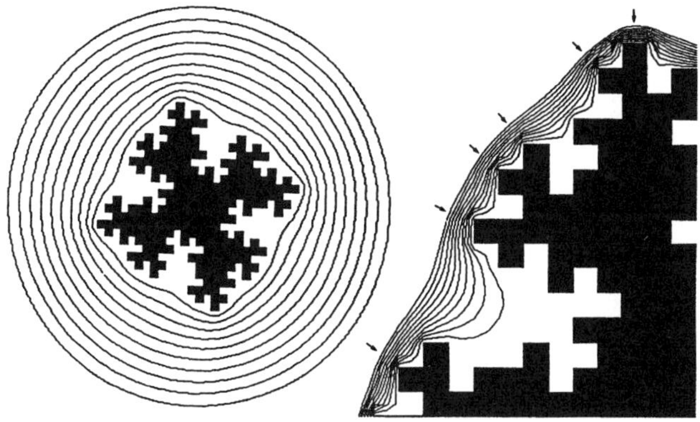

Figure 13: Potential map around an island with a fractal boundary. The isopotential curves (in an electrical problem) or the curves of equal probability of presence of random walkers (in a diffusion problem) are much smoother than the island boundary itself. They are also much closer to each other (highest gradients) in the most exposed zones. The active part of the interface (the zones where the diffusion flux is the highest) is restricted to the outer edges of the island. More than 80% of the diffusion flux reaches less than 10% of the interface. Fjords are screened and essentially inactive.

other. The outer zones are much more active than the others, whereas the deep zones are almost totally screened (Evertsz and Mandelbrot, 1992; Gutfraind and Sapoval, 1993). This is illustrated in Figure 13.

A different situation is encountered when diffusion occurs *in* a fractal space. According to what we saw in the previous sections, this might be close to the real situation in clayey lumps, whether diffusion takes place in the pore space or on the interface (in the adsorbed state). In that case, diffusion is governed by the so-called spectral exponent of the fractal space, d_s, which characterizes the way the number of distinct sites, N_{d_s}, visited by a random walker increases with time (Havlin, 1989; Van Damme *et al.*, 1986). In general, N_{d_s} increases as $t^{d_s/2}$. In a Euclidean three-dimensional space, $d_s = 2$ and N_{d_s} increases linearly with time. In a fractal space, $d_s < 2$ and the number of new sites visited increases more slowly than in a solution, for instance. Exploration is less and less efficient as time goes on (*cf.* Figure 14).

VIII Conclusion

There is ample experimental evidence that some of the most active components of soils, smectite clays, may generate fractal structures at length scales between a few nanometers and a few millimeters and, probably, beyond that. Particularly important as far as soil dynamics is concerned is the fact that the fractal space is precisely that which is involved in chemical and physical exchanges with

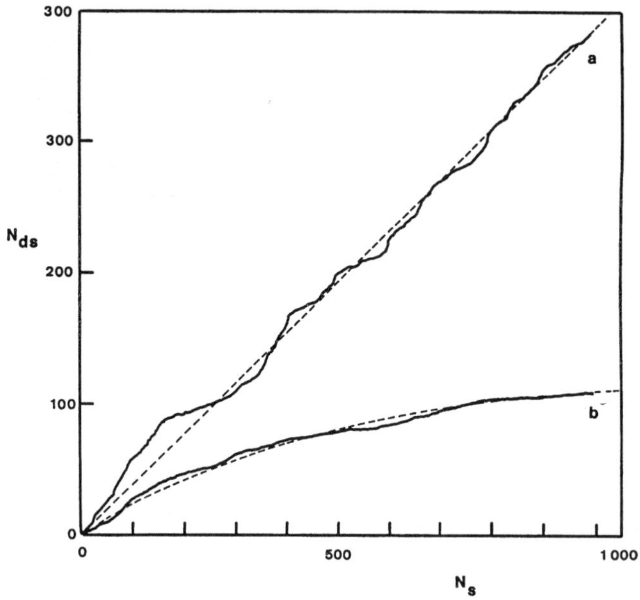

Figure 14: Time evolution of the number of distinct sites visited by a random walker on a regular lattice (a) and on a fractal network (b). Exploration of new sites slows down considerably at long times in the latter case.

the environment: the accessible interface and the open void space. The scaling exponents characterizing the geometry of this fractal space are strongly related to the interactions between the elementary clay sheets at the molecular scale, as controlled by adsorbed ions or organic compounds. Adsorbed ions or molecules that induce a strong local stacking order of the elementary clay sheets lead, at the macroscopic scale, to aggregates that are fragile and extensively microcracked, with a crack-network geometry that approaches space filling ($D = 3$). On the other hand, ions or molecules that cannot impose strong local order lead to more plastic and much less extensively microcracked aggregates of entangled sheets, in which the microcrack network geometry is non-trivial ($D < 3$).

The fractal character of the accessible interface and open void space of many smectite aggregates has far-reaching consequences as far as diffusion-reaction processes are concerned. One expects non-classical kinetic laws and time-dependent diffusion coefficients. Among the anticipated characteristic features is the occurrence of very long kinetic "tails" (the process seems to never end) in adsorption and desorption. This should be taken into account when modeling controlled release of pesticides or pollutant migration, for instance.

Fractal geometry is in itself a source of non-linearity in the kinetics of exchange processes (*i.e.*, Fick's first law with a time-dependent diffusion coefficient). In the case of smectite clays and many soils, the situation is even more complex since the geometry of the fractal space (its fractal exponents) may be modified by the textural action of the compounds that are adsorbed or desorbed. Thus, there is a double source of non-linearity. Diffusion-adsorption-reaction processes occur in a fractal space, but the geometry of this fractal space is modified by the migrating species themselves. Analyzing such processes properly is a major challenge that is at the heart of our understanding of soil chemistry.

References

Avnir, D., D. Farin and P. Pfeifer. 1984. Molecular fractal surfaces. *Nature* 308:261-263.

Avnir, D., D. Farin and P. Pfeifer. 1985. Surface geometric irregularity of particulate materials: The fractal approach. *J. Coll. Interf. Sci.* 103:112-123.

Aylmore, L.A.G. and J.P. Quirk. 1967. Micropore size distributions of clay mineral systems. *J. Soil Sci.* 18:1-17.

Ben Ohoud, M., F. Obrecht, L. Gatineau, P. Levitz and H. Van Damme. 1988. Surface area, mass fractal dimension, and apparent density of powders. *J. Coll. Interf. Sci.* 124:156-161.

Ben Ohoud, M. and H. Van Damme. 1990. The fractal texture of swelling clays. *C.R. Acad. Sc., Paris*, 311, series II:665-670.

Evertsz, C.J. and B.B. Mandelbrot. 1992. Harmonic measure around a linearly self-similar tree. *J. Phys. A* 25:1781-1797.

Farin, D. and D. Avnir. 1987. Reactive fractal surfaces. *J. Phys. Chem.* 91:5517-5521.

Farin, D. and D. Avnir. 1989. The fractal nature of molecule-surface interactions and reactions. p. 271-293. In D. Avnir (ed.) *The fractal approach to heterogeneous chemistry.* Wiley, Chichester.

Gregg, S.J. and K.S.W. Sing. 1982. *Adsorption, surface area and porosity.* Academic Press, London.

Guinier, A. 1964. *Théorie et technique de la radiocristallographie.* Dunod, Paris.

Gutfraind, R. and B. Sapoval. 1993. Active surface and adaptability of fractal membranes and electrodes. *J. Phys. (Paris)* I 3:1801-1818.

Havlin, S. 1989. Molecular diffusion and reactions. p. 251-269. In D. Avnir (ed.) *The fractal approach to heterogeneous chemistry.* Wiley, Chichester.

Ignatzek, E., P.J. Plath and U. Hundorf. 1987. The fractal character of zeolites. Part I. The fractal dimension of cobalt(II) phtalocyanine loaded faujasite. *Z. Phys. Chem.* (Leipzig) 268:859-873.

Murray, R.S., K.J. Coughlan and J.P. Quirk. 1985. Nitrogen sorption isotherms and the microstructure of vertisols. *Aust. J. Soil Res.* 23:137-149.

Murray, R.S. and J.P. Quirk. 1990. Surface area of clays. *Langmuir* 6:122-124.

Rieu, M. and G. Sposito. 1991a. Fractal fragmentation, soil porosity and soil water properties: I. Theory. *Soil Sci. Soc. Am. J.* 55:1231-1238.

Rieu, M. and G. Sposito. 1991b. Fractal fragmentation, soil porosity and soil water properties: II. Applications. *Soil Sci. Soc. Am. J.* 55:1239-1244.

Skjeltorp, A.T. and P. Meakin. 1988. Fracture in microsphere monolayers studied by experiment and computers simulation. *Nature* 335:424-426.

Stul, M.S. and L. Van Leemput. 1982. *Surf. Technol.* 16:101.

Tessier, D. 1990. Behaviour and microstructure of clay minerals. p. 387-415. In M. De Boodt, M. Hayes and A. Herbillon (eds.) *Soil Colloids and their Association in Aggregates.* Plenum Press, New York.

Van Damme, H., P. Levitz and L. Gatineau. 1986. Energetical and geometrical constraints on adsorption and reaction kinetics on clay surfaces. p. 283-304. In R. Setton (ed.) *Chemical reactions in organic and organic constrained systems.* Reidel, Dordrecht.

Van Damme, H., P. Levitz, L. Gatineau, J.F. Alcover and J.J. Fripiat. 1988. On the determination of the surface fractal dimension of powders by granulometric analysis. *J. Coll. Interf. Sci.* 122:1-8.

van Olphen, H. 1963. *An introduction to clay colloid chemistry.* Wiley Interscience, New York.

Vicsek, T. 1992. *Fractal growth phenomena.* World Scientific, Singapore.

Fractal Probes of Humic Aggregation: Scattering Techniques for Fractal Dimension Determinations

V.J. Homer

Contents

I Introduction

Humic acids comprise an important fraction of the organic materials found in natural water systems and in soils. These hydrophobic macromolecules are defined operationally by the manner in which they are isolated: as the component of natural organic matter soluble in base but insoluble in acid. Despite their low water solubility, humics have high sorption capacities for a variety of

ISBN 1-56670-105-8

trace materials. As a result, they are considered important to the transport and eventual deposition of numerous environmentally significant species, including pesticides and toxic heavy metals. These sorption and transport capabilities are expected to be strongly affected by humic aggregation characteristics, including intramolecular functional group interactions, and associations between distinct humic clusters as they form macromolecular complexes. The roles of humics in non-pathological environmental situations–the recycling of nutrients and minerals, maintaining redox activities especially in estuarine systems, interacting with clays and other particles in the shaping of sediment or soil textures–are also affected by their patterns of aggregation. Consequently, along with efforts to characterize the chemical composition of humic acids, considerable attention has been focussed over the past several years on understanding their physical structure and aggregation behavior.

Various strategies have been explored in the effort to understand humic size and shape characteristics, as will be outlined in section II. The findings from these studies have shed considerable light on humic morphology, and form a useful complement to the new experimental approach presented here: testing for fractal characteristics. This new approach to the study of humic aggregation derives naturally from the random walk manner in which these aggregates are likely to be formed, for example in percolating through soils, or sorbing onto inorganic species in surface or subsurface waters. In the past decade, several random walk colloidal systems (usually hard-sphere silica colloids) have been found to form fractal structures in solution or suspension (Schaefer *et al.*, 1984; Chen and Teixeira 1986; Weitz *et al.*, 1987; Teixeira 1988; Schmidt, 1991). The concept of fractal dimension (D_f) is now proving to be a valuable tool in understanding humic aggregation morphology, as indicated in the results of the small set of experiments presented to the scientific community thus far and reviewed here (Rice and Lin, 1992, 1993, 1994; Ostenberg and Mortensen, 1992; Ostenberg *et al.*, 1994; Homer *et al.*, 1992a, 1992b, 1994; Senesi *et al.*, 1994). These results demonstrate that D_f is a useful parameter for quantifying variations in humic aggregation in response to several distinct causes. A précis of the relevant scattering theory is presented in section III.

One of the side benefits of interest in fractal analysis is that it has unquestionably improved the access of researchers in humic materials to small-angle scattering instrumentation. This experimental technique is uniquely suited to evaluate the morphology of macromolecules, whether or not the configuration is determined to be fractal. For example, results of scattering experiments with humic acids (Homer *et al.*, 1992a) indicate that even at high concentrations these amorphous molecules do not organize as micelles or random coils in the manner that those morphologies are usually depicted, although these descriptions are useful in a generalized sense as described below. By contrast the fractal aggregation model, consisting of the random-walk organization of small amorphous humic entities into larger clusters in a self-similar scaling process, is intuitively attractive as well as being justified by experiment.

The series of fractal dimensions observed thus far for humic acids (see Table 1) suggests that there are limits to the extent that observed aggregation patterns can be neatly organized into previously specified categories (*e.g.*, cluster-cluster, reaction-limited or diffusion-limited) which have been largely developed from computer simulations. Nonetheless, these categories do help conceptualize the different ways in which small species organize into larger ones, and to understand how the "effective porosity" description inherent in the fractal dimension can vary with circumstance: be it pH (Senesi), source and/or physical form (Rice and Lin, 1992, 1993, 1994), concentration, source, and/or availability of metal ions (Homer *et al.*, 1992a, 1992b, 1994), or temperature (Osterberg *et al.*, 1994).

II Exploring Humic Aggregation Patterns

II.A Pre-Fractal Experiments and Models

Several approaches have been developed to describe the configuration of humic materials. Those listed below are of interest, to help interpret and/or to be themselves put into perspective, in comparison with the fractal description of aggregation.

II.A.a Gaussian Random Coil Model, Based on Ultracentrifugation Experiments

Cameron *et al.* (1972) extended the initial ultracentrifugation studies conducted by Stevenson *et al.* (1953) on unfractionated humics, by applying several fractionation techniques to separate their humic material into a series of distinct size classes, which were then subjected to ultracentrifugation. [An overview of the theory and application of this technique, including the diffusion/sedimentation equation used to yield diffusion and sedimentation parameters, is given by R.S. Swift, 1989.] The molecular weights of each of these fractions were estimated from the diffusion coefficients obtained in distinct sedimentation experiments; frictional coefficients were acquired similarly from the data. The set of frictional coefficients were then plotted against the molecular weight values. That plot was shown to give a reasonable fit (for the 7 lightest of the 10 fractions evaluated) to the functional dependence expected for flexible random coils, where the shape of the plot of mass *versus* radial distance from the center of the coil is Gaussian. While acknowledging that their data were also consistent with the model of condensed oblate ellipsoids, plausibility arguments led to the selection of the random coil model as being more appropriate for humic acids.

An attractive feature of this model is that its apparent lack of structure seems consistent with the amorphous chemical nature of humic materials. However, the word random in this instance can be misleading. In standard random coil models (Kratky, 1982; Cotton, 1992; and especially Kirste, 1965), the central portion of the small-angle scattering curve reflects that the overall coil has a mass fractal dimension of 2 or 5/3, depending on the interaction with the solvent. The tail of the scattering curve displays the long rod geometry of the individual segments of the coiled structures. The position of the crossover from overall coil scattering to long rod scattering specifies the persistence length, or degree of curving, of the coil. The ability to polymerize into such relatively well organized structures as Gaussian coils indicates a high degree of similarity for the comprising monomers (D.W. Schaefer, private communication, 1992). None of the spectra of the humics reviewed here, nor to my knowledge any spectra of humics as yet unpublished, fits neatly into this pattern. Nonetheless, the small-angle scattering patterns of these random coils do include features which may be roughly analogous with those of some recently examined humic acids, where some of the scattering spectra also appear to be bimodal (Homer, 1995a).

II.A.b Initial Small-Angle Scattering Experiments, Micelle/Membrane Model

In small-angle scattering experiments, the intensity of the probing beam is monitored as a function of the angle through which it is scattered by the material of interest. As detailed in section III, the shape of this scattering spectrum is determined by a combination of the morphology of individual objects, the aggregates they form, and at times the interactions between these objects. Wershaw *et al.* (1969) conducted the first small-angle scattering experiments with humic materials, which were continued into the 1980's (*e.g.*, Wershaw and Pinckney, 1973, Wershaw *et al.*, 1977, Wershaw, 1986). Samples were fractionated on Sephadex columns, and then subjected to different treatments, especially varying pH or oxidation state.

In the interpretation of their results, Warshaw's group at USGS posits the importance of intra- and inter-molecular hydrogen bonding to aggregation processes. [The proton dependence of aggregate formation has since been elegantly confirmed in fractal aggregation experiments conducted by Osterberg *et al.* (1994) discussed below.] By examining a series of distinct humic materials, the USGS group was able to capture variable responses to a changes in pH, thereby underscoring humic acid complexity. Some fractions dissociated measurably with increasing pH (above 3.5) and others did not, while at least one fraction first dissociated and then reassociated as pH increased above 7. [These latter experiments were the primary basis for the first small angle neutron scattering experiments conducted by Homer *et al.*(1992a,b) discussed below.]

Wershaw's group conducted the bulk of their experiments before the concept of fractal dimension was created in 1975 (Mandelbrot, 1983). By using small angle X-rays, rather than the not yet available small angle neutron scattering, they were limited to a system that involved substantial manipulation (desmearing) of the data. Although this consequently hampered the interpretion of their data, they were extremely careful in this process, using state-of-the-art desmearing methods. In the scattering plots expected to yield the effective average radius of the humic macromolecules (the radius of gyration), they pointed out that their scattering curves consisted of a series of such radii. They interpreted this as an indication of the polydispersity of their samples. However, these data are not inconsistent with the scattering curves from a fractal system, when plotted in such a fashion. It would be fascinating to re-evaluate their data, plotted in the manner appropriate to test for fractal behavior.

By the mid-1980's, Wershaw (1986) developed a model of humic materials similar to micelles of membrane-like structures. More recently (Hayter, 1988) it has been demonstrated that a standard feature of the small-angle scattering of ionic micelles involves the presence of an interaction peak in the scattering spectrum. This peak has not been observed with any of the humic materials examined by small-angle scattering to date, suggesting that the micellar model may over-simplify the character of humic aggregation. However the concept of hydrophilic components tending to cohere and to avoid hydrophobic components is plausible, potentially useful, and amenable to further study with the use of appropriate solvents in small-angle scattering and/or other related experiments.

II.A.c Microscopic Imaging of Humic and Fulvic Materials

Chen and Schnitzer (1976) presented scanning electron microscopy (SEM) images of humics freeze-frozen onto slides. In these, and later electron microscopy images (Chen *et al.*, 1976; Stevenson and Schnitzer, 1982), the humic and/or fulvic material appeared at times as blotches, or as rods, or as thin sheets. More recently, Tan (1985, 1988) has published highly resolved SEM images of soil humic aggregates; these are several orders of magnitude larger than estimates of the size of aqueous humic acids.

Small-angle scattering theory (and experimental results) can be used to resolve the applicability of configurations displayed in the electron microscopic imaging process. As described in section III, simple morphologies such as rods and sheets have straightforward scattering curves (*e.g.*, Guinier and Fournet, 1955; Hjelm, 1985; Hjelm *et al.*, 1990). The slopes of these curves in $\log(I)$ *versus* $\log(\mathbf{Q})$ plots (discussed in section III) are not consistent with the slopes of scattering curves of the humic materials examined thus far (Homer *et al.*, 1992a, 1994; Osterberg *et al.*, 1994; Rice and Lin, 1993). It is thus possible that the rod and sheet configurations that emerge in the SEM imaging are artifacts due to the process of sample preparation; equally likely, the soil humic materials examined in the electron microscopy studies may differ in morphology from the humics studied to date in small-angle scattering experiments. Comparison of the same humic and/or fulvic materials with both these techniques should help resolve this possible disparity.

II.A.d Quasi-Elastic Light-Scattering Experiments

Caceci and Moulin (1991) performed quasi-elastic light scattering experiments with humic and fulvic acids, using a powerful 0.5W laser source. In a demonstration of inducing humic aggregation by adding calcium, they added from 5-4 M to 1.6-2 M Ca(II) to 40 mg/l purified Aldrich humic acid. With small amounts of Ca added (0 to 4-4 M) the unimodal size distribution peaked at about 20 Angstroms, with the peak width at half-height ~30 Angstroms. Once the amount of Ca added exceeded 6.4-3 M, the size distribution (still unimodal) peaked at ~200 Angstroms. They observed no evidence of the presence of aggregates of intermediate size classes. This apparant lack of size class polydispersity, for this particular humic material, was an important consideration in the interpretation of the small-angle scattering results of similar humic materials by Homer *et al.* (1992b).

II.B Development of Fractal Scattering Approaches

II.B.a Small-Angle Scattering Approaches

[For the samples described below, fast kinetics and extensive growth were promoted by the high concentrations used. It is not unreasonable to hypothesize that the formation dynamics shaping subclusters in earlier stages of aggregation will similarly apply to the less concentrated samples found in natural settings, where cluster shapes form under conditions of slower kinetics and low growth. Under this hypothesis, findings on aggregate forms for concentrated materials should apply to those environments where humics are less extensively concentrated. Nonetheless, the concentration dependence of the the fractal dimension displayed in the humics examined by Homer *et al.* (1992a) demonstrates that formation dynamics are at least to some extent concentration dependent, and that care should correspondingly be taken in extrapolating these results to lower concentrations.]

Small-angle scattering is a well established probe of the morphology of organic and inorganic macromolecules. The scattering curve, plotting the intensity of the scattered beam as a function of scattering angle, yields information about the shapes of the aggregated macromolecules (especially those having simple geometries), and about the subparticles from which those aggregates are formed. Recent theoretical developments in this field now allow experimentalists to deduce useful information about the morphology of amorphous species, especially fractal aggregates (*e.g.*, Schaefer *et al.*, 1984; Teixeira, 1988; Schmidt, 1991). Recently, three groups working independently (Rice and Lin, 1992, 1993, 1994; Osterberg and Mortensen, 1992; Osterberg *et al.*, 1994; Homer *et al.*, 1992a,b, 1994) have used small-angle scattering to study humic acid morphology. Their results indicate that humic acids in solution aggregate as mass fractals: quasi-open structures whose degree of openness is characterized by the mass fractal dimension D_f.

Rice and Lin (1993, 1994) conducted small-angle X-ray scattering experiments and found the mass fractal dimension to vary from 1.6 to 2.5 for the two humic materials they examined in solution, whereas the surface fractal dimension of the powdered humic and fulvic acids they examined ranged from 2.2 to 2.8. By contrast, Osterberg and Mortensen (1992) found in small angle neutron scattering experiments that both humic acids they examined in solution had the same mass fractal dimension: 2.3. In more recent work, Osterberg *et al.* (1994) showed the fractal dimension to be strongly temperature dependent, rising from 1.8 to 2.35 as the temperature increased from 4°C to 22°C. By also monitoring changes in emf which were attributed to proton availability, they were able to demonstrate that protons were initially quickly released, and then slowly consumed, as aggregation and/or compaction proceeded with the rise in temperature.

Homer *et al.* (1992a,b, 1994) examined four humic acids in solution in a series of small angle neutron scattering experiments, finding the humics aggregated as mass fractals with different fractal dimensions. NMR spectra of the same materials permitted a comparison of the fractal dimension D_f of these humic acids with their chemical functional group properties. The humic acids with relatively high polysaccharide content had lower D_f values, indicative of less condensed aggregation configurations. Concentration normalized absolute intensities (scattering strengths) were monitored, to track changes in aggregation states. The scattering strength did increase with increasing aliphatic content: whether that was caused by aggregation processes, or simply indicates that increased non-substitution hydrogens may have modified the scattering length density, has not as yet been determined.

Fe^{3+} was added to some of the humic samples and was found to contribute significantly to humic aggregation. Similarly, stripping away heavy metal content through cation exchange chromatography led to sharply decreased scattering strengths. Changes in scattering strength due to changes in metallic content will not have been an artifact due to changes in scattering length densities (Equation (2)), since in neutron scattering most of these elements have similar scattering lengths. The increase in absolute scattering intensity must therefore have been caused by changes in aggregate morphology. The most probable changes would be an increase in cluster size with presence of higher metal ion content (Equation (17) and surrounding discussion), or new clusters forming from small fragments previously unaggregated. However, another possibility is that the degree of flexibility may vary for some humics, and this modification could affect the scattering intensity, for example by reducing the effective radius of gyration (the average mass radius) for highly flexible samples.

Relative openness is specified by the fractal dimension D_f throughout the range of length scales included in the fractal characterization. This feature of the aggregate geometry is thus specified down to subclusters whose characteristic lengths fall within this fractal range. In some cases, small subclusters or cluster segments which may be non-fractal, or which have a fractal dimension different from the overall aggregates, have been shown to be present in the scattering spectra (Homer, 1995a,b). Most striking were the effects of variations in Fe content, or in chemical functional group ratios.

II.B.b Turbidity Measurements of Fractal Dimension

Senesi *et al.* (1992) used lightwave transmission spectroscopy to monitor turbidity of humic acid suspensions. They explored the pH dependence of the fractal dimension of soil humic acids suspended in aqueous solutions, varying pH from 3 to 6, and found that D_f varied from 2.7 to 1.5 as pH was increased. These turbidity studies measure the total intensity of light waves impinging on the sample, minus the intensity of the diffracted waves integrated over angle diffracted (*i.e.*, minus the intensities measured in a scattering experiment). In order to recapture information about the morphology of the particles under study, the turbidity measurement is taken over a range of wavelengths. For each wavelength probed, all the spectral information contained in a scattering experiment is thereby compressed to a single data point in the transmission study. Light scattering experiments are based on the same theory as small-angle scattering experiments (see section III), the primary difference being that light waves are much larger than X-ray or neutron waves, so that larger angles must be probed to obtain the same values of the scattering vector $\mathbf{Q}$ (see Equation (1) below). Understanding the turbidity transmission studies thus entails combining small-angle scattering theory with interpretation of the wavelength dependence of the scattering spectra.

III Scattering Theory of Macromolecules

[For full appreciation of the theory of small-angle scattering, and its experimental verification, the best references are still Guinier and Fournet (1955) and Glatter and Kratky (1982); Lindner and Zemb (1991) present state of the art applications by a series of authors; the book edited by Safran and Clark (1987) is equally state of the art, and goes into great depths presenting the small-angle scattering of colloids, micelles, and fractals. The theory presented below applies to light scattering (where as explained above the angles are not necessarily small) as well as to small-angle X-ray scattering (SAXS), and to small-angle neutron scattering (SANS). Only the individual "cross section" terms vary: for light scattering this is the index of refraction, and for X-rays the electron scattering factor. The corresponding neutron scattering term, the scattering length density, will be the specific cross section parameter used in the following.]

Small-angle scattering can be used to characterize the morphology of macromolecules ranging in size from ~10 to ~2000 Angstroms. The relationship between particle morphology and the corresponding scattering curve is well understood for a variety of geometries, and continues to develop as more complex structures are considered. The basic strategy is to explore how patterns in the structural geometry generate patterns in the phase shifts of the neutron waves, and ultimately in the scattering curve itself.

The scattering intensity (I) is a function of the momentum transfer vector $\mathbf{Q}$: the change in a neutron's momentum after it interacts with the macromolecule. [The scattering described here is assumed to be elastic: there is no gain or loss of kinetic energy or magnitude of neutron momentum upon interacting with the mass in the sample, merely a change in direction of the neutron wave and its momentum vector $\mathbf{k}$.] The magnitude of the neutron wave momentum vector $\mathbf{k}$ is $2\pi/\lambda$, where λ is the neutron wavelength. Technically, $\mathbf{Q} = \mathbf{k}_i - \mathbf{k}_o$, with $\mathbf{k}_o$ the momentum of the incoming neutron wave, and $\mathbf{k}_i$ that of the scattered wave; in practice, $\mathbf{Q}$ is frequently (and unambiguously) identified with the scattering vector $\mathbf{k}_i$. The magnitude and direction of $\mathbf{Q}$ are given by the scattering angle and the neutron wavelength :

$$|\mathbf{Q}| = (4\pi/\lambda)\sin(\theta/2), = 2|k|\sin(\theta/2) \tag{1}$$

The scattering amplitude $A(\mathbf{Q})$ represents the sum of the neutron waves, scattered from all positions $\mathbf{R}_i$ in the sample where mass of the macromolecules is located:

$$A(\mathbf{Q}) = \phi_i \rho_i \exp(i\mathbf{Q} \cdot \mathbf{R}_i) \tag{2}$$

The scattering length density ρ_i measures the strength of interaction between a neutron wave and the ith subcomponent in a structure, and depends on that component's atomic composition and subsumed volume. The intensity $I(\mathbf{Q})$ of the scattering curve is the ensemble average of the square of the magnitude of the scattering amplitude:

$$I(\mathbf{Q}) =< |A(\mathbf{Q})|^2 >= \phi_i \phi_j \rho_i \rho_j \cos(\mathbf{Q} \cdot \mathbf{R}_{ij}) \tag{3}$$

where $\mathbf{R}_{ij} = (\mathbf{R}_i - \mathbf{R}_j)$ is the spacing (or chord) between mass components. Each of the terms in Equation (2) is the component of the wave shifted in phase by its difference in pathlength from a neutron scattering at an arbitrarily chosen origin. Only these phase-shifted components contribute to the net scattering intensity. Individual phase shifts are given by the inner product $\mathbf{Q} \cdot \mathbf{R}_i$ in Equation (2). Phase shifts between objects in solution (or between mass components within an object), separated by the relative position chords $\mathbf{R}_{ij} = (\mathbf{R}_i - \mathbf{R}_j)$, are given by the inner product terms $\mathbf{Q} \cdot \mathbf{R}_{ij}$ in Equation (3). The scattering intensity in a given direction $\mathbf{Q}_o$ is thus generated by the set of terms $(\mathbf{Q}_o \cdot \mathbf{R}_{ij})$. This set is determined by the distribution of the set of chords $\{\mathbf{R}_{ij}\}$, which in turn is a function of an object's morphology and of the spacing between objects.

Certain specific features of sample geometry, such as the well-defined interparticle spacing between ionic micelles, will have relatively high populations of a given chord $\mathbf{R}_{ij}$. For each Q, $I(Q)$ contains a correspondingly high population of cosine terms with arguments $\mathbf{Q} \cdot \mathbf{R}_{ij}$. In particular, for values of Q approaching $Q_1 = k\pi/\mathbf{R}_{ij}$ (for integer k), this set of cosine terms becomes more dominant, thus tending to strengthen the scattering intensity at $I(Q_1)$. This quasi-inverse relation between Q space and the sample space is a general feature in all forms of scattering, but can sometimes tend to be oversimplified. Rigorously, the scattering intensity $I(Q)$ is the Fourier transform of the correlation function $c(\mathbf{R}_{ij})$ of the distribution of relative position chords $\mathbf{R}_{ij}$.

III.A Form Factor, Structure Factor

The contribution to the scattering intensity dictated by geometry of the objects themselves (averaged over all possible orientations) is known as the form factor $P(\mathbf{Q})$. Where objects spaced close together have sufficiently uniform interparticle spacings, the correlations that develop give rise to an interparticle component of the scattering intensity known as the structure factor $S(\mathbf{Q})$. For a system of identical objects in solution, the position vectors R_i can be rewritten:

$$\mathbf{R}_i = \mathbf{l}_k + \mathbf{L}_j \tag{4}$$

where $\mathbf{l}_k$ is the distance from the center of mass of the jth object to its kth mass element, and $\mathbf{L}_j$ is the distance from an arbitrary origin to the jth object's center of mass. The scattering intensity can now be regrouped as:

$$I(\mathbf{Q}) = [\phi_m \phi_{m'} \rho_m \rho_{m'} \cos(\mathbf{Q}.\mathbf{l}_{mm'})] \cdot [\phi_n \phi_{n'} \cos(\mathbf{Q}.\mathbf{L}_{nn'})] \tag{5a}$$

or:

$$I(\mathbf{Q}) = P(\mathbf{Q}).S(\mathbf{Q}) \tag{5b}$$

where the indices $\{m, m\prime\}$ are summed over the distinct mass components within a representative object having total mass M, the indices $\{n, n\prime\}$ are summed over the total number of objects N in the solution, and the relative distances are given by $\mathbf{l}_{mm'} = (l_m - l_{m'})$ and $\mathbf{L}_{nn'} = (l_n - l_{n'})$. The $\mathbf{l}_{mm'}$'s thus denote the chords between mass components *within* a representative object, whereas the $\mathbf{L}_{nn'}$'s denote the center-of-mass spacings *between* these objects.

The first term in Equation (5b) is the *intra*particle structure factor $P(\mathbf{Q})$, known as the form factor. Because the terms in $P(\mathbf{Q})$ apply to mass components in close proximity to one another the scattering is coherent, allowing cross-terms to contribute to the net scattering intensity. As the scattering angle goes to zero, waves scattered from these mass components will tend increasingly to be in phase with one another, so that the destructive interferences that reduce net scattering intensity will disappear. [These destructive interferences, which come from variations in relative path length, lead to the relative phase shifts in Equation 5 being out of phase. The affected cosine terms subsequently tend to cancel one another, thereby decreasing the scattering intensity.] For an object of mass M having uniform scattering length density, $P(Q)$ is proportional at $Q \sim 0$ to the square of its total number (M) of mass components:

$$\begin{aligned} P(\mathbf{Q} &= \mathbf{0}) = \phi_m \phi_{m'} \rho_m \rho_{m'} \cos(\mathbf{0}.\mathbf{l}_{mm'}) \qquad \text{(m,m}'\text{=1 to M)} \\ &= \rho^2 (\phi_m 1 \phi_{m'} 1) = \rho^2 M^2 \end{aligned} \tag{6a}$$

whereas, in general:

$$P(\mathbf{Q}) = \rho^2 \phi_m \phi_{m'} \cos(\mathbf{Q}.\mathbf{l}_{mm'}) = \rho^2 P'(\mathbf{Q}) \tag{6b}$$

Destructive interferences emerge as the scattering angle increases from zero. For values of Q much less than the inverse of an object's radius of gyration R_g (its average mass radius), the scattering is Gaussian whatever the object's specific geometric configuration. This holds because the arguments of the cosine terms in $P(Q)$ are small enough in the range of Q values $QR_g << 1$ (called the Guinier region, section III.B. below) to minimize the effects of interferences generated by all but the broadest features (such as R_g) of the object's distinct geometry. As Q increases beyond the Guinier region, interferences determined by the object's morphology dominate the scattering intensity $P(Q)$. Mathematically, the relation between an object's geometry and its form factor is expressed by the fact that $P(Q)$ is generated by the Fourier transform of the object's intraparticle correlation function, which is the distribution function of its set of relative position chords $\{\mathbf{l}_{mm'}\}$.

The second factor in Equation (5b) is the *inter*particle structure factor $S(\mathbf{Q})$, usually referred to as the structure factor, and is generated by the Fourier transform of the interparticle correlation function (the distribution function of the set of relative position chords $\{\mathbf{L}_{nn'}\}$ between objects in a sample). $S(\mathbf{Q})$ becomes non-trivial when objects close to one another develop correlations in their interparticle spacing L_o (frequently the minimum distance between objects). Such correlations require more than physical proximity: the separation distances must not fluctuate too greatly over time. As Q goes towards zero ($Q << L_o$), these correlations disappear, and phase shifts from distinct objects will on average be out of phase with one another. The cross-terms in $S(\mathbf{Q} << L_o)$ will thus tend to cancel one another, only the $n = n'$ terms will remain, and $S(\mathbf{Q})$ will simply count the number of objects N in solution. For objects lacking such correlations, the cross-terms in $S(\mathbf{Q})$ will on average cancel one another throughout the full range of Q space. Thus, whether or not non-trivial interparticle correlations exist in the sample, one has:

$$\begin{aligned} S(\mathbf{Q} &\sim 0) = \phi_n \cos(\mathbf{Q}.(\mathbf{L}_n - \mathbf{L}_n))^2 \qquad (n = 1 \text{ to } N) \\ &= \phi_n 1^2 = N \end{aligned} \tag{7a}$$

while for samples lacking interparticle correlations:

$$\begin{aligned} S(\mathbf{Q}) &= \phi_n \cos(\mathbf{Q}.(\mathbf{L}_n - \mathbf{L}_n))^2 \qquad (n = 1 \text{ to } N) \\ &= = \phi_n 1^2 = N \qquad \text{for all } \mathbf{Q} \end{aligned} \tag{7b}$$

[Note that, for values of Q less than the length scale $L_{overall}$ of the overall sample, the same logic that demonstrated $P(Q \sim 0) = M^2$ (Equation (6a)) similarly shows that $S(Q << L_{overall}) = N^2$. However, such small values of Q are well below the range of small-angle scattering experiments, and so are not relevant to this discussion.]

The full expression for $I(\mathbf{Q} = 0)$ is now seen to be:

$$I(0) = P(0) \cdot S(0) = \rho^2 M^2 N \tag{8a}$$

Moreover, in the absence of interparticle correlations, one has:

$$I(Q) = P(Q) \cdot S(Q) = \rho^2 N P'(\mathbf{Q}) \tag{8b}$$

A system with sufficiently well-defined interparticle spacing will generate an interaction peak in $S(\mathbf{Q})$. The position of that peak is usually determined by the spacing $L_o \sim 1/Q_o$. In complicated geometries the form of $S(Q)$ becomes correspondingly complex, and may no longer generate a simple peak in the scattering profiles. Moreover, scattering from both inter- and intra-particle effects often overlap in the scattering curve, because they frequently involve similar length scales. However, even when it is difficult to isolate the effects of $P(Q)$ from $S(Q)$, the presence of a peak in the scattering spectrum suggests the existence of a high degree of interparticle ordering. In particular, such an interaction peak is characteristic of the small-angle scattering of ionic micelles.

III.B Characteristic Scattering Curves

The following lists common scattering curves for samples with well-characterized geometric configurations:

1) **Guinier region:** In the absence of interactions with other objects modifying the scattering intensity, at low Q the scattering from **all** structures appears to be Gaussian:

$$I(Q) = I(0)\exp(-(R_g^2 Q^2)/3) \tag{9a}$$

where R_g is the radius of gyration. The plot of $\ln(I)$ *versus* Q^2 will be linear in this Guinier region, roughly delimited by the relation $QR_g < 1$; the value of $R_g^2/3$ is obtained from the slope of the line. In particular, for $Q_G = c_G/R_g$ (where $(1/R_g) < (c_G/R_g) < (2\pi/R_g)$ indicates the upper bound in Q space of Guinier scattering), one has:

$$I(Q_G) = \exp(-c_G^2/3)I(0) = k_G M^2 N \tag{9b}$$

This model is based on analysis of the first term in the series approximation of the form factor, and has been verified by experiment countless times. For a dense but rigorous analysis of the relevant mathematics, see Guinier and Fournet (1955).

2) **Simple geometries:** For non-interacting spheres, the scattering curve is essentially Gaussian (Guinier) up to the first minima in scattering intensity. For long rods, analysis of the form factor shows that the scattering intensity decreases as Q^{-1}; for sheets or membranes, the intensity decreases as Q^{-2}. The mathematical derivation of these and related form factors is presented in Guinier and Fournet (1955); excellent examples of fine-tuning of that analysis as a function of details of geometric configuration are given in Hjelm (1985) and Hjelm *et al.* (1990).

3) **Non-fractal interfaces, surfaces:** At length scales that probe surface interfaces, the scattering intensity for hard-sphere structures (*e.g.*, those with smooth,"solid" surfaces) decreases as Q^{-4}. This usually takes place at relatively large Q values (the size range probed being relatively small) known as the *Porod region.* For objects comprised of sufficiently hard-sphere substructures (such as clusters of colloidal silica) the crossover Q_c between the Q^{-4} scattering at large Q, and the form factor scattering $P(Q)$ observed at lower Q values, gives a good estimate of the length scale **l** of the primary substructures themselves ($Q_c \sim 1/\mathrm{l}$). Those other curve shapes represent the configurations observed at larger size scales. The pre-fractal mathematical derivation for the Q^{-4} behavior for hard spheres is given by Porod in Glatter and Kratky (1982). Recent work probing behavior when the interface is fuzzy, or otherwise challenging to analyse, is found in Lindner and Zemb (1990).

4) **Micelles:** When ionic micellar structures are formed at sufficiently high concentrations from amphiphilic substructures, electrostatic interparticle repulsions lead to an interaction peak in the structure factor $S(Q)$ at low Q, due to positive correlations between neighboring objects. Many micelles form into characteristic geometric shapes, such as long rods or ellipsoids. The scattering associated with those shapes is obtained in the region of Q space that (roughly) corresponds inversely to the corresponding length scales in the object's geometry. Lucid presentations of micellar theory and experiment are outlined by Hayter (1988) and presented in depth in Hayter (1987).

III.C Scattering of Fractals

[Notation follows Weitz and Huang (1984); fractal scattering concepts found in Teixeira (1988), in Schaefer (1987), and in Schmidt (1991) are also valuable.]

For mass fractal aggregates the form of the scattering curve can be derived (Sinha *et al.*, 1984) from the correlation function *c(Rij)* of the distribution of relative position chords $\{R_{ij}\}$, which conveys the mass density correlations. By definition, for mass fractals the mass scales as:

$$M(R) = m_o R^D \qquad \text{for all } R \text{ in } \{R_{ij}\} \tag{10}$$

where M is the mass contained in the volume of length scale R in terms of the unit m_o, and $0 < D < 3$ denotes the fractal dimension. The correlation function $c(R_{ij})$ for this mass density rule is:

$$c(R_{ij}) \propto R_{ij}^{(D-3)} \tag{11a}$$

$$\Rightarrow \log(c(R_{ij})) \propto (D-3)\log(R_{ij}) \tag{11b}$$

for which the Fourier transform yielding the form of the scattering intensity is:

$$I(Q) = c_o Q^{-D} \tag{12a}$$

$$\Rightarrow \log(I(Q)) = \log(c_o) - D\log(Q) \tag{12b}$$

where c_o is a constant of proportionality, and the fractal dimension D is the slope of the line corresponding to Equation 12b. As aggregates become more condensed and the fractal dimension D increases towards 3, the relative mass density $M/V = R^D/R^3 = R^{(D-3)} = c(R)$ increases. The term $(D-3)$ in the correlation function (Equation (11a)) subsequently becomes less negative, making the slope in $\log(c(R_{ij}))$ more shallow. The slope in the corresponding scattering intensity term $\log(I(Q))$ (Equation (12b)) becomes steeper for the more highly aggregated fractal clusters. This pattern displays again the quasi-inverse relationship between the correlations of the mass position chords R_{ij} and the scattering intensity, which is formalized mathematically by their being Fourier transforms of one another.

This theory describes the scattering spectra for Q values $L_1^{-1} < Q < L_2^{-1}$, where L_1 roughly corresponds to total aggregate size ($L_1 \sim R_g$), and L_2 corresponds to the length scales of its nonfractal substructures. In the absence of other morphological correlations affecting the scattering intensity, the scattering will be primarily Guinier (Equation (9a)) in the Q-range below $Q_1 \sim L_1^{-1}$; above $Q_2 \sim L_2^{-1}$, the scattering will be dominated by spectra reflecting the morphology of these substructures. These latter spectra will be well defined to the extent that the morphology of the corresponding substructures is relatively uniform. The earlier the stage in cluster formation that fractal aggregation patterns emerge, the smaller will be the non-fractal constituents, and hence the further out in Q-space will $Q_2 = L_2^{-1}$ be found. Caution must be taken in interpreting the spectral results: in the region of overlap between fractal and non-fractal contributions to the scattering curve, non-fractal components may partially obscure the fractal scattering, especially if the fractal regime ($L_1^{-1} < Q < L_2^{-1}$) is somewhat narrow.

The spectra of fractal clusters of hard sphere silica colloids incorporate non-fractal Q^{-4} Porod interfacial behavior at large Q (see section above, and Schaefer, 1984). By contrast, the spectra of fractal clusters of numerous humic acid solutions do not (down to length scales of ~20-30 Angstroms, see Table I), indicating that the humic substructures of which they are comprised are not tightly collapsed bundles. This observation is consistent with Rice's finding that even in highly condensed freeze-dried powder form, humic and fulvic acids form surface fractals (Rice and Lin, 1993). Under certain conditions, such as low Fe content or at low concentrations, the slopes of some humic acids in solution become quite shallow at large Q, suggesting that the spectra in this region represent Guinier scattering of quite small unaggregated humic material ($R_g <$ 20 Angstroms) (J.S. Lin (private communication, 1992)); this hypothesis is consistent with some unpublished results of Rice (private communication, 1994). In some cases the unaggregated species (or aggregate subcomponents) are large enough to observe scattering indicative of their morphology (Homer, 1995). More data will need to be collected at both extremes of Q values, for different types of humic material, before the variety of morphologies characteristic of humic clusters can be established.

III.C.a Monitoring Humic Fractal Aggregation Levels

[Based on similarities in overall elemental composition, it is assumed here that the scattering length densities r are roughly the same for different humic samples and so are not a factor in the following calculations.]

In small angle neutron scattering the scattering intensity $I(Q)$, when corrected for background and sensitivity effects, represents a well-defined scattering cross section known as the absolute intensity (Wignall & Bates, 1987). This allows for unambiguous comparisons between scattering curves from different samples, and for combining scattering curves from a given sample, taken at different sample-to-detector geometries (corresponding to different Q ranges). Along with radii of gyration and fractal dimension calculations, comparisons of absolute intensities from different humic samples, or for a given sample over a period of time, can also help evaluate changes in aggregation states. The extent of aggregation can vary as a function of humic chemical composition (including metal ion content) and/or sample treatment. The possible ways in which aggregation may proceed under varying circumstances include the following, each of which will be considered in turn:

- the number N of aggregates of given size and shape may increase, and/or
- the type of cluster geometry may be modified, and/or
- aggregate masses M may increase.

In the absence of aggregation processes or morphological variations modifying different humic samples, clearly the number of aggregates N will be proportional to total concentration C. As long as no interparticle correlations modify the scattering spectra, the absolute scattering intensities for such samples will be identical to one another, varying only in the factor multiplying C (and hence N). Therefore it is useful to employ concentration-normalized absolute intensities in comparing aggregation states of different humic samples, in order to isolate the different processes that may be involved, even when aggregate geometry is unaffected. In particular, where the concentration-normed number N of aggregates does vary (but not their size or shape), the concentration-normed absolute intensities will vary by that same factor in both the Guinier and fractal portions of the scattering curves. This situation will be found in cases where previously unaggregated species begin to aggregate, whether due to increases in concentration, or to other variations in sample condition such as metal ion content or temperature changes. If larger aggregates are formed as well in these processes, then the range in Q space where the scattering is fractal will be increased (as discussed below). If smaller aggregates are formed, interferences from their Guinier scattering will reduce the apparent range of fractal scattering. The concentration and/or mass of the unaggregated species of length scale L_{small} may be large enough to generate an observable Guinier contribution to the overall scattering intensity, which will resemble a gently sloping background for $Q < 1/L_{small}$. Thus, variations in scattering curve intensities that do not modify the fractal range, and that multiply the fractal and Guinier sections of the scattering curve by the same factor, can safely be attributed to increases only in the numbers of aggregates. [This may or may not include a reduction of Guinier scattering from the previously disaggregated species, depending on their size and concentration.] Such behavior would indicate the existence of a characteristic aggregate size, as is the case with the formation of micelles. As more data is collected from different humic samples, it will be interesting to learn to what extent this type of behavior, predicted by Wershaw (1986), is observed.

When the shape of aggregate geometry is modified, the previous fractal scattering curve will be replaced by a new spectral curve that reflects the changes in cluster morphology. (This may be fractal as well, but with a different fractal dimension.) One plausible scenario would be for fractal clusters to collapse under increased aggregation, but it would be no less conceivable for lightly bound subparticles to collect loosely on previously formed clusters, especially if the mechanism of aggregation for these subparticles involves weaker bonding energies. Comparisons of absolute intensities in such cases will only be meaningful when the scattering curves can be described in terms of parameters reflecting aggregate morphology.

If the constituent mass is already fully aggregated, then increasing the size of the aggregates by a given factor (k) involves nothing more than a redistribution of the N fractal clusters of mass M and mass fractal dimension D. The effect on the scattering intensity can be seen by considering two samples (concentration normalized) whose average cluster masses are, respectively, M_1 and $M_2 = \mathrm{k}M_1$; their (normalized) numbers of aggregates are N_1 and $N_2 = (1/k)N_1$; and the the radii of gyration are $R_{1g} \sim M_1^{1/D}$ and $R_{2g} \sim M_2^{1/D} = (k^{1/D})M_1^{1/D}$. [The radius of gyration is proportional to the aggregate's actual radius R_o, where the constant of proportionality is a function of its geometry; for spheres $R_g^2 = 3R_o^2/5$. For the aggregation comparison purposes here, R_g can be safely substituted for R_o.]

The redistribution of the masses will only increase the (concentration normalized) absolute intensity below the lower bound of fractal scattering for the less aggregated sample: $Q_{1f} = c_f/R_{1g}$, where as before $((1/R_{1g}) < (c_f/R_{1g}) < (2\pi/R_{1g}))$. [In general there may be a region of crossover between the upper bound Q_G of Guinier scattering and the lower bound Q_f of fractal scattering, so that c_G may be less than c_f. However, as long as the shapes of the aggregates being compared are essentially the same, the shape of the crossover scattering should be as well, and thus not interfere with the reasoning below. For simplicity then, in the following it is assumed that $Q_G = Q_f = Q_{G-f}$.] Both fractal absolute intensities can be shown to overlap above Q_{1G-f}, as is shown below. Remembering that $Q_{1G-f} = c_G/R_g = c_G/M^{1/D}$, and that $I(Q_{G-f}) = k_G M^2 N$ (Equation (9b)), then $I(Q)$ above the crossover position Q_{G-f} can be expressed in terms of the parameters M, N, and k_G, as well as by the fractal dimension D. By linearity of $\log(I)$ *versus* $\log(Q)$ plots (Equation (12b)), one has:

$$\begin{aligned}\log I(Q) &= \log I(Q_{G-f}) - D(\log(Q) - \log(Q_{G-f}))\\ &= \log(kG) + 2\log(M) + \log(N)\\ &\qquad -D\log(Q) - \log(M) + D\log(c_G)\\ &= \log(kG) + D\log(c_G) + \log(M) + \log(N) - D\log(Q)\end{aligned}$$

or

$$I(Q) = (k_G c_G^D) M N Q^{-D} \qquad \text{for } Q > Q_{G-f} \tag{13}$$

throughout the fractal regime, where k_G and c_G are functions of overall aggregate shape independent of fractal dimension and aggregate size. At $Q = Q_{G-f}$,

$$I_{fractal}(Q) = k_G M^2 N = I_{Guinier}(Q) \tag{14}$$

in this simplified scenario, positing that Guinier scattering meshes seamlessly into fractal scattering. [In general, there may well be an intermediate crossover region in the scattering spectra between pure Guinier and pure fractal scattering, requiring the introduction of another constant of proportionality into Equation (13). This constant will also depend on overall geometry, but would be expected to be independent of D and R_g.] From Equation (13), it is clear that when absolute intensities are concentration normalized, the fractal scattering intensities will overlap for fixed values of the fractal dimension D, so long as the samples being compared are fully aggregated. This expression can also be used to assess aggregation in situations where D does vary.

Equation (13) is consistent with an analytic approximation to the scattering spectra which encompasses both the Guinier and fractal ranges (Cabane, 1991, modified here to eliminate a term describing hard sphere scattering at high Q values; Osterburg and Mortensen, 1992):

$$I(Q) = \rho^2 M^2 N \cdot (1 + \xi^2 Q^2/3)^{-D/2} \tag{15}$$

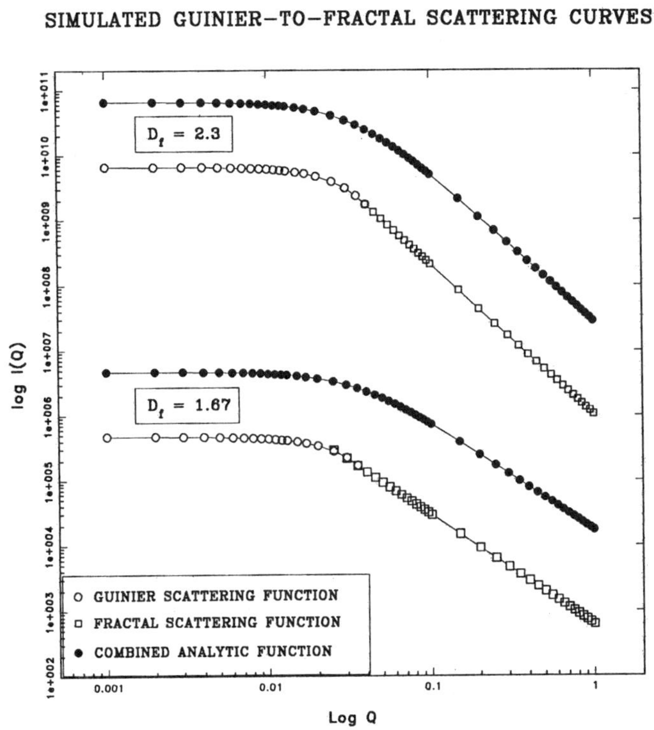

Figure 1: Simulated fractal aggregate spectra: single analytic function combining Guinier and fractal scattering (Equation (15)) [filled symbols]; Guinier scattering (Equation (9)) overlapped with fractal scattering at Q_{G-f} (Equation (13)) [open symbols].

where ξ approximates a renormalized R_g. This expression yields a crossover from Guinier to fractal scattering where it would naturally be expected (at $Q \sim 1/\xi$), and when concentration-normed absolute intensities are used, yields overlapping intensities in the fractal scattering range similar to Equation (13). Figure 1 gives examples of simulated scattering curves from Equation (15) (filled circles) and from Equation (13) (open squares) combined with Equation (9) (open circles). The curves generated by Equation (15) have a smoother quasi-Guinier crossover region between the Guinier and fractal scattering. Neither of these expressions has been tested extensively against scattering from humic acids, which might be expected to vary in their crossover properties from scattering from hard sphere colloids. Limited results with humic acids (Homer, 1995a) suggest the actual crossover scattering may follow Guinier scattering well out in Q space, that is to say that $c_G = Q_G R_g$ may be relatively large.

It should be clear from the discussion above that when concentration normalized absolute intensities of samples **do** differ in the fractal regime, it indicates that previously non-aggregated species are now aggregating. This contribution to the aggregation may be caused by increases in the masses M of the clusters and/or in the numbers N of these clusters. Whether or not the fractal dimension also varies, comparisons of intensities at $Q \sim 0$, and at Q in the fractal regime, along with radii of gyration ratios, can help distinguish which of these processes is taking place. Scattering intensity at $Q \sim 0$ is proportional to the square of aggregate mass, while in the fractal regime it is directly proportional with mass. As previously noted, scattering intensity is directly proportional to N in both places. Thus, the ratios of intensities at $Q \sim 0$ and for Q values within the fractal regime will both be the factor multiplying N, when only the number of aggregates N increases. When only the masses of the aggregates differ, that factor will be given by the ratio of intensities in the fractal regime; the value of that ratio will be doubled at $I(Q) \sim 0$.

In intermediate situations where aggregation generates (or increases) size polydispersity by involving increases in both M and N, the ratio at $Q \sim 0$ will be less than twice that in the fractal regime. Further information about the sizes of the largest clusters can be read from the radii of gyration obtained from Guinier plots (plots of $ln(I)$ *versus* Q^2). The existence of much polydispersity will interfere with the linearity of the fractal regime in $\log(I)$ *versus* $\log(Q)$ plots at Q values below the Guinier-to-fractal crossover Q_{G-f} of the group of smallest aggregates. A nonlinear curve will in general be obtained in both Guinier and fractal plots, in the range of Q values between Q_{2G-f} of the group of largest aggregates, and Q_{1G-f} of the group of smallest aggregates, and thus should give some estimate of the extent of polydispersity.

IV Investigations of Fractal Humic Aggregation

Rice and Lin (1992,1993,1994) have used SAXS to examine the fractal properties of a series of freeze-dried humic and fulvic acids. In powder form these materials do not form mass fractals, indicating that they have become deeply entangled in a volume-filling manner. However, they do form surface fractals, where the fractal dimension increases above the Euclidean surface dimension of 2 with increasing convolutions of these surfaces. Curiously, fulvic acids appear to have more convoluted surfaces than do humic acids from the same source.

Fulvic acid is the component of base-soluble natural organic matter that is also soluble in acid. Compared with humic acids, which have high ratios of hydrophobic groups (such as aliphatics and aromatics), these materials have high ratios of chemically reactive hydrophilic functional groups (such as carboxyls). Fulvic acids are smaller in size than humic acids; they do not aggregate well in solution, and so are difficult to observe in small-angle scattering experiments, even in concentrations as high or higher than are needed for the small-angle scattering of humic acids. Neither do they tend to adhere nearly as well as humic acids to clay particles (Murphy *et al.*, 1990). It is possible that the characteristics that prevent them from aggregating successfully in solution and onto surfaces also contribute to the lack of intermeshing at their surfaces upon freeze-drying, which produces a high surface D_f when they are in powder form.

Rice and Lin have also compared the surface and volume fractal properties of two humic materials: a stream sediment humic acid and a groundwater humus (where humus refers to the combined humic and fulvic fractions of base-soluble natural organic material.) In both cases, the volume D_f differs farther from the corresponding Euclidean dimension than does the surface D_f. This suggests that in solution the mass becomes more open than the surface is in powdered form. The groundwater material is particularly interesting: its surface and volume dimensions are both sharply distinct from non-fractal space-filling morphology, with a highly porous interior in solution, and a deeply convoluted surface in powder form. Perhaps even more noteworthy is the concentration used to collect this data: about 150 times less material than was needed for the stream sediment humic acid, yet the analysis time was only on the order of 5 times longer (Rice, personal communication, 1994). All other factors being equal (a large and perhaps unfair assumption), this ~30 greater quasi-normalized sensitivity suggests that the aggregate concentration exceeds sample concentration by the same factor, although the effect is somewhat diminished by their lower fractal dimension. It would be tempting to read overmuch into this single datum: Rice and Lin have examined other groundwater humus material not having the same striking sensitivity. Nonetheless, it will be interesting to observe as they continue with their work how common or rare is groundwater humic material found to be so highly aggregated.

Osterberg *et al.* (1994) have examined how differences in temperature can affect the degree of openness of humic aggregates. Humic acid maintained in solution for an extended period of time at 4°C was then raised to 22°C, whereupon SANS data collection began. As shown in Table I, the fractal dimension increased from a starting value of 1.85 to 2.35 at the end of the experiment, demonstrating that the increase in temperature somehow allowed a compaction of the humic material. One possible explanation is that the increase in kinetic energy was sufficient to allow functional group interactions to overcome some activation energy barrier. In a replicate experiment, changes in pH were monitored via the solution emf, as protons were emitted from or consumed by the humic acid. Initially, protons were quickly ejected from the humic acid in the span of a few minutes; thereupon they were slowly consumed by the humic material, raising the pH of the solution. The time constant for this experiment, 22 hours, is consistant with the time frame of the neutron scattering experiment, suggesting that the processes controlling pH either led to or were caused by the processes controlling the compaction of the humic acid. Osterberg *et al.* suggest that as the temperature was raised, ordered waters clinging to the hydrophilic interiors of the humic material at lower temperatures were released. With the continuing increase in kinetic energy allowing restructuring of the humic material to take place, more compact energy-favored configurations could be formed.

Senesi *et al.* (1994) looked at the pH dependence of the configuration of a soil humic acid suspension by monitoring the fractal dimension in turbidity experiments. Their results (Table 1), that D_f dropped substantially from 2.7 to 1.5 as the pH was raised from 3 to 6, are of interest in part because of

Technique	Researchers	Material/Treatment /Source	D_f	Comments
Small angle	Rice and Lin	stream sediment HA	2.2	powdered materials
X-ray		stream sediment FA	2.3	surface D_f's
Scattering		peat HA	2.7	"
(SAXS)		soil FA	2.5	"
		Lignite HA	2.3	"
		Lignite F	2.8	"
		Groundwater humus	2.5	"
		stream sediment HA, 6.7 mgC/ml	2.5	materials in solution
		groundwater humus, 45 mgC/L	1.6	mass D_f's
Small angle	Homer et al.	All samples in solution		
Neutron		GTHA 10 mgC/ml, pH=5	2.20	CEX
Scattering		" " 15 " " pH=5	2.14	no CEX, higher abs. int.
(SANS)		" " 20 " " pH=5	2.02	CEX
		" " 7.5 " " CEX, pH=7		higher absolute intensity
		" " 7.5 " " CEX, pH=7, Fe^{3+} added		after Fe added
		(sampling analysis incomplete)		
		Aldrich HA, 0.4 mgC/ml, pH=5	2.41	
		" " 2 " " pH=5	2.45	
		" " 10 " " pH=5	2.56	
		" " 10 " " pH=11	2.66	
		" " 10 " " pH=11, Ca^{2+} added	2.66	
		" " 2 " " CEX, pH=5	2.67	
		" " 10 " " CEX, pH=5	2.66	
		" " above, 2 weeks later	2.45	higher abs. int. now
		North Carolina HA from R.L. Wershaw		
		#2, 15mgC/ml, pH=7	2.29	
		#4, " " , pH=7	2.46	
Small Angle	Osterberg	IHSS HA 3.6 mg/ml	2.3	all samples in sol.
Neutron	et al.	A1 horizon soil humic acid, 3.0 mg/ml		
Scattering		T=22C, after 1 hr.	1.85	
(SANS)		T=22C, after 11 hrs.	2.15	
		T=22C, after 34 hrs.	2.30	
		T=22C, after 60 hrs.	2.35	
Turbidity	Senesi et al.	Soil humic acid, pH=3	2.7	all samples in susp.
		" " pH=4	2.6	
		" " pH=5	2.4	
		" " pH=6	1.5	

Table 1: Fractal behavior of humic materials

the negative results in other tests of the pH dependence of humic materials (Goldberg and Weiner, 1989; Caceci and Moulin, 1991; Homer *et al.*, 1992a). By the very definition of humic acids they are less soluble in acid than in base. Therefore it would not be surprising for them to become more open as the aqueous solution becomes more basic, yet such results of disaggregation with increasing pH are not often reported. It is quite possible that the aquatic humic materials sampled by Homer *et al.* (1992a, 1994) were too high in concentration, which promotes aggregation, to observe changes in D_f as a function of pH. The fluorescence polarization experiments of Goldberg and Weiner, and the photon correlation experiments by Caceci and Moulin, may not have been sensitive to changes in internal compactness, but only to gross changes in aggregate size. As ratios in intensity were not presented, it is not known whether the strong decrease in fractal dimension for the humic material studied by Senesi *et al.* was correlated with other aggregation/dissagregation processes.

Homer *et al.* (1992a,b, 1994) calibrated absolute intensity as specified in Wignall and Bates, with hydrogen incoherent background corrections made in accordance with Homer *et al.* (1994). The corrected absolute intensities could then be concentration normalized, to allow comparisons of aggregation levels between different samples. [In all cases, concentrations were normalized to the same standard: $C_o = 10$ mgC/ml.] All data were collected at the SANS facility at the High Flux Isotope Reactor at ORNL. A series of humic samples were examined, including the ubiquitous Aldrich humic acid, humic acid extracted at different times from the organic carbon in a marshy wetlands pond in Georgetown SC, and humic acid fractions from Hyde County NC, extracted at USGS and generously donated by R.L. Wershaw. The following summarizes some of the more interesting results. [A detailed study of curious bimodal fractal character of the USGS humic material is discussed in depth in Homer, 1995.] Scattering from Aldrich humic acid is shown in Figure 2, from Georgetown humic acid is shown in Figure 3, and from the USGS North Carolina humic acid is shown in Figure 4.

IV.A Sensitivity of Absolute Intensity in the Scattering Experiment to Changes in Fractal Dimension

The density in mass fractals decreases with increasing length scale being probed, so that relatively small differences in D_f can signify relatively large differences in aggregate mass density, and hence in concentration normed scattering intensity, if the overall cluster sizes are sufficiently large. For example, suppose the radius of gyration of the unaggregated subspecies forming the fractal clusters is R_o for each of the Aldrich and Georgetown humic acids, and that these units aggregate to form clusters having a radius of gyration $R_g = 25R_o$. (Our SANS data at low Q indicate that 500 Angstroms is a lower bound to the overall cluster sizes for several of these humics, while $R_o \sim 20$ Angstroms is a typical estimate of the length scale for the unaggregated material.) Consistent with our SANS data (Table 1), assume that D_2 for the Aldrich humic acid is 2.65, and D_1 for the Georgetown humic acid is 2.15. In terms of the unit $[R_o]$, the mass

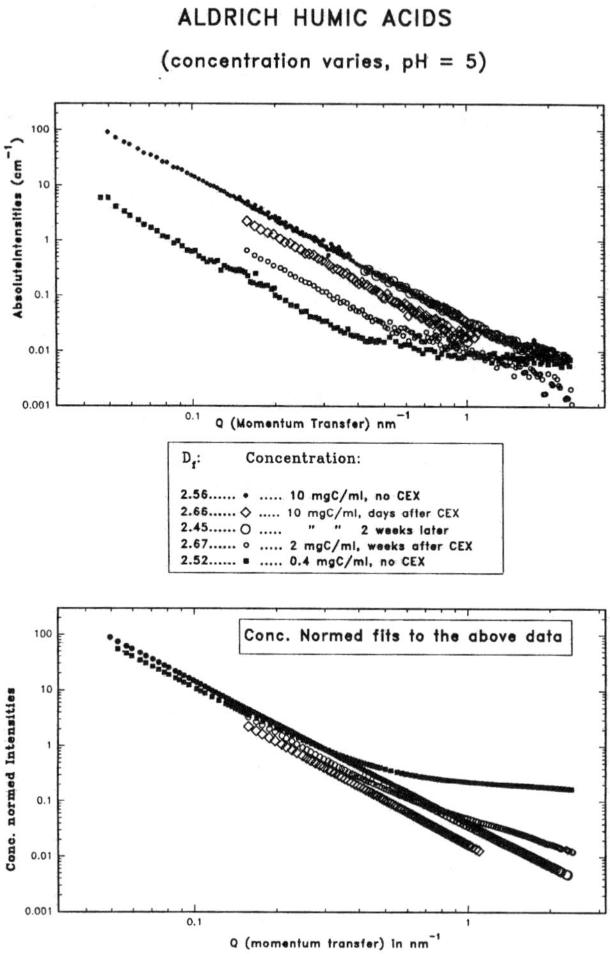

Figure 2: Small-angle scattering of Aldrich humic acid, thought to be high in lignin material. The normed intensities (as concentrations range from 0.4 mgC/ml to 10 mgC/ml) are similar in magnitude, whether or not the humic material was subjected to cation exchange columns (CEX).

per aggregate M is:

$$M_{(Georgetown)} = (25[R_o])^{2.15}$$

whereas

$$\begin{aligned} M_{(Aldrich)} &= (25[R_o])^{2.65} = (25[R_o]^{2.15}(25[R_o])^{0.40} \\ &= M_{(Georgetown)}(25[R_o])^{0.40} \end{aligned} \quad (16a)$$

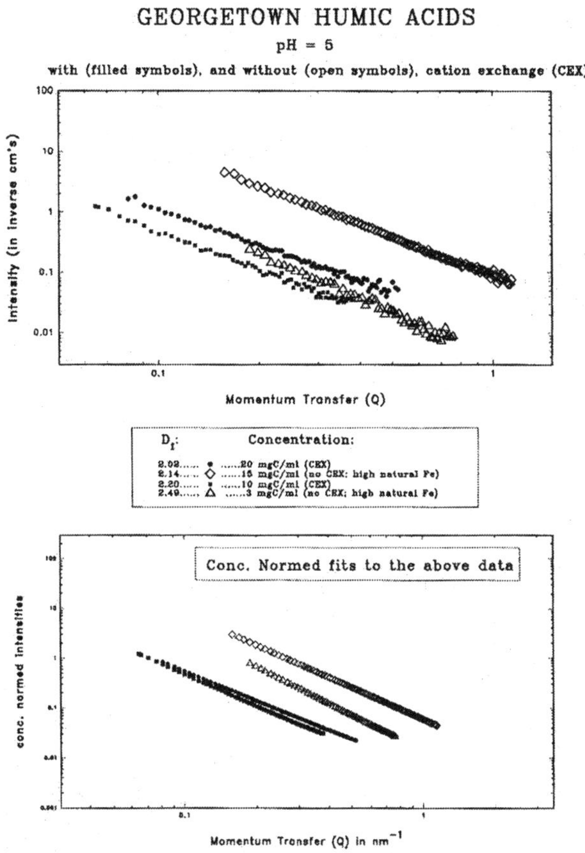

Figure 3: Small-angle scattering of humic acids from an estuarine pond in Georgetown, SC. Previous research (Homer and Bidleman, 1982) has shown this material to be highly effective in reductive dechlorination processes when spiked with Fe^{3+}.

so the mass of the 500-Angstrom Aldrich cluster is $25^{0.40} = 3.6$ times greater than the mass of the Georgetown humic cluster of the same size, since for same size clusters:

$$(R_g/R_o)^{D_2 - D_1} = k = M_2/M_1 \tag{16b}$$

Since both samples have the same radius of gyration, the Guinier-to-fractal crossover position will be the same for both (Equation (9b)). Positing the humic mass to be fully aggregated in both samples, then (by Equation (13b)) in the Guinier portion of the scattering curve the concentration normed intensity for the Aldrich humic acid is 3.6 times as high as that for the Georgetown humic acid. In the fractal regime beyond the Guinier-to-fractal crossover, the scattering goes from being proportional to the square of the mass M to being directly

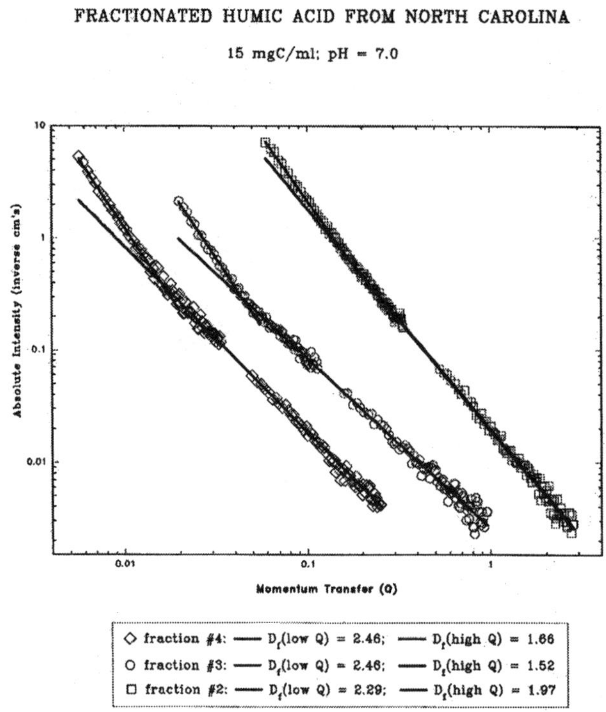

Figure 4: Small-angle scattering of humic acid from North Carolina, extracted and fractionated at USGS (Wershaw and Pinckney, 1973). The bimodel spectra of fractions #4 and #3 could not be readily fitted as being superposed, hence are thought to represent intra-aggregate morphology properties, perhaps like the bimodal scattering spectra from polymer coils (*e.g.*, Kirste, 1967).

proportional to M (Equation (13)). Indeed, the ratio of these scattering curves at any given point Q in the fractal regime gives the ratio k' (where $k' < k$) of the masses M' in the clusters enclosed by the length scale $R = 1/Q = M'^{1/D}$. The value of k (the difference in scattering intensity at the crossover point) increases with the ratio R_g/R_o, which can be viewed effectively as the number of self-similarity iterations the clusters incorporate.

In point of fact, non-cation exchanged Aldrich humic acid is similar in normalized fractal intensity (Figure 2b) to non-cation exchanged Georgetown humic acid (Figure 3b), suggesting that in the absence of cation exchange they have somewhat similar levels of total aggregated material. When both humic acids are subjected to cation-exchange columns, the intensity of the Aldrich humic acid (at higher concentrations) drops initially, but then is regained after several days are allowed to pass, giving time for any reaggregation processes to come to equilibrium (Figure 2). The loss of the cations does not appear to be important to the aggregation mechanism(s) employed by this material. The Georgetown humic acid behaves quite differently (at similarly high concentrations): after cation exchange has removed much of the metallic ions, its normalized intensity is about an order of magnitude lower than that of the same humic acid beforehand (Figure 3), showing that the Georgetown material has become considerably disaggregated (Equation (13) and surrounding discussion). Moreover, months later the same cation-exchanged samples decrease even slightly further in intensity, showing that for this material the cations are crucial for the aggregation mechanism(s) employed. Yet even this disaggregated material is still fractal in ranges of Q corresponding to cluster lengths on the order of 500 Angstroms. Both the Aldrich and Georgetown humic acids are forming aggregates larger than can be directly measured with the SANS instrument used to date.

IV.B Concentration Dependencies within Given Humic Source

As the concentration of Aldrich humic acid at pH = 5 (not subjected to cation exchange) increased from 0.4 to 10 mgC/ml, the absolute intensity increased by a factor of $\sim$1.5 at the lowest Q values available for comparison, signifying that aggregation was more extensive at the higher concentration (Figure 2). D_f was larger for the more concentrated sample, so some of the increase in aggregation involved compression of the clusters. A nearly flat portion of the scattering curve visible at low concentrations was attributed to previously unaggregated species which appeared to be aggregated at higher concentrations. This contribution to the scattering intensity, which presumably represents Guinier scattering from small humic subspecies, disappeared at higher concentrations.

When the concentration for Georgetown humic acid depleted in iron (by having undergone cation exchange) doubled from 10 to 20 mgC/ml, the concentration normed scattering intensities were identical at low Q (Figure 3), while the value for D_f decreased with increasing concentration, from 2.20 to 2.02. These data suggest that in the absence of metallic ions, even though aggregation concentration effects were minimal (at these high concentrations), other concentration-dependent morphological processes were present. Compare with Georgetown humic acid naturally high in Fe^{3+} at 15 mgC/ml: the concentration-normed intensity is about 20 times higher than the 10 mgC/ml sample, and is about 15 times higher than the 20 mgC/ml sample. Its fractal dimension is 2.14, intermediate between the others. Clearly extensive aggregation from previously disaggregated materials has taken place with this cation-rich sample. Other SANS experiments, where iron was added to cation-exchanged humic material, establish that the addition of this cation by itself generates aggregate morphology quite similar to that of the Georgetown material before cation exchange was conducted (Homer, 1994). It will be of interest in future work to examine the impact on morphology of adding varying amounts of other cations. [The cation exchange process reduced the Fe content of this humic acid by two orders of magnitude. Previous experiments with natural humic material from Georgetown by Homer and Bidleman (1982) suggested that increasing iron content sharply increased the ability of the humic material to interact with and subsequently degrade the chlorinated hydrocarbon pesticide toxaphene in aqueous solutions, whereas the addition of Fe alone did nothing to enhance the degradation of the pesticide. This study of the synergism between humic material and Fe led to interest in probing the impact of iron on humic morphology and activity (Homer, 1983).]

IV.C Sensitivity of Cluster Size to Presence of Cations

The plausible cause for the leap in absolute intensity for the non-cation exchanged material, on the order of 15 times higher than expected from the differences in concentration alone, is that the presence of iron (and/or other aggregating cations removed in the exchange columns) causes extensive further aggregation for this humic acid. This trend in the data gives insights into the causes and patterns of aggregation that would be difficult to obtain experimentally otherwise. The fact that the value of D_f for the 15 mgC/ml Georgetown humic acid falls between the D_f's for the 10 and 20 mgC/ml samples (Table 1) does not appear to be consistent with the widely accepted idea that the iron causes the clusters to collapse into compact structures. The presence of the iron greatly increases total mass of the aggregated clusters, either by creating huge networks from extensive interparticle bridging, and/or by stimulating nucleation of small subclusters into fractal aggregates. In either case, the net effect is that the aggregates formed still maintain their "porous" fractal character. Because the extended aggregation made possible by the presence of iron for this humic material was so pronounced, it is worth examining the implications of whether it was caused by increasing the mass M, or the number of clusters N, or something in between. [At the lowest range of Q values the SANS instrument at HFIR was capable of examining, the scattering of the cation-exchanged Georgetown humic acid was still fractal (corresponding to aggregate radii of gyrations of over 1000 Angstroms), so the Guinier data could not be obtained with this instrument. Beamtime has not as yet been alloted for examination of the same Georgetown humic acid before cation exchange, at similarly low values of Q.]

i) assuming the average mass M per aggregate was the parameter increased:
In this case, the mass per cluster can be approximated to be increased by a factor of about 15, since the concentration normed intensity is about 15 times greater than would be expected for an iron-depleted Georgetown humic acid. Positing that these humics are mass fractals with the same fractal dimension ($D \sim 2.2$) for the pre-iron cluster mass M_1 (with aggregate length scale L_1) as for the post-iron cluster mass M_2 (with length scale L_2) gives:

$$\begin{aligned} M_1 &= (L_1)^D \qquad & M_2 = (L_2)^D = 15M_1 \\ L_2^D &= 15(L_1)^D \qquad & L_2 = 15^{1/D}L_1 = 3.42L_1 \end{aligned} \tag{17}$$

The net effect under this scenario is that the iron generates extensive interparticle bridging, creating large cluster networks that still maintain their porous character. The volume of the larger clusters is $(L_2)^3 = 3.42^3 L_1^3 \sim 40L_1^3$, *i.e.*, about 40 times the volume of the smaller clusters. One can imagine several environmental consequences based on this model of loose floating pillows of humic material, rather than (but quite possibly in addition to) the compact pellets usually thought to be formed in the presence of excess iron. This postulated aggregation effect is consistent with other unpublished results collected by Homer with these humic acids and added iron, and with unpublished small angle X-ray scattering results collected by Rice (1994) upon adding cations to humic materials.

ii) Assuming the number N of aggregates was the parameter increased:

In this case, the number of clusters would be increased by a factor of about 15. While the clusters so formed would be smaller, they would have to be large enough not to interfere with the observed fractal scattering range; their length scales would thus have to be at least over 300-400 Angstroms. Given the low fractal dimension of iron-rich Georgetown humics (2.14 for the sample in Figure 3; 1.91 for other Georgetown samples), it is well within reason that they would contain "pores" on the order of tens of Angstroms. In the presence of high Fe, even relatively low concentrations of humic material are shown to be capable of developing "porous" clusters, large enough to enfold or cohere with other interesting quasi-hydrophobic particles in solution (Figure 3). The iron (and possibly other cations) clearly triggers a powerful aggregation mechanism with this material. More studies with humic materials relatively high in functional groups such as polysaccharides or carboxyls will help specify the correlation between functional group organization and aggregation effectiveness.

IV.D Variations in Concentration Dependencies between Different Humic Sources

As discussed above, the fractal dimension of the Aldrich humic acid increases with increasing concentration, whereas that of the Georgetown humic acid decreases with increasing concentration. The Aldrich material seems to become more densely intertangled as the concentration is raised, a reasonable response to the increasing aggregating pressures. The response of the Georgetown material is more complex. One explanation for the decreasing fractal dimension with increasing concentration for this humic acid may be that at higher concentrations, these materials might form "pre-aggregate clusters" that have more exposed repelling functional groups (Figure 5). The faster rate of formation, increased at higher concentrations, may simply allow the clusters too little time to organize into energetically favored, more compact structures having lower surface energies, before the next stage of aggregation begins. This conjecture is consistent with the temperature dependence observed by Osterberg *et al.* (1994).

IV.E NMR-Based Conjecture for Chemical Basis of Formation of Compact versus Loose Aggregates

The sparse data currently available comparing fractal dimension with functional group properties (Homer *et al.*, 1992b) are nonetheless suggestive that humic

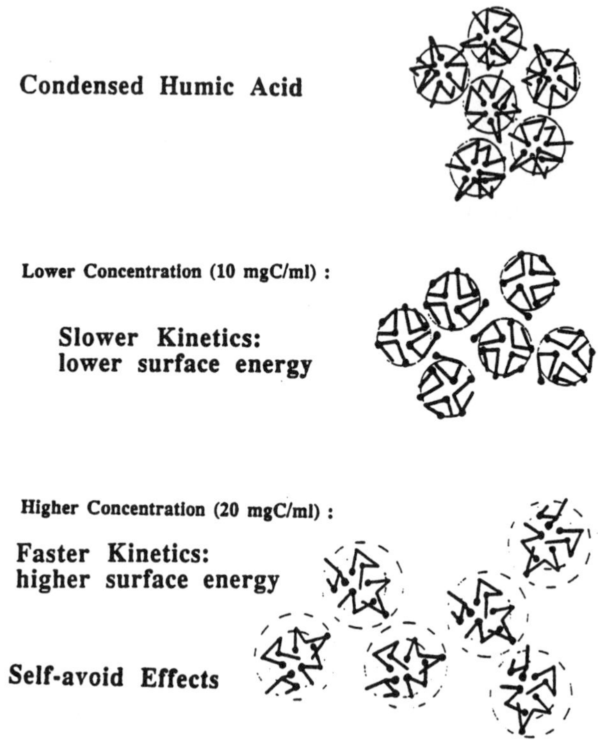

Figure 5: Schematic diagram of proposed aggregation formation kinetics for Georgetown humic acids: higher concentrations ⇒ faster kinetics ⇒ being further from equilibrium ⇒ higher surface energy ⇒ self-avoid surface effects, which lead to lower fractal dimensions.

acids with relative excesses of polysaccharide to aromatic functional groups may tend to have lower fractal dimensions (Figure 6, Table 2). CP-MAS NMR data were collected by Z-H Chen on a Bruker 400 MSL NMR instrument in the Chemistry Division at ORNL. Standard functional group analysis was performed for humic acid from Aldrich, for two samples from Georgetown, and for the fractionated humic acid from North Carolina via USGS (labelled NC#2, NC#3, and NC#4). Both Aldrich and NC#4 humic acids (Figure 4), with larger aromatic than polysaccharide content, have higher values for D_f, indicating that the networks they form are relatively dense, with substantial interpenetration. Humics collected from the Georgetown SC wetlands humic acid and the less hydrophobic fraction from North Carolina (NC#2) had larger polysaccharide than aromatic content; they formed much more open networks. Although clearly more data sets are needed to see which chemical functional groups most strongly affect the relative compactness of humic aggregates, the variations in D_f already observed show this aggregation to be sensitive to the relative proportions of functional groups contained.

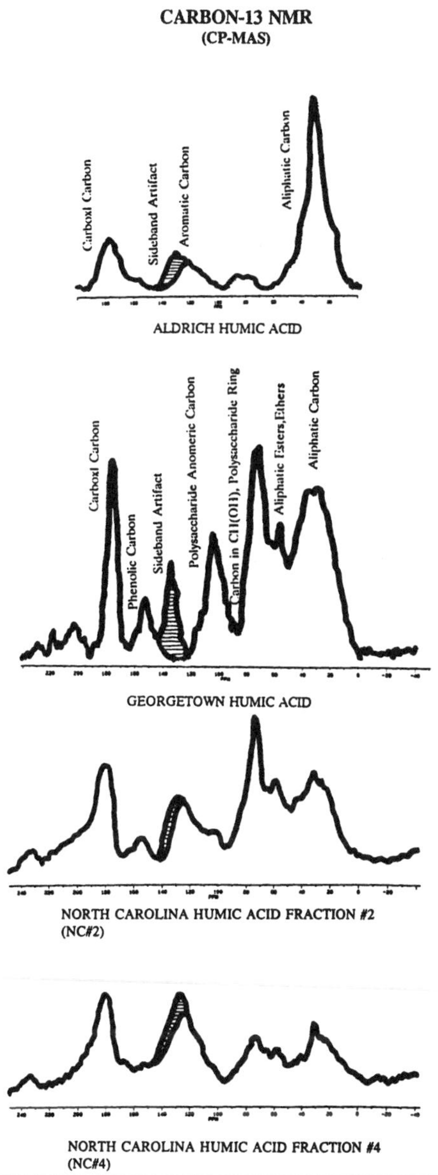

Figure 6: NMR functional group analysis of several humic acids. Functional group ratios from these data are presented in Table 2. [Shadowed areas at $\sim$ 130 ppm are side band artifacts, isolated by varying the spinning rates.]

D_f	Humic Acid	Carbo -xyl	Pheno -lic C	Non phenolic Aromatics	Poly- saccharides	Ali- phatic C	Poly- saccharides less aromatics
2.3	GTHA	20%	7.5%	0	37.5%	35%	+ 30
2.25	NC#2	25%	0	20%	30%	25%	+ 10
2.65	NC#4	30%	0	30%	20%	20%	- 10
2.6	AldHA	20%	0	15%	5%	60%	- 10

Table 2: Fractal dimensions and chemical characteristics of humic acids

V Summary

Fractal aggregation of objects in solution is frequently thought of as a form of random walk process. Steps in the random walk are the collisions of objects with one another; when there is sufficient attractive force between them, they combine and aggregation continues. The type of aggregation will vary with internal properties including flexibility, stochiometry, and attractive forces based on chemical makeup, and will also depend on external properties including temperature, concentration, and availability of other cosorbing species such as cations. Perhaps even the presence of inorganic colloids might serve as platforms catalyzing aggregation processes. Models of aggregation have been developed which provide well-defined endpoints of the ways in which aggregation is likely to proceed. For example, in the cluster-cluster aggregation model, colliding groups of subclusters (whose size corresponds to the current stage of aggregation) cohere in a series of aggregation steps. In the ballistic aggregation model, small isolated species sorb onto a few nucleated clusters. Many more aggregation steps (slower kinetics) are involved as the clusters incorporate these small fragments, a process that allows the aggregates to become more condensed.

Internal conditions can be important to aggregation kinetics. The greater the attractive forces between separate species, the higher the probability that they will "sense" each other's presence from a distance and combine. The reaction rates of these attractive forces will then control the rates of aggregation. Conversely, a self-avoiding random walk results when repulsive forces exist between neighboring species at the outset of aggregation. The aggregates formed in this latter process will tend to be less interconnected, and hence more open, than those formed from species having attractive forces. Other factors being equal, species neutral to each other are expected to form aggregates having intermediate degrees of openness. Since the fractal dimension is a reliable measure of the degree of openness of aggregate structures, it is often used to try to understand reaction kinetics for complex aggregating processes (Family and Landau, 1984).

Small-angle scattering can be used to track these aggregation processes and stages. Spectra of fractal samples not fully aggregated will display Guinier scattering from unaggregated species at high Q. As aggregation proceeds the absolute intensity will vary with the concentration and sizes of the subclusters or nonaggregates: in pure cluster-cluster processes the Guinier scattering will be modified by well-defined increments in radii of gyration as well as by changes in concentration, while in pure ballistic processes only the concentration of the nonaggregates will be affected. Variations in fractal dimension D_f are monitored in the plots of $\log(I)$ *versus* $\log(Q)$. Osterberg *et al.* (1994) have demonstrated that reaction kinetics can be successfully probed directly, and that monitoring changes in D_f indicates the type of aggregation taking place. The time-frame of their emf measurements of the changes in proton release were completely consistent with the time-frame of the changes in fractal dimension as their samples became more compact. Homer *et al.* (1992b, 1994) have shown that the removal of cations can significantly alter the extent of humic aggregation. For Aldrich humic acid at pH = 5, removing iron and other cations with cation exchange columns leads to an initial reduction in overall mass of aggregated material, (lowering of concentration-normed absolute intensity), whereas, if anything, it increases the compactness of the remaining clusters (raising the fractal dimension). Within two weeks after the initial spectrum was collected, the absolute intensity regained the same value as for non-cation exchanged Aldrich HA at the same pH and concentration (Figure 2), while D_f became lower, indicating that the clusters had reaggregated now in less compact form. For Georgetown humic acid, cation exchange did not appear to affect the fractal dimension, but dramatically reduced absolute intensity, and by inference the overall mass of aggregated material (Figure 3). This latter disaggregation configuration was stable over time, unlike the case for Aldrich humics. In these experiments, the combined fractal dimension and concentration-normed intensity values were used to infer the conditions under which increasing aggregation corresponded with altering the degree of compaction. For the humic acid from Georgetown, differences in cation content (where for example the iron content varied by two orders of magnitude) led to dramatic differences in aggregation levels. Since the morphology of these aggregates is fractal, any larger aggregates formed are also more "porous"; such increased bouyancy would tend to counteract flocculation.

Rice and Lin (1993) have demonstrated that humic and fulvic materials from different sources have different aggregate morphologies, indirectly underscoring the importance of specific chemistry to humic material geometry. Their observation of a highly aggregated groundwater humus in solution with low fractal dimension raises the question of how that aggregation process proceeded: were large soil humic aggregates dissolved and then transported to the groundwater? As the water carrying the humic material percolated through the soil, did interaction with metallic cations on soil interfaces help modify the form and degree of aggregation?

Senesi *et al.*'s work (1992a,b) with soil humics demonstrates that at least some humics are indeed strongly pH sensitive, in contrast with studies of other humic material, where far less pH sensitivity has been observed. This pH dependence of humic acid configuration may have been allowed by the degree of concentration of the suspension; another very real possibility is that the pH configuration dependence is a function of the chemistry of the humic acids involved. NMR has been found to be a sensitive tool for determining the chemical functional group content of humic acids (*e.g.*, Malcolm, 1989, 1991; Lobartini *et al.*, 1991; Homer *et al.*, 1992a,b, 1994). Although other factors might also account for the observed differences in pH dependence, perhaps the key to specifying the properties that lead to strong pH aggregation dependence lies in correlating the pH effects with NMR data for given humic materials.

VI Environmental Implications and Speculations

The advent of high resolution techniques such as electron microscopy and NMR have provided mechanisms to conduct highly detailed studies of the interconnection between humic acid form and function. The technique of small-angle scattering, capable of elucidating numerous aspects of macromolecular morphology, continues to be the primary experimental tool for determining D_f for a variety of colloidal systems. Using scattering and related techniques to investigate the fractal behavior of natural organic matter continues the process of developing means to quantify important distinguishing features of these materials. Initial experiments have shown that this can be a powerful tool for interpreting humic responses in a variety of situations. The following conjectures attempt to extract from this information possible future directions suggested by these early results. In particular, an attempt is made to extrapolate from results with humic materials in solution to implications about humic interactions with inorganic surfaces.

One of the earliest observations applying small-angle scattering to humics was that, for the same total carbon concentration, humic acids were up to 2 orders of magnitude stronger scatterers than fulvic acids (Ramsey, 1991, personal communication; Homer, 1991, unpublished results). Given their similar scattering length densities, their cluster masses must be quite small (Equations (8a) and (15)), showing that fulvic acids resist aggregation even in high concentrations. This suggests that even under these favorable conditions they must have strong self-repelling energies that allow them to resist combining together. This suggestion of self-avoidance may factor in the higher surface fractal dimension for fulvic as opposed to humic acids observed by Rice and Lin (1993): even when forced by freeze-drying to collapse together, their surfaces maintain separated, loosely organized structures. Considered with experimental observations of the weak coherence of fulvic materials to inorganic surfaces (*e.g.*, Murphy *et al.*, 1990, and schematic diagram in Figure 7), fulvic lack of aggregation suggests that the presence of carboxyl groups alone is an insufficient predictor of the probability of bridging to such surfaces (compare with Spos-

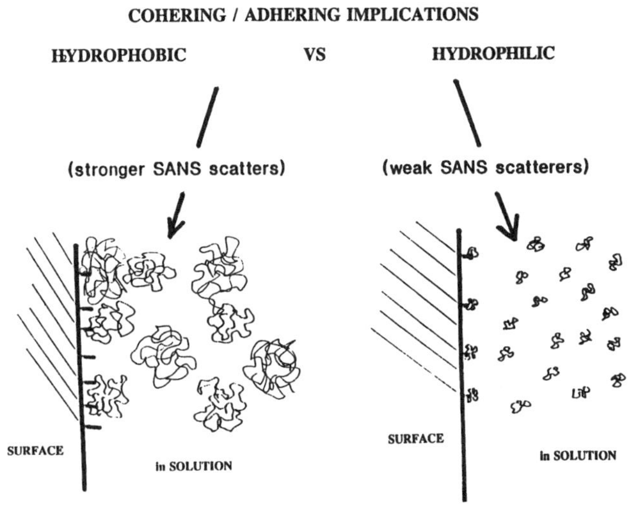

Figure 7: Quasi-hydrophilic materials such as fulvic acids, with high carboxyl levels, do not appear to aggregate as extensively in aqueous solution or on inorganic surfaces (Murphy *et al.*, 1990) as do the more hydrophobic humic acids.

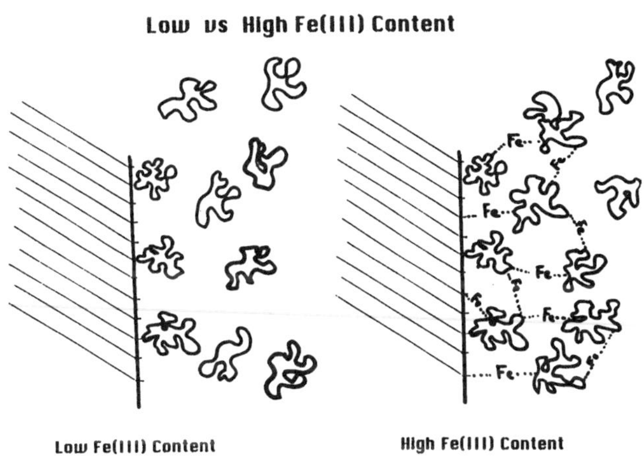

Figure 8: Some humic acids with high Fe content aggregate extensively in solution, while keeping open cluster morphology (Figure (3)). It remains to be learned whether this openness is maintained when humics high in Fe cohere with other species in solution and/or to inorganic surfaces.

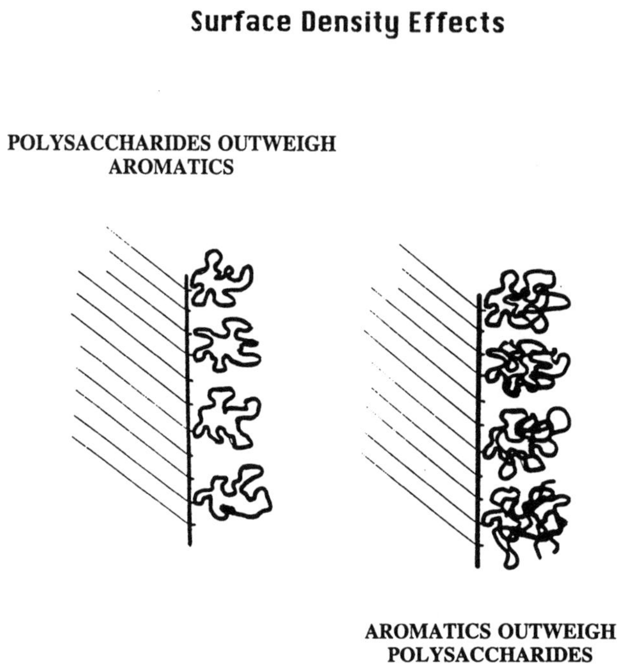

Figure 9: Humic acids have been observed to vary sharply in their ability to adhere to ionrganic surfaces (Murphy *et al.*, 1990); they vary equally sharply in their degree of openness (fractal dimension D_f; see Table 1). The possibility that these two observations are correlated should prove a rich topic for future experimental work.

ito, 1984 and references therein), since fulvic acids are substantially richer in this functional group than are humic acids. This speculation requires subjection to further experimentation before it can be given serious credence, but is offered here as an example of the kind of information morphological analyses can elicit. Since carboxyl groups are known to be important for the actual binding of natural organic to inorganic materials, perhaps shielding caused by aliphatic or other functional groups is equally important to the overall energy balance in organic acid to inorganic interface sorption processes. Whatever the functional group basis, humic acids do organize into highly aggregated complexes, and this ability increases with higher metal cation content (Homer *et al.*, 1992b; Rice, personal communication, 1994). The surprising suggestion from the fractal analysis is that the cation-effected complexes need not be collapsed structures. It will be of interest to determine whether this openness in solution is translated to the formation of relatively open networks upon sorbing onto inorganic surfaces (schematic diagram in Figure 8). In a similar vein, the correlation of NMR functional group properties with the openness of the humic aggregates, in terms of their fractal dimensions (Table 2), suggests that related correlations might exist between the functional group fingerprint and the thickness of adherence to inorganic surfaces (schematic diagram in Figure 9). Humic acids with higher fractal dimensions might be expected to form denser surface coatings; similarly, those forming larger clusters in solution may form deeper coatings on surfaces.

Relative quantities of chemical functional groups and of cations such as Fe^{3+} are two of several factors contributing to aggregation variations. The tendency in the past has been to think in terms of small *versus* large aggregates. The mass fractal model encourages thinking in terms of open *versus* compact, as well. Variations in source material, temperature, and/or pH can all affect aggregating properties, which in turn affect adsorbing and cosorbing behaviors. The groundwater humus from the Big Sioux aquifer extracted and analyzed by Rice and Lin (1993) is an example of the kind of large, loose microsponges that could readily transport smaller bodies far downstream from their source. The prevalence of such humic transporters, and the conditions that favor/inhibit their generation, can now be studied and quantified in a manner newly available: by assessing their morphologic responses in depth, and in particular their fractal dimension.

Acknowledgements

It is a pleasure to thank R.L. Wershaw for his insights in several discussions, his groundbreaking work applying small-angle scattering to the aggregation of humic and fulvic materials, and his generous provision of carefully cleaned and separated humic material from North Carolina. Conversations with J.A. Rice, who conducted the first small-angle scattering experiments with humic acids here at ORNL, have been very helpful. G.D. Wignall and R. Triolo were generous in their time teaching the basics of SANS experimentation. Z.-H. Chen graciously performed the NMR analysis despite considerable backseat driving by V.J.H., as we both learned how to isolate sideband spinning peaks. This research was supported in part in part by the Division of Materials Sciences, in part by a program managed jointly by ORAU and ORNL, and in part by the Laboratory Directed Research and Development Program of Oak Ridge National Laboratory, managed for the United States Department of Energy by Martin Marietta Energy Systems, Inc. under contract No. DE-ACO5-840R21400.

References

Cameron, R.S., B.K. Thornton, R.S. Swift and A.M. Posner. 1972. Molecular weight and shape of humic acid from sedimentation and diffusion measurements on fractionated extracts. *J. Soil Sci.* 23:394-408.

Caceci, M.S. and V. Moulin. 1991. Investigation of humic acid samples from different sources by photon correlation spectroscopy. p. 97-105. In Bhattacharji, S., Friedman, G.M., Neugebauer, H.J. and Seilacher, A. (eds.) *Hu-*

mic Substances in the Aquatic and Terrestrial Environment. Springer-Verlag, Berlin.

Chen, S-H. and J. Teixeira. 1986. Structure and fractal dimension of protein-detergent complexes. *Phys. Rev. Lett.* 57:2583-2586.

Chen, S.-H. and T.-L. Lin. 1987. Colloidal solutions. *Methods of Experimental Physics* 23B:489-543.

Chen, Y. and M. Schnitzer. 1976. Scanning electron microscopy of a humic acid and a fulvic acid and its metal and clay complexes. *Soil Sci. Soc. Am. J.* 40:866-872.

Chen, Y., A. Banin and M. Schnitzer. 1976. Use of the scanning electron microscope for structural studies on soils and soil components. p. 425-432. In O. Johari (ed.) Proc. Inter. Symp. on SEM.

Cotton, J.P. 1991. Introduction to scattering experiments. In P. Lindner and T. Zemb (eds.) *Neutron, X-Ray, and Light Scattering: Introduction to an Investigative Tool for Colloidal and Polymeric Systems*. North-Holland, Amsterdam.

Glatter, O. and O. Kratky (editors). 1984. *Small Angle X-Ray Scattering*. Academic Press, New York.

Goldberg, M.S. and E.R. Weiner. 1955. Fluorescence measurements of the volume, shape, and fluorophore composition of fulvic acid from the Suwannee River. In R.C. Averett, A. Guinier and G. Fournet (eds.) *Small Angle Scattering of X rays*. John Wiley, New York.

Hayes, M.B.H., P. MacCarthy, R.L. Malcolm and R.S. Swift. 1989. *Humic Substances II: In Search of Structure*. John Wiley, New York.

Hayter, J.B. 1987. SANS studies of micellar and magnetic fluids. In S.A. Safran and N.A. Clark (eds.) *Physics of Complex and Supermolecular Fluids*. Wiley & Sons, New York.

Hayter, J.B. 1988. Interparticle interactions and polarization effects in colloids. *J. Appl. Cryst.* 21:737-743.

Hjelm, R.P. 1985. The small-angle approximation of X-ray and neutron scatter from rigid rods of non-uniform cross section and finite length. *J. Appl. Cryst.* 18:452-460.

Hjelm, R.P., P. Thiyagaragan, D.S. Sivia, P. Lindner, H. Alkan and D. Schwahn. 1990. Small-angle neutron scattering from aqueous mixed colloids of lecithin and bile salt. *Progr. Colloid Polym. Sci.* 81:45-51.

Homer, V.J., J.F. McCarthy and Z.-H. Chen. 1992a. The effect of concentration and pH on the size and aggregate structure of humic acid, presented at Inter. Conf. on Organic Subst. in Soil and Water, Lancaster University, UK.

Homer, V.J., J.F. McCarthy and Z.-H. Chen. 1992b. Fractal dimension: A probe of humic aggregation properties, presented at 1992 Clay Minerals Soc. Meeting in Minneapolis.

Homer, V.J. 1994. Aggregation patterns from humics of different chemical compositions. In: Abstracts of Invited Papers of the 7th IHSS Inter. Meeting. (Proceedings published in 1996.)

Homer, V.J. and Z.-H. Chen. 1995. Fractal character of humic acid aggregation: Chemical underpinnings, environmental implications. Unpublished manuscript.

Lindner, P. and T.H. Zemb (eds.). 1991. *Neutron, X-Ray, and light scattering: Introduction to an investigative tool for colloidal and polymeric systems.* North-Holland, Amsterdam.

Lobartini, J.C., K.H. Tan, L.E. Asmussen, R.A. Leonard, D. Himmelsbach and A.R. Gingle. 1991. Chemical and spectral differences in humic matter from swamps, streams and soils in the Southeastern United States. *Geoderma* 49:241-254.

Malcolm, R.L. 1989. Applications of solid-state ^{13}C NMR spectroscopy to geochemical studies of humic substances. p. 310-338. In M.B.H. Hayes, P. MacCarthy, R.L. Malcolm and R.S. Swift (eds.) *Humic Substances II: In Search of Structure.* John Wiley, New York.

Malcolm, R.L. 1991. Factors to be considered in the isolation and characterization of humic substances. p. 9-37. In S. Bhattacharji, G.M. Friedman, H.J. Neugebauer and A. Seilacher (eds.) *Humic Substances in the Aquatic and Terrestrial Environment.* Springer-Verlag, Berlin.

Mandelbrot, B.B. 1983. *The Fractal Geometry of Nature.* W.H. Freeman and Co., New York.

McKnight, D.M., K.E. Bencala, G.W. Zellweger, G.R. Aiken, G.L. Feder and K.A. Thorn. 1992. Sorption of dissolved organic carbon by hydrous aluminum and iron oxides ocurring at the confluence of Deer Creek with the Snake River, Summit County, Colorado. *Env. Sci. Technol.* 26:1388-1396.

Murphy, E.M., J.M. Zachara and S.C. Smith. 1990. Influence of mineral-bound humic substances on the sorption of hydrophobic organic compounds. *Environ. Sci. Technol.* 24:1507-1516.

Olausson, E. and I. Cato. 1980. *Chemistry and Biogeochemistry of Estuaries.* John Wiley, New York.

Osterberg, R. and K. Mortensen. 1992. Fractal dimension of humic acids. *Eur. Biophys. J.* 21:163-167.

Osterberg, R., L. Szajdak and K. Mortensen. 1994. Temperature-dependent restructuring of fractal humic acids: A proton-dependent process. *Environ. International* 20:77-80.

Rice, J.A. and J.S. Lin. 1993. Fractal nature of humic materials. *Env. Sci. and Tech.* 27:413-414. [Note: the label for the humic material reported with a mass fractal dimension of 1.6 in this paper is correctly labelled in this chapter as a groundwater humus, as verified by J.A. Rice 1994.]

Rice, J.A. and J.S. Lin. 1992. Fractal nature of humic materials. Extended abstract for the A.S.C. Meeting, San Fransisco, CA.

Rice, J.A. and J.S. Lin. 1994. Fractal dimensions of humic materials. In Proceedings of the 6th IHSS International Meeting.

Safran, S.A. and N.A. Clark. 1987. *Physics of Complex and Supermolecular Fluids.* Wiley & Sons, New York.

Schaefer, D.W. 1987. Small-angle scattering from disordered systems. In: G.D. Wignall, B. Crist, T.P. Russell and E.L. Thomas (eds.) *Scattering, Deformation, and Fracture in Polymers*, Materials Research Society, Pittsburgh, Pennsylvania.

Schaefer, D.W., J.E. Martin, P.W. Wiltzius and D.S. Cannell. 1984. Fractal geometry of colloidal aggregates. *Phys. Rev. Lett.* 52:2371-2374.

Schaefer, D.W. 1986. Small-angle scattering of silica colloid fractals. *Phys. Rev. Lett.* 56:2199.

Schmidt, P.W. 1991. Small-angle scattering studies of disordered, porous and fractal systems. *J. Appl. Cryst.* 24:414-435.

Senesi, N., G.F. Lorusso, T.M. Miano, C. Maggipinto, F. Rizzi and V. Capozzi. 1994. The fractal dimension of humic substances as a function of pH by turbidity measurements. In Abstracts of Invited and Volunteered Papers of the 6th IHSS Inter. Meeting.

Sinha, S.K., T. Freltoft and J. Kjems. 1984. Observation of power-law correlations in silica-particle aggregates by small-angle neutron scattering. In: T. Family and D.F. Landau (eds.) *Kinetics of Aggregation and Gelation.* Horth Holland, New York.

Sposito, G. 1984. *The Surface Chemistry of Soils.* Oxford University Press, New York.

Stevenson, F.J., Q. Van Winkle and W.P. Martin. 1953. Physicochemical investigations of clay-adsorbed organic colloids: II. *Soil Sci. Soc. Am. Proc.* 17:31-34.

Stevenson, I.L. and M. Schnitzer. 1982. Transmission electron microscopy of extracted Fulvic and humic acids. *Soil Sci.* 133:179-185.

Suffet, I.H. and P. MacCarthy (eds.) 1989. *Humic Substances, Influence on Fate and Treatment of Pollutants.* American Chemical Society, Washington, D.C.

H.K. Tan. 1985. Scanning electron microscopy of humic matter as influenced by method of preparation. *Soil Sci. Soc. Am. J.* 49:1185-1191.

Tanford, C. 1973. *The Hydrophobic Effect.* Wiley & Sons, New York.

Teixeira, J. 1988. Small-angle scattering by fractal systems. *J. Appl. Cryst.* 21:781-785.

Weitz, D., M.Y. Lin and J.S. Huang. 1987. Fractals and scaling in kinetic colloid Aggregation. In S.A. Safran and N.A. Clark (eds.) *Physics of Complex and Supermolecular Fluids.* Wiley & Sons, New York.

Wershaw, R.L. 1986. A new model for humic materials and their interactions with hydrophobic organic chemicals in soil-water or sediment-water systems. *J. Contam. Hydrol.* 1: 29-45.

Wershaw, R.L., D.J. Pinckney and S.E. Booker. 1977. Chemical structure of humic acids– Part I, A generalized structural model. *US Geol. Surv., J. Res.* 5: 565-569.

Wershaw, R.L. and D.J. Pinckney. 1973. Determination of the association and dissociation of humic acid fractions by small angle X-ray scattering. *US Geol. Surv., J. Res.* 1:701-707.

Wershaw, R.L., P.J. Burcar and M.C. Goldberg. 1969. Interaction of pesticides with natural organic materials. *Environ. Sci. Technol.* 3:271-273.

Wignall, G.D. and F.S. Bates. 1987. Absolute calibration of small angle neutron scattering data. *J. Appl. Cryst.* 20: 28-40.

Fractal Geometry and the Description of Plant Root Systems: Current Perspectives and Future Applications

G.M. Berntson, J. Lynch and S. Snapp

Contents

I Introduction

Roots are the most enigmatic of plant organs. Despite their fundamental role in water and nutrient acquisition, anchorage, and resource storage, we know little of their structure and function in soil compared to what we know about above-

ISBN 1-56670-105-8

ground structures such as leaves, flowers, and shoots. In recent years there has been a growing interest in the fundamental biology of plant root systems and the role they play in plant growth, soil formation and ecosystem processes (*e.g.*, Waisel *et al.*, 1991). Studies of the form and function of roots in their native environment are faced with daunting methodological obstacles. Most obviously, roots grow in soil, an opaque medium from which they cannot easily be extricated or even observed without introducing artifacts, destroying the native root architecture, or precluding subsequent observation of the same individual (*e.g.*, Böhm, 1979). Root systems themselves are exceedingly complex structures, typically being composed of thousands of individual root tips and different classes of roots that vary developmentally, physiologically and morphologically (Waisel and Eshel, 1991). Furthermore, root growth and architecture is very plastic and interacts dynamically with a wide array of physical, chemical, and biological factors in the soil that vary in time and space.

Several researchers have proposed and developed techniques for quantifying various aspects of root system structure and function. Examples of such techniques include the topological characterization of root system branching patterns (*e.g.*, Fitter, 1985, 1987), *in situ* spatial and temporal dynamics of root growth and death (*e.g.*, Hendrick and Pregitzer, 1993; Berntson *et al.*, 1995), use of tracers to quantify dynamics and distribution of ion uptake within root systems (*e.g.*, Scott Russell and Clarkson, 1976; Jackson *et al.*, 1990), and numerous models for describing the acquisition of nutrients from soils (*e.g.*, Nye and Tinker, 1977; Barber, 1984), the spatial distribution of roots within the soil (*e.g.*, Lungley, 1973; Diggle, 1988; Pagés *et al.*, 1989; Davis, 1994) and the role the spatial distribution of roots plays in the acquisition of nutrients from soils (Fitter *et al.*, 1991; Berntson, 1994a). Each of these techniques has resulted in a better understanding of root system growth and function. Taken together, however, these studies reinforce the observation that the plant root system development and function is exceedingly complex. In this chapter we explore the potential utility of fractal geometry and scaling relationships as a means of providing alternative and potentially useful tools for making sense of this complexity.

The unifying concepts underlying fractals and power laws is self-similarity and self-affinity. An object is said to be self-similar when a small portion of that object is an exact replica of the object as a whole. For simple constructs from Euclidean geometry this concept is easily understood - a fragment of a one-dimensional line is a one-dimensional line. Throughout nature many objects (and processes) show a more complex type of scale invariance. When looking at a cloud it is hard to tell how far away the cloud is by its shape or texture alone. A cloud is similar in form when it is near as well as when it is far. But a cloud is not a line, a box or sphere. It is a complex shape that shows simple scaling of apparent form at different scales of observation. The complexity of the shape of this cloud - how its bumps, folds and texture fill space - can be quantified using fractal geometry. We refer the reader to the first chapter of this book (Baveye and Boast, 1997), and to Mandelbrot's seminal work (1983), for a more detailed presentation and discussion of these ideas.

Fractal geometry has been successfully used to develop relatively simple algorithms that can reasonably approximate many complex plant forms (usually above ground) observed in nature (Prusinkiewicz and Lindenmayer, 1990). The root systems of plants also appear have fractal properties (*e.g.*, Tatsumi *et al.*, 1989). On first principles, it should be obvious that fractal geometry should be a useful tool for describing plant root systems. Plant root systems grow through the hierarchical, iterated production of new root tips - a process that suggests nested complexity and thus fractal properties. Inasmuch, fractal geometry may provide new perspectives on root function as well as root structure. Several aspects of root architecture have been shown to be highly correlated with the fractal dimensions of plant root systems (*e.g.*, Fitter and Stickland, 1992a; Berntson, 1994b). But, to date there have been no applications of fractal geometry to plant root system structure or function beyond simple demonstrations that fractal dimensions can be calculated and can show high correlations with well understood aspects of root architecture. In this chapter we review existing literature (up to the date of the writing of this manuscript) on the application of fractal geometry to plant root systems, present new information and ideas on the applicability of fractal geometry and scaling relationships to the space-filling and branching structures of roots, and consider issues and prospects for future work in this area.

II Summary of Research to Date

Fractal geometry was not applied to plant root systems until 1989, when Tatsumi and colleagues proposed to use the fractal dimension, D, to characterize whole root system forms (Tatsumi *et al.*, 1989). Since this work was published there have been several studies applying fractal geometry to plant root systems, averaging about one published study a year, with more studies appearing in recent years (see Table 1). The techniques used to calculate the fractal dimensions of plant root systems and the potential applications of these measurements will be discussed in detail through the rest of this chapter (especially see "Methodological Considerations" for an overview of techniques applied in previous studies). In this section of the chapter we present a basic overview of how fractal geometry has been applied to plant root systems (Table 1).

Several studies have demonstrated that as root systems grow and become larger (greater total root length or mass), D increases. This has been observed by sequential measurements of root systems through time (Fitter and Stickland, 1992a; Lynch and van Beem, 1994) and comparisons of root systems of plants of equal age but different size (Eghball *et al.*, 1993; Berntson, 1994b; Lynch and van Beem, 1994). These observations need to be interpreted with caution. Estimates of D through ontogeny increase during early growth and then level off (Fitter and Stickland, 1992a; Snapp, unpublished; Figure 1). This suggests that consideration of D as an estimate of root system size is sound only during initial growth. Independent of the total size of a root system, increasing density of roots (length or mass per unit area within which roots are distributed) is highly correlated with increasing fractal dimensions (Berntson, 1994b). Topology and

Ref	Species/Genotypes	Comp	Meas	Prep	D	R_{min}*
1	*Panicum miliaceum* L., cv. Saitama local	0	1	2, 3	1.48 - 1.58	0.28 - 0.70
	Zea mays L., cv. Golden Cross Bantam					
	Secale cereale L., cv. Haruichiban					
	Sorghum bicolor Moench, cv. Saitama local					
	Triticum aestivum L., cv. Norin 26					
	Pisum sativum L., cv. Shirokinusaya					
	Arachis hypogaea L., cv. unknown					
2	*Betula populifolia* L.	1	1	1, 2	0.87 - 1.75	0.17 - 0.34
	Betula alleghaniensis L.					
3	*Phaseolus vulgaris* L., cv. Tostado	0, 3	1	3	1.20 - 1.59	?
	", cv. Porrillo Sintetico					
	", cv. Carioca					
	", HAB 229					
4	*Zea mays* L., B73xLH105	1	1	2	1.35 - 1.57	6.35
	", N74xMo17					
5	*Pinus taeda* L.	2	1	3	1.27 - 1.45	0.05
6	*Trifolium pratense* L.	1, 3	2	1	1.24-1.43	?
	Plantago lanceolata L.					
	Festuca ovina L.					
	Festuca rubra ssp. *commutata* L.					

*Ref (References): 1 - Tatsumi *et al.* 1989; 2 - Berntson 1994; 3 - Lynch & van Beem 1994; 4 - Eghball *et al.* 1993; 5 - Diebell & Feret 1993; 6 - Fitter & Stickland 1992

Comp (Additional Comparison); 0 - none; 1 - nutrient supply; 2 - morphological grade; 3 - ontogeny

Meas (Method of Measurement of D); 1 - Grid Intercept; 2 - Dividers

Prep (Method of Image Preparation); 1 - *in situ* tracing; 2 - spatial distribution preserved (*e.g.* pin-board); 3 - spatial distribution not preserved

R_{min} (Minimum Image Resolution) - mm

Table 1: Summary of published studies examining fractal dimensions of plant root systems.

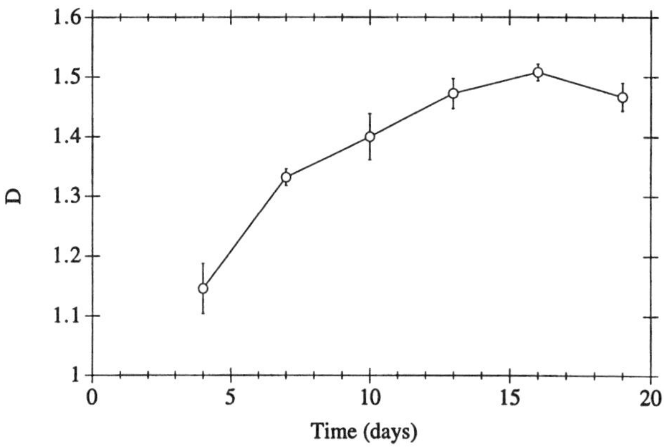

Figure 1: Fractal dimensions of sequentially measured *Arabidopsis* root systems grown in root bags. Bars represent one standard error of the mean. (Data from Snapp, unpublished.)

distributions of link length (which can directly affect root density) are also highly correlated with D (Fitter and Stickland, 1992a), though it is not clear if these architectural features of root systems influence fractal dimensions through their effect on root density. In general, differences in fractal dimensions appear to reflect changes in root system size and architecture. In all but one published study (Tatsumi *et al.*, 1989), fractal dimensions varied significantly between different species and/or genotypes. Two of the published studies examined the effect of variations in nutrient supply (either N or combined N:P:K, Eghball *et al.*, 1993; Berntson, 1994b, respectively). In both cases, increases in the supply of nutrients resulted in significant increases in fractal dimensions.

III Methodological Considerations

III.A Sample Preparation

Quantifying root deployment through space is vital to understanding root system ability to capture resources present in a 3-dimensional, heterogeneous matrix. Assessing the costs associated with resource capture is also contingent on the ability to measure root distribution in 3 dimensions. Measuring three-dimensional objects is no easy task. Quantifying root system space-filling properties presses the limits of commonly available experimental methodology. Researchers have attempted to measure 3-dimensional root systems primarily by artificially imposing a 2-dimensional, or approaching 2-dimensional, view. Methods to measure root system architecture have depended on three main approaches to sample preparation: 1) non-destructive tracing, 2) removal of soil with attention to spatial arrangement, and 3) removal from soil with spatial arrangement lost.

III.A.a Non-Destructive Root System Imaging

Root system distribution can be observed and traced at a transparent wall surface. For this technique, roots are grown between two planes, one of which is glass or plastic, in a narrow space filled with soil, imposing an approximation to a 2-dimensional root system distribution (Berntson and Woodward, 1992; Berntson, 1994b; Fitter and Stickland, 1992a). If root systems are not grown in a confined space, much of the root system will not be observable at the interface (Snapp and Shennan, 1994). This is a severe constraint, as root system branching angles are altered when plants are grown in a confined environment. Not all roots will be visible using this technique, as some roots remain hidden in the soil matrix (Gross *et al.*, 1992; Berntson *et al.*, 1995). In a variation of this technique, plants can be grown in pouches that are buried in the soil. The pouches are constructed of a porous membrane that separates the roots from the soil but allows water and nutrients from the soil to freely move to the roots (*e.g.*, Drew, 1993). Images of root systems growing in these pouches can be obtained non-destructively by retrieving the bag, opening it, obtaining an image of the root system, and returning the bag to the soil (Snapp, unpublished). These techniques are advantageous because the spatial arrangement of the root system is preserved (Berntson, 1994b; Fitter and Stickland, 1992a), and root system development over time for the same individuals can be monitored (Figure 1). It remains to be tested whether confining a root system to approximately 2-dimensional space will alter the essential space-filling properties of the root system. Anatomical data on lateral root distribution suggest that it may be possible to extrapolate from 2-dimensional to 3-dimensional if the effects of confinement on circumference positioning of lateral branching is determined. Root branching from the central axis occurs along protoxylem-based ranks in a wide range of monocotyledonous and dicotyledonous species (Charlton, 1975, 1982, 1983; Pulgarin, 1988). Circumferential orientation of lateral roots will be altered by the 2-dimensional confinement of a root system between plates, but longitudinal spacing of branching and branching angle will not necessarily be altered (Charlton, 1982, 1983). Modeling based on geometric simulation could test the extent to which 2-dimensional information can help predict 3-dimensional root system deployment (*e.g.*, Berntson, 1994a).

The ability to make repeated measures of the same root system is a substantial advantage of this system. Root system deployment through space is a dynamic process (Berntson and Woodward, 1992). Due to the methodological difficulties of measuring root systems, studies often do not address changes with ontogeny (*e.g.*, Eghball *et al.*, 1993; Tatsumi *et al.*, 1989). Alteration of root distribution as resources are encountered, acquired and depleted is one of the most under-studied aspects of below-ground ecology (Körner, 1991). Non-destructive tracing reduces the number of replicates needed for developmental studies. It allows root system development over time to be monitored without the variation encountered with destructive measurements that confound plant-to-plant variation with ontogenetic changes in root system architectures.

III.A.b Removal from Soil with Attention to Spatial Arrangement

A 2-dimensional approximation of the 3-D deployment of roots in space can be preserved by careful removal of soil and placement of the root system on a planar surface (Eghball *et al.*, 1993; Lynch and van Beem, 1994). This will necessarily mean some loss of spatial information as root systems are subjected to loss of the supporting soil matrix. A closer approximation of the *in situ* spatial arrangement of roots can be achieved by using a pin-board technique to support roots as soil is removed (Berntson, 1994b; Tatsumi *et al.*, 1989). The entire root system is exposed by this technique. This is in contrast to non-destructive tracing, which will not record roots hidden from view by the growth media. Spatial arrangement preservation is approximate, but the resolution appears to be sufficient to detect differences in root system architecture induced by nutrient supply (Eghball *et al.*, 1993; Berntson, 1994b).

Problems with this method include its destructive nature. It is a particularly time-consuming means of sample preparation. Also, the process of recording a 3-D system in 2 D will incur a loss of information. This is due to root overlap in 2 D that does not occur in 3 D, changes included in root system branching angles to force a root system into approximately 2 D, and alteration of space-filling properties. The high density of roots present in a root system that grew in 3 D and then is confined to 2 D for measurement purposes can also pose problems, as there are practical limits to image resolution.

III.A.c Removal from Soil without Spatial Arrangement Preserved

Removal of roots from soil allows measurement of the topological branching structure of a root system - but this capability requires the separation of overlapping branches (Berntson, 1992; Lynch and van Beem, 1994). The approach is similar to the technique already discussed and has similar advantages and disadvantages. The main difference with this technique is that after soil is removed the root system, or root system subsection, it is placed on a surface and branches are separated before the image is recorded. This allows quantification of the entire root system length, and is a standard root extraction method. However, *in situ* spatial arrangement of the root system is lost by this technique. Root system deployment in space is intricately related to the fractal dimension of root systems (in contrast to the topological dimension of the root system), and it is expected that not preserving the spatial arrangement of roots will bias the calculation of D.

III.B Calculation of Fractal Dimensions

To date, published reports have calculated fractal dimensions on 2-dimensional images of 3-dimensional root systems. Sample preparation to obtain a 2-dimensional representative image can have a large effect on the calculated fractal dimension (Berntson, 1994b). However, the minimum resolution used in calculating the fractal dimension of a root system appears to have little effect on the calculated fractal dimension (Table 1). Calculating the fractal dimension from a 2-dimensional image, on the other hand, is a straightforward process. There are two calculation methods that have been used successfully to calculate fractal dimension for root systems, the box counting method and the dividers method.

The box-counting method is a widely used method to calculate fractal dimension. It is straightforward to automate this technique using a computer. This method consists of superimposing a grid over an image and counting the number of boxes (N) which cover or overlap any portion of the root system. This is repeated for the same image using a range of grid sizes (s, length of side of grid). The power law relating the N with s is of the form $N \propto s^{-D}$, where D is the fractal dimension (more precisely, the fractal box-counting dimension). Usually this power function is estimated by logarithmic regression of N versus s (Tatsumi *et al.*, 1989). We have used the images presented by Mandelbrot (1983) with mathematically derived D to calibrate computer automated estimations of D using the box-counting method and found less than a 2% variation between actual and estimated D values.

The dividers method has also been used to calculate the fractal dimension of 2-dimensional images. This method consists of stepping along the perimeter of a curve and counting the number of steps (N) required to traverse the curve. This is repeated for the same curve using a range of step sizes (s, length of step). The power law relating the N with s is of the form $N = s^{-e}$, where D is equal to $1 + e$ and is the fractal dimension (more precisely, the compass or divider dimension). A substantial disadvantage of this method is the requirement that the root system image not contain any overlapping segments. The perimeter of an object is the basis for calculating a fractal dimension by this method, thus root sections cannot overlap. This technique will only work with particular types of images of root systems – those that have no overlapping portions. This method has been used by Fitter and Stickland (1992a). We recommend the box counting method, as it does not have the serious constraint of the dividers method and it can be used for a wide range of root system images, from subsections to entire tree root systems (Table 1).

An important consideration in calculating fractal dimension is the resolution of the image. Large root systems can only be digitized, given image capturing hardware limitations, with a sacrifice. The resolution required by a large image does not allow small features within root systems to be resolved. If 0.5-mm diameter roots can not be resolved by an image capture operation, then the question must be raised whether the image accurately represents the space filling properties of a root system. The fine roots of a system can represent the overwhelming majority of the surface area, indicating the important role they play in resource capture and probably in root system costs (Eissenstat, 1992). The space-filling and resource capture ability of root hairs may be adequately represented by the roots on which they appear. Functionally, root hairs have been successfully modeled as increases in the diameter of the roots that support them (*e.g.*, Nye and Tinker, 1977). Thus, a biological reasonable limit for high resolution image capture to calculate fractal dimension may be one that resolves fine roots but does not resolve root hairs, for a given root system. If this is not practical given equipment limitations, then images of portions of a root system should be obtained (allowing higher resolution to be maintained) and the relationship of fractal dimension for a portion of a root system to the fractal dimension of an entire root system should be determined (but see "Subsampling" section), or a system developed to join together and analyze multiple smaller images (*e.g.*, Tatsumi *et al.*, 1989).

IV Dimensionality of Root Systems and The Calculation of Fractal Dimensions

One obstacle to our understanding of root architecture is that although root systems grow in 3-dimensional space, most analytical approaches do not adequately represent the *in-situ* deployment of roots in 3 dimensions. Roots are most commonly studied by soil coring or its modern derivative, minirhizotron observation tubes, which provide a rather one-dimensional perspective on root density with depth (Mackie-Dawson and Atkinson, 1991, and references therein). Another technique is to expose roots at the surface of a soil profile revealed by a rhizotron or trench, which provides thereby a two-dimensional perspective on root distribution with depth (Bhm, 1979). Root architecture in 3 dimensions can be obtained by serially sectioning roots embedded in resin followed by image analysis (Commons *et al.*, 1991). This process is, however, laborious and destructive of the root system under observation. Recently, some noninvasive techniques have been developed, such as MRI, for the visualization of root systems in their *in situ* 3-dimensional configuration. Such methods are expensive, requiring access to highly specialized equipment, and have limitations in terms of size of root systems that can be examined and resolution of the smallest roots that can be discerned (Bottomley *et al.*, 1986, 1993; Brown *et al.*, 1991; Rogers and Bottomley, 1987).

Just as the self-scaling nature of some fractals suggests a potential utility of fractals in describing the architecture of a root system from some subset of the system, fractal geometry also suggests an approach to the three-dimensional architecture of roots from the one- or two-dimensional measurement techniques most root biologists employ. A set of results in fractal theory called *projection theorems* state that the fractal dimension of a set in 3-D space should bear predictable relationships to the fractal dimension of the set projected onto 2 dimensions (such as through an X-ray), or the dimension of the set projected onto one dimension. The fractal behavior of root systems in 1D, 2D, 3D, and 3D projected onto 2D, are therefore of interest. As yet only fractal values of roots in 3D projected onto 2D (*i.e.*, of plants restricted to an artificially flattened root growth zone or of plants excavated, then laid out and flattened) have been reported in the literature.

Figures 2 - 5 show the results of deriving fractal dimensions from geometric models of bean root systems as simulated by *SimRoot* (Davis, 1994). In Figure 2 the actual 3-dimensional values of D are compared with values of D for a root system growing in 3 dimensions projected onto 2 dimensions.

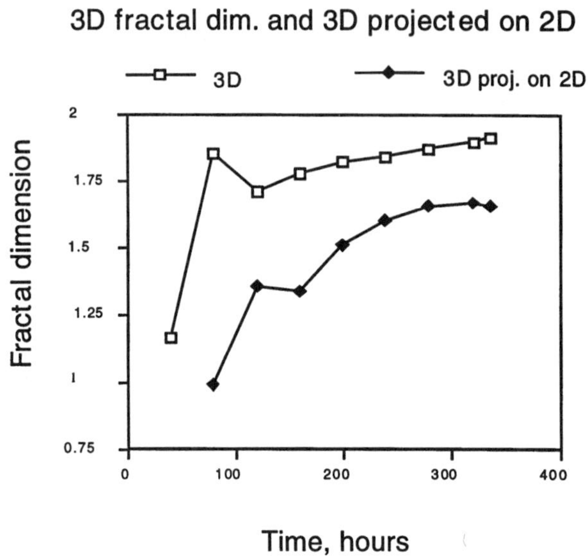

Figure 2: Fractal dimensions of *Phaseolus* root systems simulated using *SimRoot*. Values were calculated from simulated 3-dimensional roots and 2-dimensional projections of modeled root structure.

The projected values are similar to published values for actual flattened root systems (see Table 1). Fractal dimensions calculated from the 3-dimensional structure and 2-dimensional projection both showed asymptotic increases in fractal dimensions through time (Figure 2). However, D in 3 dimensions for the same (simulated) root system was higher than D projected onto 2 dimensions, and, more importantly, showed somewhat different trends over time, especially during early growth before values in both contexts approached their asymptote. Since actual root systems live in 3-dimensional space, the nonlinear relation of D in 3 dimensions and D projected onto 2 dimensions for precisely the same simulated root system suggests that present methods of estimating D in 2 dimensions may be problematic.

Figures 3 and 4 show true 2-dimensional values (rather than values from 3-dimensional objects projected onto 2 dimensions) of D for a root system 'sliced' vertically (Figure 3) or horizontally (Figure 4). The vertical slices correspond to roots as observed from the side by means of a rhizotron or soil trench. The value of about 1.2 is fairly constant throughout most of the width of the root system, rising slightly in the center near the root base and dropping or becoming erratic at the root periphery, which is consistent with expectations that root branching complexity would increase towards the center of the root system. Horizontal slices show roughly the same pattern, with a mean value approaching 1.2, but increasing near the base of the root and decreasing distally (Figure 4). In an actual soil it may be interesting to compare the D of 2-dimensional slices

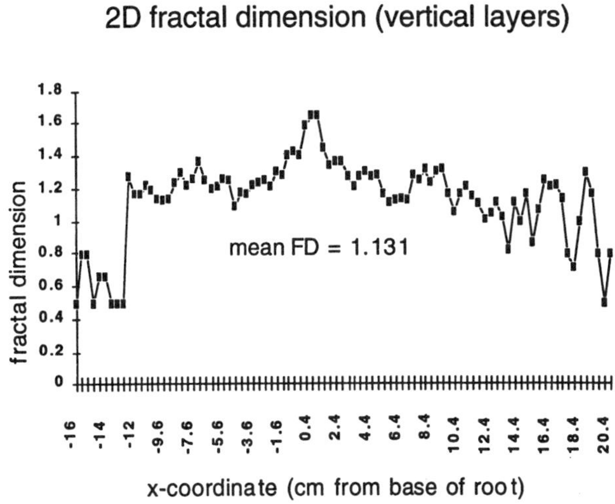

Figure 3: Fractal dimensions of *Phaseolus* root systems simulated using *SimRoot*. Fractal dimensional calculated from vertical slices of simulated root systems.

or root systems with the D of soil pores, especially in heavy-textured soils where soil pores and channels may be associated with avenues of root growth. Figure 5 shows true 1-dimensional values of D for a root system sampled vertically in a grid of x and y values. The values of D average 0.46 and increase with proximity to the base of the root system, as with 2-dimensional slices.

This preliminary analysis suggests that values of D for a simulated root system do show some regularity: in 3 dimensions D is about 1.9, D for the same root in 2 dimensions is about 1.2, about 0.7 less, and for the same root in 1 dimension about 0.5, again 0.7 less than the next higher dimension. This apparently regular relationship between values of D in progressive dimensions suggests a possible utility of 1-D (cores) or 2-D (soil faces) analyses in estimating D of roots in 3 dimensions. This subject, and the issue of whether the progression of D in different dimensions itself has any relevance, deserves further study.

V Fractal Properties of Root System Branching Structures

The previous discussion of root systems and fractals was restricted to characterizing the space filling properties of plant root systems. These techniques are valuable because they allow us to characterize static, empirical space-filling properties of plant root systems. However, these techniques are currently of limited value for two reasons: 1) they do not characterize the processes of

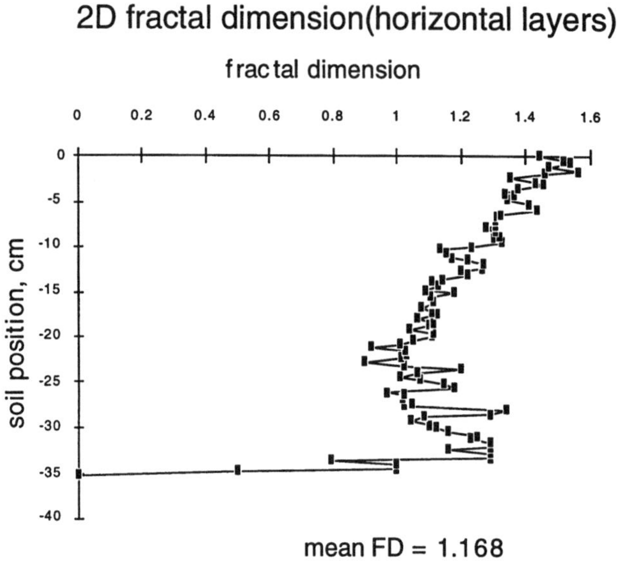

Figure 4: Fractal dimensions of *Phaseolus* root systems simulated using *SimRoot*. Fractal dimensional calculated from vertical slices of simulated root systems.

development that lead to the observed space-filling properties; and 2) the application of these techniques is fraught with methodological difficulties due to the difficulty in observing 3D *in situ* spatial distributions of roots within the soil.

The scaling behavior of the branching patterns within individual plant root systems, independent of the *in situ* spatial distribution of the roots within the soil, may allow us to address some of the issues not possible with the dividers and box-counting estimations of D. For example, characterization of the scaling properties or root production (branching) within individual root systems in relation to developmental hierarchies may allow us to use simple scaling relationships for describing the development of plant root systems. Also, if we can characterize the scaling properties of the branching patterns within individual plant root systems in relation to functional hierarchies, then we may be able to use these scaling properties to relate branching patterns and processes to physiological activity (including nutrient uptake). We make the distinction between *developmental* and *functional* hierarchies in order to distinguish between hierarchical systems for classifying different portions of root systems on the basis of developmental (*sensu* Rose, 1983) or functional (*sensu* Fitter, 1985, 1986, 1987) criteria. Both of these approaches can be explored for the branching structures of plant root systems by use of different types of ordering systems. The exploration of these possibilities is not methodologically difficult. As will be discussed later in this section, techniques have been developed and used extensively for rapidly quantifying root system branching patterns.

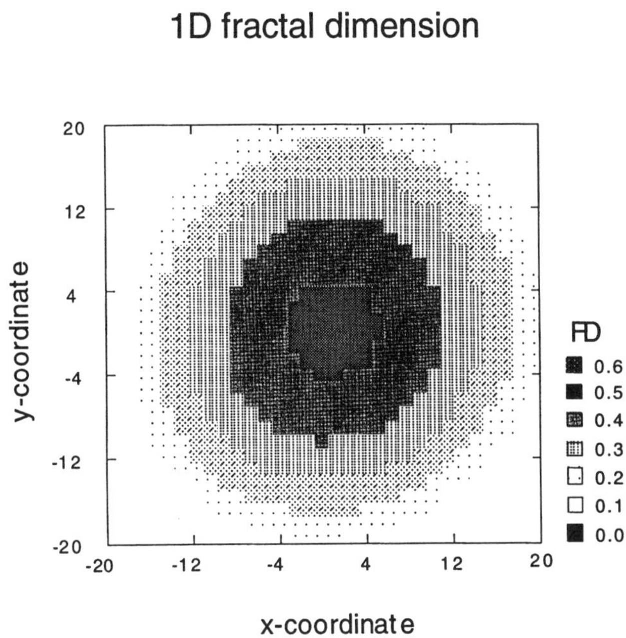

Figure 5: Fractal dimensions of *Phaseolus* root systems simulated using *SimRoot*. Fractal dimensional calculated from grids placed within vertical slices of simulated root system.

V.A Terminology

Topologically, it is convenient to regard plant root systems as rooted binary trees (*sensu* Clapham, 1990). For purposes of clarity and consistency with previously published descriptions of plant root systems (*e.g.*, Berntson and Woodward, 1992), we will refer to the root, vertices and leaves of a rooted binary tree as they refer to components of a plant root system as the base, links and root tips.

V.B Ordering and Scale

Characterizing topological and geometric features of branching structures in biology has its roots in the quantitative description of stream drainage networks. Fitter has applied techniques for describing stream drainage networks developed by Strahler (1952; Fitter, 1982) and Werner and Smart (1973; Fitter, 1985, 1986, 1987). The interesting thing about these techniques is that, depending on how they are applied, they are analogous to a fractal analysis of plant root geometry and topology by characterizing the power laws governing root size-metrics in relation to branching order. To date, however, there have been no comprehensive reviews of the application and applicability of the power laws relating root system topology and geometry to different ordering systems.

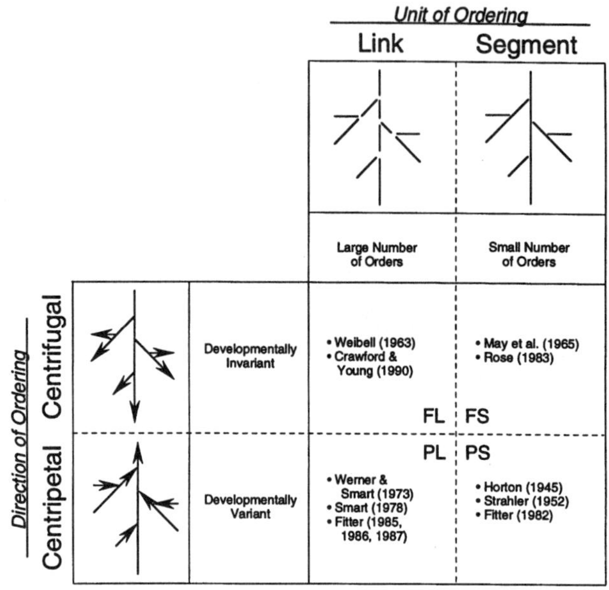

Figure 6: Schematic diagram of the *DIRECTION* and *UNIT* based components of the systems used in this paper for assigning different orders (scales) within root systems. The diagram illustrates the basic features of each variation in ordering methodology and gives citations of the introduction and/or application of the particular ordering scheme. See text for additional information and Figure 7 for an application of each of the ordering methods.

In this section we compare and contrast four different systems for ordering portions within rooted binary trees which represent two orthogonal ordering criteria (Figures 6 and 7). The two ordering criteria are the *direction* of ordering and *unit* of ordering. The direction of ordering can be either centrifugal ("F") or centripetal ("P"). A centrifugal ordering system starts at the base of the system (or root of the binary tree, following Clapham) and orders are assigned in increasing magnitude *away* from the base. This direction of ordering is "developmental" in that it reflects the direction of growth of root systems. A centripetal ordering system starts at the root tips and orders are assigned increasing magnitude *toward* the base. This direction of ordering is "functional" in that equal orders are functionally equivalent insofar as the order reflects topological distance from the root tips or root age. Distance from root tips has been shown to be an important determinant of the morphological state and physiological capacity of plant roots (Clarkson *et al.*, 1968; Harrison-Murray and Clarkson, 1973; Scott Russell and Clarkson, 1976). Units of ordering are either individual *links* ("L", also known as "nodes") or contiguous, linear groups of links

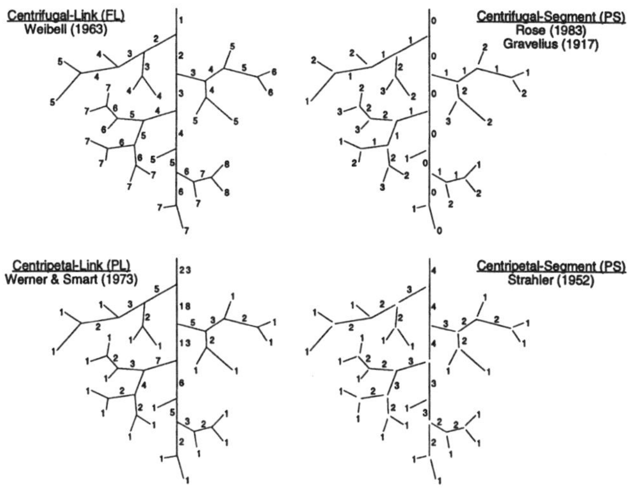

Figure 7: Schematic diagram illustrating the application of each topological ordering scheme used in this chapter. For the two segment based ordering schemes, separate segments are drawn as unconnected lines.

(*segments*, "S"; *sensu* Fitter, 1986). When the ordering units are links, demarcation between adjacent orders is unambiguous. Demarcation between adjacent segmental orders can be ambiguous or more complicated, depending on the criteria used. For example, the developmental ordering system (*e.g.*, May *et al.*, 1965, 1967) often used for describing different components of growth of plant root and shoot systems can be ambiguous when clear patterns of dominance or differential growth are not observed between adjacent developmental orders. A review of these ordering systems and others is provided by MacDonald (1983).

Fitter first introduced the centripetal link-based ordering system ("PL"; AKA "topological" ordering; Fitter, 1985, 1986, 1987) and centripetal segment-based ordering system ("PS"; also known as "morphometric" ordering; Fitter, 1982, 1985) to plant root systems. Both of these ordering systems were first introduced in the geomorphology literature for describing the topological properties of stream drainage networks. The PL system was first introduced by Werner and Smart (1973). In this system, root tips are assigned an order of one (Figure 7). Each proximal link (traversing in the direction of the base) is assigned an order equal to the sum of the orders of the two distal links. The order of the base of the root system is equal to the number of root tips in the system. The PS system was first introduced by Horton (1945) and then revised by Strahler (1952). This system assigns all root tips an order of one (Figure 7). Each proximal link is assigned an order equal to the larger order of the two distal links, or one greater than the order of the distal links if they are of equal order. Contiguous groups of links of equal order are regarded as a single segment.

The only centrifugal-segment based ordering scheme ("FS") that has been applied to plant root systems is the developmental ordering scheme. This scheme has been used extensively in describing patterns of root growth (*e.g.*, Hackett, 1968; May *et al.*, 1965, 1967) and in modeling the growth of plant root systems (*e.g.*, Lungley, 1973; Rose, 1983; Diggle, 1988; Pags *et al.*, 1989). In this scheme, different orders reflect the hierarchical nesting of multiple growth axes in a root system (Figure 7). In a simple tap-rooted plant where apical dominance is clearly expressed, there is a single axis (*e.g.*, the growth of a single root tip) where the length between the base of the root system and the tip of this axis is significantly longer than any other potential axis. This axis is referred to as the developmental axis. Each root tip emerging from this developmental axis constitutes a subordinate developmental axis. Usually the developmental axis is referred to as order zero, subordinate axes to the developmental axis are order one, and so on. In order to express these orders on log scales we have offset this ordering system by one so that the developmental axis is of order one.

The remaining ordering system, the centrifugal-link based scheme ("FL"), has never been applied to plant root systems. Recently this ordering scheme was applied to the shoot systems of mature deciduous trees (Crawford and Young, 1990), but its primary application has been in describing the structures of bronchial trees (Weibell, 1963). This system of ordering assigns order one to the base of the root system (Figure 7). Each adjacent distal link (away from the base) is assigned an order one greater than its adjacent proximal link. Thus, a link's order is its topological distance from the base of the system.

V.C Topological Analysis of Plant Root Systems

Unlike the measurement of *in situ* spatial distributions of roots within the soil, reliable and moderately rapid techniques have been developed for the complete measurement of plant root branching patterns. These techniques involve carefully laying out intact root systems removed from the soil on a flat surface so that no two non-adjacent links touch. These roots are then digitized into a computer and the images are processed using software designed for measuring root architecture (Berntson, 1992; Fitter and Stickland, 1991, 1992a). This method of preparation preserves the topological structure and every link length within the root system, but retains no information about the original *in situ* pattern of deployment of root systems.

Root systems can be stored in computer memory as link data vectors (*sensu* Smart, 1978) or binary trees (Berntson, 1992). Storing the root systems in computer memory as binary trees greatly facilitates the implementation of simple algorithms for applying all of the ordering schemes discussed above. All of the above ordering schemes can be applied from a knowledge of the topological structure of the root system independent of root geometry, except for the FS scheme (developmental). In this paper, FS orders were calculated by assuming that different orders could be distinguished by length. Initially, the longest single axis within the root system (longest distance between any given root tip and the base of the root system) was identified and this was designated the developmental axis. Each sub-tree (branch) attached to the developmental axis was processed similarly.

This process was repeated until all segments within a root system had been assigned an order.

V.D Fractal Power Laws of Topological Systems

The calculation of fractal dimensions using the box-counting or dividers techniques involves looking at the relationship between size (area or length) and scale (grid size or divider length). Looking at these relationships on log-log plots allows the power-scaling factor between observer scale and apparent size to be determined. The general form of the relationship relating observer scale and apparent size is $y = s^D$, where y is apparent size, s is the scale of the observation and D is the scaling factor (equivalent, in previous examples, to the fractal dimension). We can apply this same technique to rooted binary trees by using the different orders within the branching structure as different scales. Clearly we cannot directly equate any of the above-described ordering schemes with geometric scales. Thus, scaling factors obtained using any of the above ordering schemes as the scale do not have direct relevance to space-filling properties. Rather, the scaling factors obtained in these analyses represent the exponents in power laws governing the processes of scaling in relation to the applied ordering system within roots (hereafter referred to as scaling exponents).

We will be characterizing these scaling exponents by examining the power-law relationships between the ordering schemes and three size components of the plant root systems. The size components we examine are: 1) number of links/segments of a given order; 2) total length of root in a given order; and 3) average length of links/segments of a given order. The exploration of size components of plant root systems in relation to topological structure advocated here is not without precedent. Most importantly is the work originally proposed by Strahler (1952). Strahler proposed expressing the scaling properties of size components in relation to his ordering system as N_w/N_{w+1}, where N_w is the size at order w, and N_{w+1} is the size at order $w+1$. The two size components Strahler proposed examining that are analogous to those proposed here are 1) number of links/segments of a given order; 2) average length of links/segments of a given order. The statistic determined from the first size component is the *Strahler bifurcation ratio* (R_b) and the second is the *Strahler stream-length ratio* (R_L). We have opted to not use Strahler's method of looking at scaling between root size components and branching structure. Rather, we calculate scaling exponents that are directly analogous to the fractal dimensions derived from the dividers and box-counting methods. The only difference between how we present scaling factors and the presentation of D for the box-counting technique is that because the two direction ordering systems we employ assign order values in opposite directions (centrifugal and centripetal), the sign of the derived scaling factors are both positive and negative.

		Number	Total Length	Average Length
FL	% Significant	6.670	3.333	6.667
	R^2	0.096	0.074	0.078
FS	% Significant	1.163	0.000	1.163
	R^2	0.387	0.174	**0.942**
PL	% Significant	**95.556**	**87.778**	26.667
	R^2	0.681	0.546	.0163
PS	% Significant	3.488	0.000	0.000
	R^2	**0.974**	**0.893**	0.594

Table 2: Summary of type-I linear regression analysis (see text for a discussion of terminology and methods). Presented are the percent of all regressions significant at the $p < 0.05$ level (p-values Bonferroni-corrected for n = 92, p-value cut-off of 0.05/92 = $5.26e^{-4}$) and the average coefficient of determination of the regression. All groups of regressions in which greater than 85% of the regressions are significant and average coefficients of determination greater than 0.85 are highlighted in bold type and underlined.

V.E Scaling Exponents of Root System Branching Patterns

To explore the above-discussed ordering systems in relation to root system branching structure, the root systems of five species of plants were examined: *Senecio vulgaris*, *Arabidopsis thaliana*, *Acer rubrum*, *Betula alleghaniensis*, and *Betula populifolia*. *Arabidopsis* and *Senecio* are both short-lived annuals. The *Arabidopsis* root systems examined were from two-week-old *Arabidopsis* (n = 5; Berntson unpublished data). Root systems of *Senecio* aged 2, 3 and 4 weeks were examined (n = 20 for each age; data from Berntson and Woodward, 1992). *Acer* (n = 8) and the two *Betula* species (*populifolia*, n = 7; *alleghaniensis*, n = 12; Berntson unpublished data) are trees. For the tree species, root systems from three-week-old seedlings were examined. Root systems of each species were prepared, digitized and analyzed as described above (Berntson, 1992). The binary tree representations were ordered using each of the four direction and unit ordering combinations and the number, total length and average length of each link/segment per order were determined. These results were analyzed by performing linear, type-I regressions where log(order) was the independent variate and log(number, length and average length) were the dependent variates. To avoid type-II errors in comparing multiple p-values, a sequential Bonferoni-correction was employed to adjust p-values (Rice, 1989).

There are several interesting trends that emerge from examining the results of the regressions (Table 2). In general, only segment based ordering systems showed high coefficients of determination and only link based ordering schemes showed high levels of significance. The high coefficients of determination observed in the segment ordering schemes suggest that these ordering schemes result in power-law relationships that accurately describe the vast majority of variance observed in the distribution of segment numbers, total length and average length of segments. The low levels of significance of these regressions is the result of few orders being used in the regressions (between 3 and 6). When so few data points are used in regression analysis the power of significance testing is significantly decreased even though the regressions explain an excess of 89% of the variation observed in the dependent variables. Conversely, the high levels of significance observed in the PL ordering scheme (for number and total length of root per order) with low coefficients of determination (0.681 and 0.546) are the result of large numbers of observations included in each regression and thus greater power of significance testing. Crawford and Young (1990) recognized that this discrepancy would result from the utilization of segment versus link ordering schemes and thus advocated the use of link based ordering. For the results presented here, we suggest that the segment based ordering schemes are more valid means of determining simple power laws regulating structure within plant root systems due to the substantially higher coefficients of determination.

The FS and PS regressions showed different though complementary patterns of coefficients of determination in relation to the size components considered in the regressions. The FS ordering scheme showed an extremely high coefficient of determination with the average length of root segments, but not with the number of segments or total length of root per order. The pattern with PS was exactly the opposite, showing high correlation with the number of segments and total length of root per order. Considering the components of root systems that the orders from these segment based ordering schemes represent, these results are not surprising. The orders within the FS ordering system represent the hierarchies of developmental axes. If rates of elongation and maximum length of each root segment are inversely related to its FS order then we would *expect* to see a consistent pattern between average segment size and order.

The reason why the FS system shows low coefficients of determination in relation to the absolute size of each order within the root systems (number of segments and total length of root per order) is due to a *delay* in production of roots between successive developmental orders. For example, the relationship between FS order and number and total length per order rises until the highest order, where it drops off substantially, while average length shows a strong linear relationship with order (Figure 8). In models of plant root growth the delay between the initiation of subsequent FS orders has been described using empirical constants (*e.g.*, Rose, 1983). Thus, the total amount of root (in terms of number or size) in a given developmental order does not scale simply for the higher developmental orders. Because the rate of growth of different FS orders is consistently different, the average length of roots of different developmental orders does show simple scaling properties between different FS orders.

The interpretation for why PS ordering shows simple scaling properties with segment number and total root length per order is not as simple as the relation-

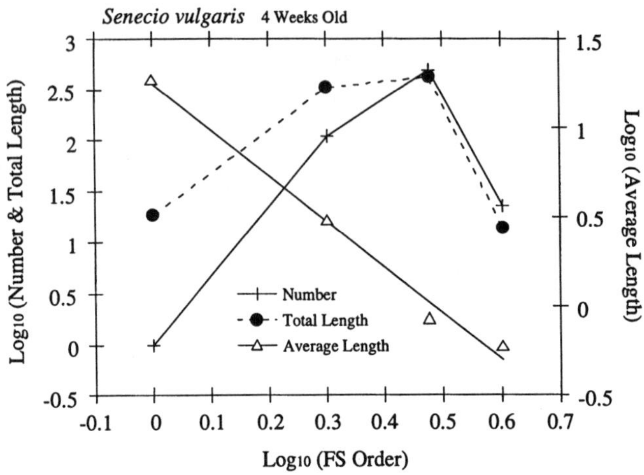

Figure 8: Centrifugal-segment (FS) ordering scheme in relation to various size components of a four-week-old *Senecio vulgaris* root system. Note that the relationship between average length and order is linear. In contrast, the number of segments and total length per order show substantial departures from linearity due to low values observed in the highest FS orders.

ship between FS ordering and average segment length. PS ordering is not developmental and thus does not represent hierarchies of developmental axes as does FS ordering. Instead, PS ordering is a system that is dynamic, continually changing as a root system grows. The application of this ordering system for describing branching structure topology (R_b) has resulted in linear relationships in every reported usage in the literature. In biological systems, this ordering system has been successfully applied to plant root systems (Fitter, 1982), the shoot system of trees (Leopold, 1971; Barker *et al.*, 1973; Whitney, 1976), fungal hyphae (Park, 1985), foraging trails of ants (Ganeshaiah and Veena, 1991), and the bronchial trees of lungs (Horsfield, 1980). Thus, it is not surprising that this ordering scheme has shown simple scaling with number of segments per order. Total length of root per order is roughly analogous to total number of segments per order, both being measures of the total amount of a root system per order. It is interesting to note that average segment length does **not** show simple scaling properties with PS order. This lack of a simple scaling relationship indicates that the elongation rates (rates of root length production per segment) are not consistent between PS orders.

V.F Ontogeny of Scaling Exponents

One of the main advantages of determining the scaling properties of branching structures of plant root systems rather than using the box-counting or dividers

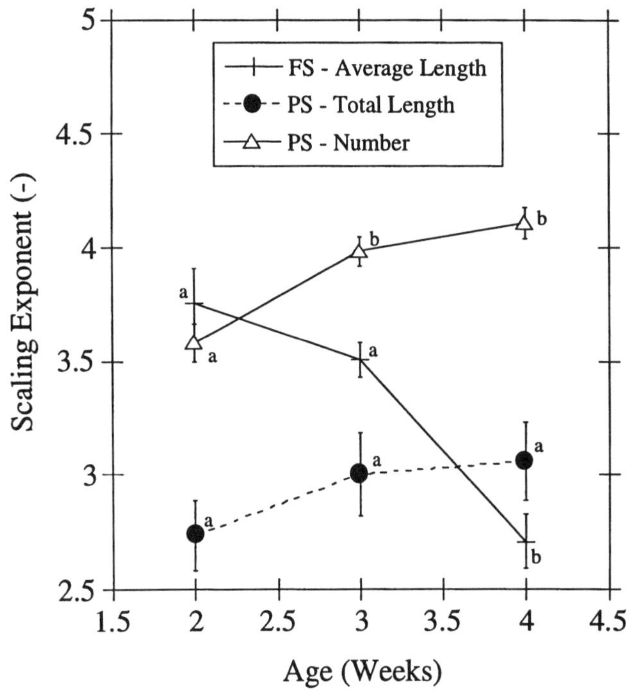

Figure 9: Ontogeny of scaling exponents for *Senecio vulgaris* root systems. If letters differ between different times for any given scaling exponent, then the values are significantly different at $P < 0.05$ (Bonferoni-corrected least squared means comparisons). Bars are single standard errors.

method to calculate D is that a knowledge of scaling properties of branching structures can be used to understand the *processes* that lead to the formation of root systems. Inasmuch, it is critical to know if these scaling properties are themselves variable through time. Figure 9 presents a summary of the three main ordering scheme-metric component regressions that showed high coefficients of determination for two-, three- and four-week-old *Senecio vulgaris.* Both the number of segments and total length per PS order showed trends of increasing scaling exponents with time - though only for number of segments per PS order was this trend significant (ANOVA, $P = 0.4351$ and $P < 0.0001$, respectively). For number of segments per PS order, the difference was due to a lower scaling exponent at week two. Average segment length by FS order showed a highly significant decrease in scaling exponent through time (ANOVA, $P < 0.0001$), where the lowest scaling exponent was at week four. These observations of the ontogeny of the scaling exponents of root system branching patterns suggest that for *Senecio* the scaling of length by PS order is relatively constant through ontogeny. However, the scaling of segment numbers by PS order and average segment length by FS order are variable through ontogeny.

V.G Complex Scaling Properties

Non-linear log-log relationships between scale and apparent size do not *a priori* imply that the structures being examined are not fractal. Mandelbrot (1983, 1988) and others have pointed out that some patterns have nonuniform fractals. Many of these fractals (called multifractals) have fractal dimensions that vary depending on the nature of the processes or parameters that regulate their formation. For structures that exhibit distinct fractal dimensions within particular scales of observation, the variable fractal dimensions may be the result of different processes operating at the different scales (Burrough, 1981; Krummel *et al.*, 1987; Sugihara and May, 1992). Thus, in cases where a fractal analysis does not show simple scaling properties, the observed complexity may result from fundamentally different processes operating between different scales. This observation has been applied in ecology as evidence of differing processes acting on different scales in the formation of landscapes (Krummel, 1987) and complex spatial structuring of vegetation allowing for diverse assemblages of habitats for invertebrates (Morse *et al.*, 1985; Gunnarson, 1992).

Fitter has suggested that the PL ordering system is valuable for describing plant root systems because it results in categorizing root links of equivalent function (proximity to a root tip) similarly. According to this logic, if complex scaling properties are observed between PL order and a size component, this would suggest that different functional components of plant root systems show different scaling properties. Such an observation would not be entirely surprising. Root sections of different age and different root members can show pronounced physiological differentiation (Scott Russell and Clarkson, 1976; Waisel and Eshel, 1991).

To interpret these non-linear scaling properties we first wanted to know the scaling properties of root systems with known growth properties. To this end we performed a series of simulations of root growth where the rules regulating growth varied from the topological extremes of dichotomous branching to herringbone branching (*sensu* Fitter, 1985). The intermediate growth rules we investigated we refer to as multi-axis (MA) and constrained dichotomous (CD) branching patterns. These simulations were performed using a topologically explicit modification of Rose's (1983) numerical model of root growth (Berntson unpublished). This model requires input for time of first appearance, elongation rate, and distance between adjacent daughter roots for each developmental order and a maximum developmental order. A summary of the input parameters used in the simulations is given in Table 3. Unlike Rose's (1983) original application of this model, the topologically explicit model creates a binary tree representation of the growing root system.

The results of these simulations are presented in Figure 10. The simulations of topological extreme root systems showed highly consistent patterns. The dichotomous simulation always resulted in perfect linear relationships between

	FS Order	D	H	CD	MA
DO_{max}		∞	2	5	3
ER	1	1	1	3	1
	2	1	1	2.5	1
	3	1		1.5	0.1
	4	1		0.4	
	5	1		0.2	
	6	1			
	∞	1			
IBD	1	1	1	1	1
	2	1	na	1	1
	3	1		1	na
	4	1		1	
	5	1		na	
	6	1			
	∞	na			
T	1	0	0	0	0
	2	0	0	1	0
	3	0		2	0
	4	0		3	
	5	0		4	
	6	0			
	∞	0			

Table 3: Input parameters to topologically explicit modification of Rose's (1983) numerical model of root growth. Refer to text for a more detailed description of model. FS Order = centrifugal segment-based ordering system (developmental ordering), D = dichotomous branching pattern, H = herringbone branching pattern, CD = constrained dichotomous branching pattern, MA = multi-axis branching pattern, DO_{max} = maximum developmental order, ER = elongation rate (length time^{-1}), IBD = interbranch distance (length), T = time (time). Length and time units arbitrary. The highest developmental order values for IBD are not applicable (na) because there is no branching on the highest developmental order.

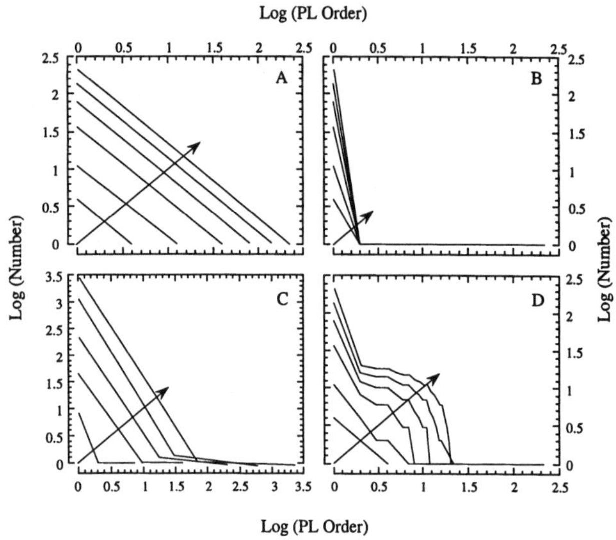

Figure 10: Summary log-log plots of number of links versus PL order for simulated root growth. The arrow in each plot indicates the ontogenetic trajectories of each simulation, with the oldest simulated root systems near the arrow-head. A = dichotomous, B = herringbone, C = constrained dichotomous, D = multi-axis.

PL order and number, with a scaling exponent of -1 (R^2 = 1). The herringbone simulation resulted in relationships with a scaling exponent of zero for PL order > 1 because there is only one link of each higher order. The scaling exponent of the tail end of these simulations was log(number links of order 1)/log(2). The larger the root system, the steeper the slope. The constrained dichotomous simulation resulted in number-order relationships consisting of two linear segments. The PL order which marked the boundary between these segments was equal to order 2 for small root systems (indicating that they were initially herringbone) but increased through ontogeny. The scaling exponent between order and number remained fairly constant for small orders through ontogeny. The magnitude of this scaling exponent was decreased with increasing maximum developmental order and increasing relative elongation rates at higher developmental orders. The multi-axis simulation showed the most complex scaling properties between PL order and number of links per order. For the larger root systems (older simulated root systems) there were three distinct regions in the relationship between order and number. Similar to the herringbone pattern, there was a small region (from order 1 to 2) which showed a steeper slope with increasing root system size. Similar to both the constrained dichotomous and the herringbone system, the scaling exponent for the largest orders was zero. For intermediate orders, there was convex hump. At the low order end of this

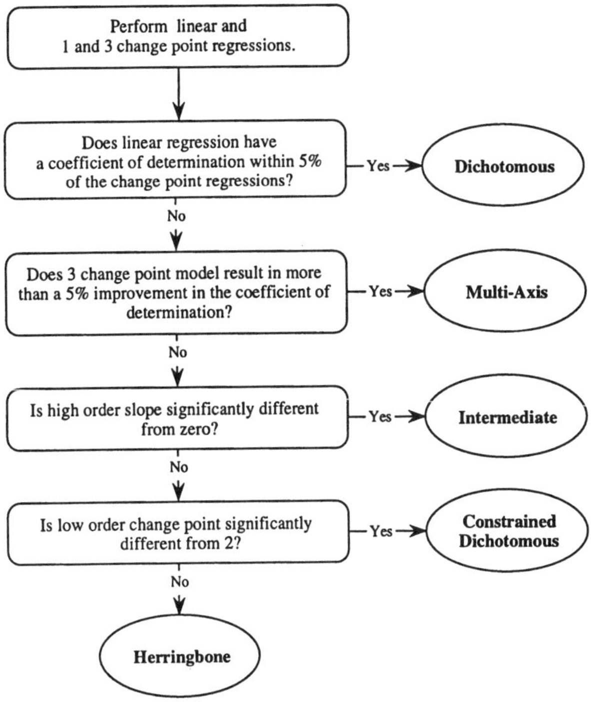

Figure 11: Schematic diagram illustrating the use of change point regressions for characterization of root systems. See text for details.

hump, the slope is nearly zero. The number of links this region corresponds to is the number of axes within the root system. The lower orders within this system behave analogously to the herringbone pattern. The difference is that in the multi-axis system there are several herringbones being ordered simultaneously. The right-hand side of the convex hump results from the multiple axes being of different sizes and thus, once a certain order is reached, the constant offset from a herringbone pattern decays.

To characterize these patterns in actual root systems we applied one and three change point linear models (*sensu* Chappell, 1989) using a least squares maximum likelihood estimation (Press *et al.*, 1989). These two models allow continuous linear segments (two in the single change point model and four in the three change point model) to be fitted to the data. The models allow us to characterize a root system in relation to the above simulations (H, CD, and MA). A schematic diagram illustrating the characterization of root systems as one of the above simulated types through the application of the change point regressions is provided in Figure 11. An additional branching pattern is introduced in Figure 11, the "intermediate" branching pattern. This pattern results from root

	Age	Branching Pattern			
		H	CD	I	MA
Senecio	2	22 (1.00)	22 (0.97)	17 (0.94)	39 (0.99)
	3	0 -	20 (0.97)	30 (0.92)	50 (0.99)
	4	0 -	15 (0.97)	10 (0.88)	75 (0.99)
Non-*Senecio*	•	3 (1.00)	72 (0.97)	16 (0.93)	9 (0.97)

Table 4: Summary of multiple change point regressions (see text for a discussion of terminology and methods). Presented is the percent of all regressions that are representative of H (herringbone), CD (constrained dichotomous), I (intermediate) or MA (multi-axis) branching patterns for each age class for *Senecio* and all the other species pooled together. The average coefficient of determination of each set of regression is noted in parentheses. Refer to the text for a description of criteria used to categorize change point regressions.

systems that show no statistically discernible zero scaling exponent region at high orders. Instead, these systems show two non-zero scaling regions. These root systems are similar to multi-axis branching patterns, but have low curvature of the "convex" region and a reduced zero scaling exponent region. These regions are pooled together in the "intermediate" branching pattern.

Using these techniques, we explored the PL ordering system in relation to number of links per order in more detail. The log-log linear regression of PL order versus number of links per order and total length of root per order showed high levels of significance but low coefficients of determination (Table 2). A summary of using the change point regressions to characterize root system branching patterns is given in Table 4. Sample plots for each of the four branching patterns is given in Figure 12. In no case did any root system exhibit dichotomous branching. *Senecio* root systems exhibited each of the observed branching patterns. The young *Senecio* roots showed a large proportion of herringbone and constrained dichotomous branching patterns. In contrast, the older root systems primarily exhibited multi-axis branching patterns. The other species' root systems primarily exhibited multi-axis branching patterns.

From these data, it is clear that the PL ordering system can be used to characterize different types of branching patterns of root branching structures. The data presented here for examining complex scaling properties of plant root systems are from small root systems. Ultimately, it is desirable to apply these techniques for characterizing complex scaling properties to larger root systems. Assessing whether the scales over which different branching processes appear to be operating correspond with physiological and morphological variation in

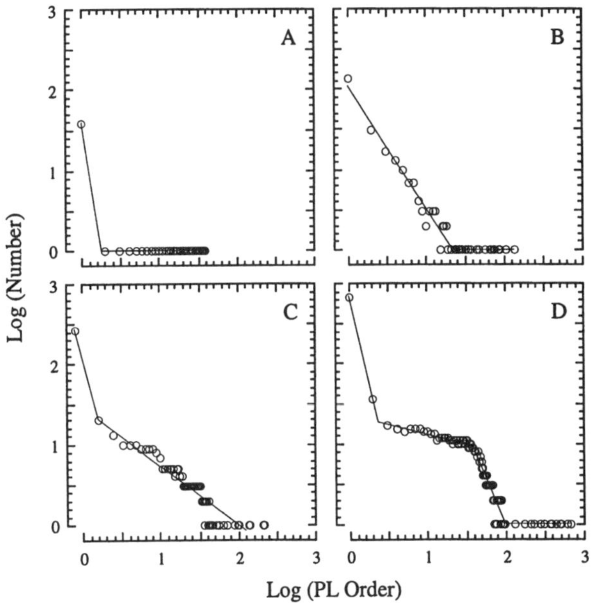

Figure 12: Sample scatterplots of PL order versus number of links per order for root systems characterized using the criteria diagrammed in Figure 5. A = herringbone (2-week-old *Senecio*), B = constrained dichotomous (*Betula alleghaniensis*), C = intermediate (3-week-old *Senecio*), D = multi-axis (4-week-old *Senecio*).

root systems is an unexplored but potentially exciting line of study. If branching processes correspond with physiological capacity within plant root systems, then the spatial distribution of roots (as influenced branching patterns) is intricately associated with physiological capacity. Another potential application of identifying the scales of branching processes in relation to the PL ordering scheme is to increase our understanding of the spatial scales that separate root production in very large plants such as trees. The root systems of trees typically contain multiple classes of roots that differ functionally and morphologically. Large, woody roots, which may contain the majority of a large tree's standing below ground mass, are long-lived and primarily responsible for structural stability and transport between fine roots and the stem (Lyford and Wilson, 1964; Lyford, 1980). In contrast, the fine roots of trees can be relatively short lived, are the primary resource uptake organs (Vogt and Bloomfield, 1991) and exhibit different branching patterns than woody roots, foraging more intensively within small volumes of soil (Lyford, 1975). For these reasons, applications of these techniques to the root systems of trees (and possibly other large, long-lived perennials) would allow us to quantify the scale over which different growth processes are taking place. This information may allow us to characterize the spatial and temporal scales over which different components of these large, long lived root systems deploy their roots.

VI Subsampling: Practical Considerations of Fractal Geometry for Plant Root Systems

Fractal objects are those whose structure is similar at any scale of observation. From this characteristic of fractal objects, it follows that subsamples of the object can be reasonable representations of the structure of the entire object. This observation has profound implications for plant root systems. Measurements of the structure of plant root systems can be a tedious process, requiring large amounts of manual labor for relatively little data (*e.g.*, Böhm, 1979). Citing some of the studies discussed in the "Summary of Research to Date" section as evidence that plant root systems are fractal and therefore self-similar, Fitter and Stickland (1992b, p. 243) have suggested that "random pruning of a root system should not alter its branching pattern". This hypothesis has very important implications to studies of root architecture on field grown plants because, if true, it suggests that subsamples of root systems may be reliable proxies for entire root systems.

In order to assess the validity of this hypothesis it is important to clearly define the methods used to characterize plant root systems as fractal, the methods used to subsample the root systems, and the subsequent inferences about the structure of the entire root system from the subsamples. There have been two published attempts of relating subsampling methods within root systems to estimates of fractal dimensions that we can explore to gain a preliminary understanding of the limitations of conceptualizing plant root systems as fractal with regard to properties of self-similarity. One method explored the effect of subsampling the images used to calculate fractal dimensions for the entire root system (hereafter referred to as the "Image Subsample"; Eghball *et al.*, 1993), the other method examined the variability in fractal dimensions of different classes of roots within individual root systems (hereafter referred to as the "Morphological Subsample"; Lynch and van Beem, 1994). In order to directly explore the limitations of a subsampling scheme that would represent a distinct methodological advantage in our ability to measure root system growth, we conclude this section by presenting the results of a simulation of the removal of root systems from field conditions in which the patterns of root truncation are explored.

VI.A Image and Morphological Subsampling

The Image Subsample method was applied by Eghball *et al.* (1993). The root systems of maize that Eghball *et al.* (1993) examined were prepared by removing them from soil profiles with their spatial orientation preserved. Two-dimensional images were then taken of the cleaned root systems. The box-counting method was used to calculate the fractal dimension of these images.

The images of the root systems were divided into nine separate sections (three columns x three rows). Fractal dimensions were calculated separately for each of these sections and Eghball *et al.* observed that fractal dimensions were significantly greater near the soil surface than in the deepest portion of the root system. Otherwise, they observed no significant variations in observed fractal dimensions.

The Morphological Subsample method was applied by Lynch and Van Beem (1994). The root systems of common bean were subsampled on the basis of root type: tap root, lateral roots and basal roots. Prior to preparing the root system for analysis, each of the different root types were separated from one another and their fractal dimensions determined. This study found that at any given root age, fractal dimensions of different root types could vary significantly. Through ontogeny the composition of the root systems (relative contribution by different root types) changed and the fractal dimensions of these individual root members tended to increase.

These two approaches are similar insofar as they examined the effect of subsampling on root fractal dimensions. These approaches differ significantly in the components of root systems that are subsampled. The Image Subsample method, when applied to root systems whose *in situ* spatial orientation is preserved, is a means of assessing the spatial variation within root system fractal dimensions. The Morphology Subsample method is a means of assessing developmental variation within root system fractal dimensions. Both of these subsampling methods demonstrated that fractal dimensions within plant root systems can be variable. Thus, the assumption that observations subsamples of root systems can be used to make inferences about whole root system structure is flawed due to variation within root system structure.

Neither the Image Subsampling method nor the Morphological subsampling method can be easily (rapidly) applied to plant root systems growing in the field. Both subsampling procedures require that the root systems be carefully excavated and prepared prior to carrying out the subsampling procedures. Thus, these methods are valuable mostly from the perspective of demonstrating that within root system structure is variable – not from the perspective that they might have represented viable subsampling procedures for rapidly assessing root system structure.

VI.B Practical Limitations to Self-Similarity: Root Collection from Field Conditions

Fitter and Stickland's (1992b) proposal that "random pruning of a root system should not alter its branching patterns" led them to "suggest that architectural analysis can be safely applied to field-grown root systems, or more accurately the pruned fragments that can be recovered by normal sampling techniques". Neither of the above-discussed subsampling schemes (Image or Morphological) can be easily applied to the type of subsampling that takes place when collecting roots from field conditions. Fitter and Stickland's proposal, were it true, represents a profoundly significant practical implication of the fractal properties of plant root systems. This proposal has not, however, been tested.

The critical assumption behind Fitter and Stickland's proposal hinges on the work of Van Pelt and Verwer (1984). Van Pelt and Verwer demonstrated that random removal of a single sub-tree within a rooted tree does not alter the probability distributions of resultant trees. Van Pelt and Verwer's analysis rests on three assumptions that present complications when applying their results to plant root systems. First, only a single sub-tree (*e.g.*, branch) is removed from the root system. Second, the removed branch is selected at random. Third, the population of root systems being considered is the product of a random growth model (either segmental or terminal, see Van Pelt and Verwer, 1983). There are several reasons why these assumptions are not necessarily applicable to plant root systems. First, pruning a root system by removal from the soil is likely to result in breakage of more than a one branch. Second, breakage of individual roots will be dependent upon the strength of the root, which is largely a function of diameter (Ennos, 1990). Diameter varies in a regular pattern with developmental and topological order (Fitter *et al.*, 1991), and thus distribution of probable points of breakage within the root system is not random. Third, the distribution of branching events within a root system is dependent on developmental order (Rose, 1983). Thus, the random growth models described by Van Pelt and Verwer (1983) are not directly applicable to root systems.

VI.B.a Simulation Model of Field Extraction

To directly explore the potential effects of subsampling by field collection, we performed a simulation representative of extraction from the soil where the plant is pulled up from the soil by pulling on the stem (Berntson, unpublished). In this simulation, root systems of 4-week-old seedlings of *Senecio vulgaris* L. were loaded into computer memory (Berntson, 1992). Two water and CO_2 treatments had been imposed on the plants, resulting in a wide range of root system sizes and architecture (see Figure 13). A more detailed description of experimental design, growth conditions and root system architecture can be found in Berntson and Woodward (1992).

Simulations of truncation due to tensile failure were based on the work of Ennos (1989, 1990). Ennos calculated the critical length (L_{crit}) of an individual root, at which a root would be broken rather than extracted from the soil without breakage, as

$$L_{crit} = \frac{\sigma R}{2\alpha\tau} \tag{1}$$

where s = breaking stress (kNm^{-2}), R = radius of root (cm), t = soil strength (kNm^{-2}) and a = relative strength of root/soil bond where 0 = no bond and 1 = root/soil bond > soil strength. To simplify the simulations, the terms describing the soil and root/soil bond were lumped together, where

$$K = \frac{\sigma}{\tau} \tag{2}$$

K is a unitless parameter representing the force of the breaking stress relative to the strength of the soil. The product of Ka^{-1} was used as an index of

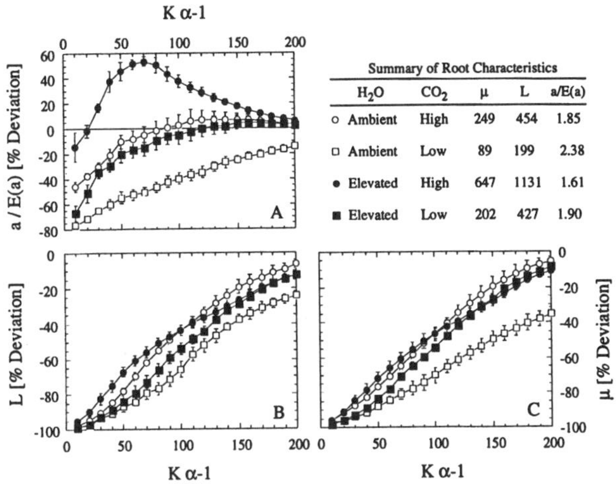

Summary of Root Characteristics

	H_2O	CO_2	μ	L	a/E(a)
○	Ambient	High	249	454	1.85
□	Ambient	Low	89	199	2.38
●	Elevated	High	647	1131	1.61
■	Elevated	Low	202	427	1.90

Figure 13: Results of simulations of removal of root systems from from the soil with varying ratios of root-soil to root tensile strength (Ka^{-1}). Effects on A - topology, B - root length and C - root tips (including broken fragments). Included in the figure are average size and topological features of root sytems used in simulations.

the integrated strength of root tissue relative to the soil/root bond and the soil strength. The radii of individual roots within the root systems were determined using the relationship proposed by Fitter (Fitter, 1987; Fitter *et al.*, 1991):

$$R = R_{ext} + (\mu - 1) * RIR \tag{3}$$

where R is the radius of the link (cm), R_{ext} is the radius of external links (cm), μ is the magnitude of the link (number of external links subtending the link) and RIR is the relative rate of increase in radius (cm μ^{-1}). To best approximate observed distributions of root diameters in *S. vulgaris,* values of 0.025 and 0.0005 were selected for R_{ext} and RIR, respectively. This resulted in a maximum root diameter for the largest root system of approximately 0.35 cm. Values for Ka^{-1} were varied from 10 to 200 in increments of 10 to simulate a wide range of values for L_{crit} over the observed range of root radii. Values of L_{crit} were calculated for each root (*sensu* centrifugal-segment ordering: see section V.B.) within the root system. L_{crit} was calculated with each root modeled as a cylinder of constant radius equal to the radius of the base link (proximal to base of root system) of the root (Ennos, 1990).

VI.B.b Extraction of Roots From the Field: Cautionary Demonstration of Non-Self-Similarity

Results of the simulations are presented in Figure 13. The effect of variations in Ka^{-1} on remaining root length (L) and number of root tips (also known as magnitude, μ) were approximately similar for all root systems. There was a tendency for the smaller, less branched root systems to show greater levels of underestimation of L and especially μ. The effect of variations in Ka^{-1} on root system topology (as measured by root system altitude over expected altitude, aE_a^{-1}; Fitter *et al.*, 1991; Fitter and Stickland, 1991, 1992a, 1992b) differed profoundly from the effects on L and μ, and varied significantly by species. Root systems of intermediate size and aE_a^{-1} (elevated CO_2, low soil moisture and ambient CO_2, high soil moisture treatment combinations) showed relatively reliable estimations of aE_a^{-1} with pruned root systems even when L and μ were still being significantly overestimated. The smallest root systems showed consistently greater underestimations in aE_a^{-1}. The largest root systems showed a non-linear, non-monotonic pattern of aE_a^{-1} over the range of simulated Ka^{-1}.

These results suggest that, depending on the soil structure relative to root system strength and the size/complexity of branching of the root system, very different patterns of pruning will occur when roots are extracted from the soil by pulling them out. Clearly, the results of this simulation have little relevance to carefully conducted excavations of root systems. However, they have condemning implications to Fitter and Stickland's (1992b) assumption that "pruned fragments that can be recovered by normal sampling techniques" are reliable proxies for whole root system architecture. There are not consistent relationships between root system size or topology and the effects of pruning by pulling them out of the soil - in some cases complexity of branching structure is overestimated and in other cases it is underestimated. We suggest that roots are *complex* fractal structures, showing spatially and developmentally dependent variations in fractal structure and thus many simple subsampling procedures cannot be used to infer whole root system structure following the assumption that root systems show simple fractal behavior.

VII Potential Utility

The newness of this area of research makes it difficult to assess the potential utility of fractal geometry for root studies. However, benefits in several areas are possible. As a geometry suited to the description and analysis of natural objects, it is reasonable to hope that fractal geometry may provide useful simple descriptors integrating much of the complex architecture of plant root systems. The 3-dimensional (or even 2-dimensional) configuration of root systems is cumbersome to summarize using standard Euclidean geometry. Fractal geometry could provide condensed, quantitative descriptions of root architecture (such as D in 2-space or 3-space) that would be amenable to wide comparisons of diverse roots or mass screening in breeding programs and for rapid assessment of the belowground adaptive plasticity. In fact, the extreme condensation of information represented by D for an entire root system may be counterproductive for such purposes, since many distinct root shapes could give rise to the same value of D. As discussed in section II, D shows high correlations with several different architectural attributes - from spatially independent patterns of topology, to direct measures of root system size and density.

Although the functional significance of D may depend on the absolute scale over which D is measured, D itself is theoretically independent of the absolute size of an object. Thus, D may be a useful integrated parameter of root architecture because it is uncoupled from absolute size. However, in this regard it must be noted that in the several studies which have reported D over time in growing root systems, it appears that D increases through ontogeny (or at least, size and D co-vary with time) until a stable value is reached (*e.g.*, Fitter and Stickland, 1992a; Snapp, unpublished data). A size-independent architectural parameter would be useful for comparisons of diverse taxa of different sizes or to facilitate comparisons within a taxa at different times or at different growth stages. Convergence of D towards some asymptotic value may render it insensitive to treatment differences, which would undermine its utility as a descriptive parameter.

A better understanding of the relationship of D to time and root system size is needed in order to resolve these issues. Just as projection theorems permit us to consider D in a one-, two-, or three-dimensional context, it is tempting to consider time the 'fourth dimension' and try to condense or concatenate spatial and temporal aspects of D, since root systems typically branch in time as well as space. The obstacle to this approach is that if comparisons were made over time intervals that were too short, root growth would be negligible, and thus the dimension of the four-dimensional fractal would just be the largest of the spatial fractal dimensions over time (Weiss and Lynch, unpublished). Hence, it is possible that we gain no new information by considering time as an extra dimension in deriving/calculating empirical values for D.

In contrast to adding an additional integrative dimension to the application of fractal geometry to plant root systems, it would be wise for future studies to refine our understanding about the multifractal behavior of plant root systems. We have seen that the fractal dimenson of plant root systems varies as a function of *location* (Eghball *et al.*, 1993) and *developmental class* (Lynch and van Beem, 1994). The methods presented in this paper for identifying variations in the scaling exponents for branching patterns within different scales (orders) of root systems is one possible way that we can derive an understanding of how these complex patterns are generated. Characterization of the fractal properties of plant root systems in relation to developmental hierarchies within plant root systems may allow us to use simple scaling properties for describing the development of plant root systems. Also, if we can characterize the fractal properties of plant root systems in relation to functional hierarchies of plant root systems (differentiation on the basis of physiological capacity, solute flow, life span) then we may be able to use simple scaling properties to branching structure in relation to function. Such approaches are difficult or impossible using standard variations of box-counting and dividers methods for determining fractal dimensions. By examining the topological/developmental structure of plant root systems, these goals are possible. The preliminary results we have presented here suggest that fractal properties of root system branching patterns can be variable through ontogeny, making the application of these scaling properties, as generators of the complex architecture of plant root systems, difficult. On a more positive note, it seems that there are regular, albeit complex, scaling properties within root systems in relation to a functional classification of within-root system structure. These scaling properties show consistent patterns with regard to ontogeny and widely varying growth patterns, and thus may prove a useful tool for studying within-root system variation in development and possibly function.

A potential advantage of fractal analysis of root systems noted by Fitter and Stickland (1992b) is that due to the inherent self-similarity of fractal objects, architectural features of a set of roots may be inferred by observation of some smaller subsystem, thereby greatly simplifying sampling procedures and analytical requirements. This benefit would be obtained only if plant root systems were, indeed, self-similar throughout their entire structure. As discussed previously, this assumption appears to not be valid due to developmental and spatial variation within single root systems (see section VI.). It should also be noted in this regard that self-similarity for a mathematical fractal may extend over an infinite range of scales, whereas for a biological entity such as a root system, the fractal behavior of interest to root architecture is only relevant within a certain range of scales, generally between several mm and several m, and comparisons of fractal behavior among roots of similar functional classification (say, roots active in nutrient foraging as opposed to structural roots) would only be valid within an even smaller range of that spectrum. Further, subsampling root systems through excavation may obscure self-similar properties by preferentially damaging or removing roots of small diameter (see section VI.B.). Therefore, even in uniform environments, the self-similarity of some mathematical fractals may be of limited utility in root analysis.

Fractal geometry may provide new perspectives on root function as well as root structure. Within the range of scales appropriate to root system studies, increasing D would be expected to be associated with increasing foraging intensity by root systems (Berntson, 1994 b), caused by the fineness of branching and intensity of root proliferation in a given volume. The adaptive significance of foraging intensity in terms of soil resource acquisition will depend upon the resource being considered. Water and mobile nutrients such as nitrate and Mg can move relatively long distances through soil to roots by mass flow of water (Barber, 1984), so foraging intensity will have less impact on the acquisition of these nutrients than on diffusion-limited resources such as P and K. Phosphorus is noteworthy in that it is a primary constraint in many terrestrial ecosystems and it is the most immobile of the principal soil resources; therefore, foraging intensity as denoted by D may have particular relevance to P acquisition. In the context of specific soil resources the actual scale of fractal behavior will be important in the interpretation of functional responses. Root-branching patterns on a mm scale may have very different implications than the same patterns on a cm scale, depending upon whether depletion of the resource in question (and therefore, intra-root competition) occurs over cm or mm. A potentially fruitful approach to this problem would be to consider the fractal nature of root structure in relation to the fractal nature of resource distribution in soil; we might expect efficient exploitation of soil resources to display some association of the two.

VIII Future Prospects

Although the utility of fractal geometry in root analyses remains to be demonstrated, at the moment it is intriguing as an architectural parameter that integrates much of the complexity of actual root systems. Ironically, this property makes direct interpretation of fractal dimensions for plant root systems difficult. In order to gain a firmer grasp of the utility of fractal geometry and its development as an analytical tool we need to continue research in several areas. 1) We need to further develop techniques for calculating fractal dimensions of actual root systems. No research to date has determined the fractal dimension of actual root systems within a 3-dimensional matrix. 2) We need to determine the relationships between root system fractal dimensions and root growth and plasticity. Such information may allow us to use these indices to summarize a large amount of complex information that may have direct relevance to root system form and function, root responses to environmental factors, and variation between species and genotypes. 3) We need to better understand the relationship between ontogenetic variation in fractal properties in relation to the dynamic properties of root development. 4) Within root systems, developmental structure and function can vary substantially. Understanding this variation in the context of fractal properties of plant root systems, in terms of both developmental and space-filling patterns, could greatly aid our ability to determine the functional implications of these patterns.

Acknowledgments

We wish to thank Kai Nielsen for helpful discussions and for providing unpublished data. Helpful comments on the manuscript were provided by D. Ackerly, D. Karpa and D. Shumway. For GMB, this research was performed under appointment to the Graduate Fellowships for Global Change Program administered by Oak Ridge Institute for Science and Education for the U.S. Department of Energy, Office of Health and Environmental Research, Atmospheric and Climate Research Division. For JPL, support was provided by USDA/NRI grant 94371000311.

References

Barber, S. A. 1984. *Soil Nutrient Bioavailability: A Mechanistic Approach.* John Wiley and Sons, New York.

Barker, S. B., G. Cumming and K. Horsefiled. 1973. Quantitative morphometry of the branching structure of trees. *J. Theor. Biol.* 40:33-43.

Baveye, P. and C.W. Boast. 1997. Fractal geometry, fragmentation processes and the physics of scale-invariance: an introduction. (This volume, Chapter 1).

Berntson, G. M. 1992. A program for characterizing root system branching patterns. *Plant and Soil* 140:145-149.

Berntson, G. M. 1994a. Modeling root architecture: Are there tradeoffs between efficiency and potential of resource capture? *New Phytologist* 127:483-493.

Berntson, G. M. 1994b. Fractal analysis of plant root systems: How reliable are calculations of fractal dimensions? *Anal. Bot.* 73:281-284.

Berntson, G. M. and F. I. Woodward. 1992. The root system architecture of *Senecio vulgaris* L. under elevated CO_2 and drought. *Functional Ecology* 6:324-333.

Berntson, G. M., E. J. Farnsworth and F. A. Bazzaz. 1995. Allocation, deployment and root turnover in two birch species. *Oecologia* 101(4):439-447.

Böhm, W. 1979. *Methods of Studying Root Systems*. Springer-Verlag, Berlin.

Bottomley, P. A., H. H. Rogers and T. H. Foster. 1986. NMR imaging shows water distribution and transport in plant root systems in situ. *Proc. Natl. Acad. Sci. USA* 83:87-89.

Bottomley, P. A., H. H. Rogers and S. A. Prior. 1993. NMR imaging of root water distribution in intact *Vicia faba* L. plants grown in elevated atmospheric CO_2. *Plant, Cell and Environment* 16:335-338.

Brown, D. P., T. K. Pratum, C. Bledsoe, E. D. Ford, J. S. Cothern and D. Perry. 1991. Noninvasive studies of conifer roots: nuclear magnetic resonance (NMR) imaging of Douglas-fir seedlings. *Canadian Journal of Forest Research* 21:1559-1566.

Burlando, B. 1990. The fractal dimension of taxonomic systems. *J. Theor. Biol.* 146:99-114.

Burrough, P. A. 1981. Fractal dimensions of landscapes and other environmental data. *Nature* 294:240-242.

Chappell, R. 1989. Fitting bent lines to data, with applications to allometry. *J. Theor. Biol.* 138:235-256.

Charlton, W. A. 1975. Distribution of lateral roots and pattern of lateral initiation in *Pontederia cordata* L. *Botanical Gazette* 136:225-235.

Charlton, W. A. 1982. Distribution of lateral root primordia in root tips of *Musa acuminata Colla. Anal. Bot.* 49:509-520.

Charlton, W. A. 1983. Patterns of distribution of lateral root primordia. *Anal. Bot.* 51:417-427.

Clapham, C. 1990. *A Concise Oxford Dictionary of Mathematics*. Oxford University Press, Oxford.

Clarkson, D. T., J. Sanderson and R. Scott Russell. 1968. Ion uptake and root age. *Nature* 220:805-806.

Commons, P. J., A. B. McBratney and A. J. Koppi. 1991. Development of a technique for the measurement of root geometry in the soil using resin-impregnated blocks and image analysis. *J. Soil Sci.* 42:237-250.

Crawford, J. W. and I. M. Young. 1990. A multiple scaled fractal tree. *J. Theor. Biol.* 145:199-206.

Davis, R. D. 1993. *Modeling and Visualization of Botanical Root Systems*. M.S.

Thesis. Pennsylvania State University, University Park, Pennsylvania.

Diebel, K. E. and P. P. Feret. 1993. Using fractal geometry to quantify loblolly pine seedling root system architecture. *Southern J. Applied Forestry* 17:130-134.

Diggle, A. J. 1988. ROOTMAP - a model in three-dimensional coordinates of the growth and structure of fibrous root systems. *Plant and Soil* 105:169-178.

Drew, A. P. 1993. Establishment of willow cuttings grown in porous membrane root envelopes. *Plant and Soil* 148:289-293.

Eghball, B., J. R. Settimi, J. W. Maranville and A. M. Parkhurst. 1993. Fractal analysis for morphological description of corn roots under nitrogen stress. *Agron. J.* 85:287-289.

Eissenstat, D. M. 1992. Costs and benefits of constructing roots of small diameter. *J. Plant Nutr.* 15:763-782.

Ennos, A. R. 1989. The mechanics of anchorage in seedlings of sunflower, *Helianthus annuus* L. *New Phytol.* 113:183-192.

Ennos, A. R. 1990. The anchorage of leek seedlings: The effect of root length and soil strength. *Anal. Bot.* 65:409-416.

Fitter, A. H. 1982. Morphometric analysis of root systems: application of the technique and influence of soil fertility on root system development in two herbaceous species. *Plant, Cell and Environment* 5:313-322.

Fitter, A. H. 1985. Functional significance of root morphology and root system architecture. p. 87-106. In A. H. Fitter, D. J. Read, D. Atkinson and M. B. Usher (eds.) *Ecological Interactions in Soil.* Blackwell Scientific Publications, Oxford.

Fitter, A. H. 1986. The topology and geometry of plant root systems: Influence of watering rate on root system topology in *Trifolium pratense*. *Anal. Bot.* 58:91-101.

Fitter, A. H. 1987. An architectural approach to the comparative ecology of plant root systems. *New Phytologist* 106:61-77.

Fitter, A. H. and T. R. Stickland. 1992a. Fractal characterization of root system architecture. *Funct. Ecol.* 6:632-635.

Fitter, A. H. and T. R. Stickland. 1992b. Architectural analysis of plant root systems. III. Studies on plants under field conditions. *New Phytol.* 121:243-248.

Fitter, A. H., T. R. Stickland, M. L. Harvey and G. W. Wilson. 1991. Architectural analysis of plant root systems. I. Architectural correlates of exploitation efficiency. *New Phytol.* 118:375-382.

Ganeshaiah, K. N. and T. Veena. 1991. Topology of the foraging trails of *Leptogenys processionalis* - why are they branched? *Behavioral Ecology and Sociobiology* 29:263-270.

Gross, K. L., D. Maruca and K. S. Pregitzer. 1992. Seedling growth and root morphology of plants with different life-histories. *New Phytologist* 120:535-542.

Gunnarsson, B. 1992. Fractal dimension of plants and body size distribution in spiders. *Funct. Ecol.* 6:636-641.

Hackett, C. 1968. A study of barley. I. Effects of nutrition on two varieties. *New Phytol.* 67:287-299.

Harrison-Murray, R. S. and D. T. Clarkson. 1973. Relationships between structural development and the absorption of ions by the root system of *Cucurbita pepo*. *Planta* 114:1-16.

Hendrick, R. L. and K. S. Pregitzer. 1993. Patterns of fine root mortality in two sugar maple forests. *Nature* 361:59-61.

Horsfield, K. 1980. Are diameter, length and branching ratios meaningful in the lung? *J. Theor. Biol.* 87:773-784.

Horton, R. E. 1945. Erosional development of streams and their drainage basins: hydrophysical approach to quantitative morhpology. *Geol. Soc. Amer. Bull.* 56:275.

Jackson, R. B., J. H. Manwaring and M. M. Caldwell. 1990. Rapid physiological adjustment of roots to localized soil enrichment. *Nature* 344:58-60.

Körner, C. H. 1991. Some often overlooked plant characteristics as determinants of plant growth: a reconsideration. *Funct. Ecol.* 5:162-173.

Krummel, J. R., R. H. Gardner, G. Sugihara, R. V. O'Neill and P. R. Coleman. 1987. Landscape patterns in a disturbed environment. *Oikos* 48:321-324.

Leopold, L. B. 1971. Trees and streams: the efficiency of branching patterns. *J. Theor. Biol.* 31:339-354.

Lungley, D. R. 1973. The growth of root systems - a numerical computer simulation model. *Plant and Soil* 38:145-159.

Lyford, W. H. 1975. Rhizography of non-woody roots of trees in the forest floor. In Torrey and Clarkson (eds.) *The Development and Function of Roots*. Academic Press, London.

Lyford, W. H. 1980. Development of the root system of Northern red oak (*Quercus rubra* L.). *Harvard Forest Paper*. No. 21.

Lyford, W. H. and B. F. Wilson. 1964. Development of the root system of *Acer rubrum* L. *Harvard Forest Paper.* No. 10.

Lynch, J. P. and J. J. van Beem. 1994. Growth and architecture of seedling roots of common bean genotypes. *Crop Science* 33:1253-1257.

MacDonald, N. 1983. *Trees and Networks in Biological Models*. John Wiley and Sons, New York.

Mackie-Dawson, L. A. and D. Atkinson. 1991. Methodology for the study of roots in field experiments and the interpretation of results. In D. Atkinson (ed.) *Plant Root Growth: An Ecological Perspective*. Blackwell, London.

Mandelbrot, B. B. 1983. *The Fractal Geometry of Nature*. W. H. Freeman and Co, New York.

Mandelbrot, B. B. 1988. An introduction to multifractal distribution functions. In H.E. Stanley and H. Ostrowsky (ed.) *Fluctuations and Pattern Formation*. Kluwer Academic, Dordrecht.

May, L. H., F. H. Chapman and D. Aspinall. 1965. Quantitative studies of root development. I. The influence of nutrient concentration. *Aust. J. Biol. Sci.* 18:23-35.

May, L. H., F. H. Randles, D. Aspinall and L. G. Paleg. 1967. Quantitative studies of root development. II. Growth in the early stages of development.

Aust. J. Biol. Sci. 20:273-283.

Morse, D. R., J. H. Lawton, M. M. Dodson and M. H. Williamson. 1985. Fractal dimension of vegetation and the distribution of arthropod body lengths. *Nature* 314:731-733.

Nye, P. H. and P. B. Tinker. 1977. *Solute Movement in the Soil-Root System.* Blackwell Scientific, Oxford.

Pagés, L., M. O. Jordan and D. Picard. 1989. A simulation model of the three-dimensional architecture of the maize root system. *Plant and Soil* 119:147-154.

Park, D. 1985. Does Horton's law of branch length apply to open branching systems? *J. Theor. Biol.* 112:299-313.

Press, W. H., B. P. Flannery, S. A. Teukolsky and W. T. Vetterling. 1989. *Numerical Recipes in Pascal.* Cambridge University Press, Cambridge.

Prusinkiewicz, P. and A. Lindenmayer. 1990. *The Algorithmic Beauty of Plants.* Springer-Verlag, Berlin.

Pulgarin, A., J. Navascus, P. J. Casero and P. G. Lloret. 1988. Branching pattern in onion adventitious roots. *Amer. J. Bot.* 75:425-432.

Rice, W. R. 1989. Analyzing tables of statistical tests. *Evolution* 43:223-225.

Rogers, H. H. and P. A. Bottomley. 1987. *In situ* nuclear magnetic resonance imaging of roots: influence of soil type, ferromagnetic partical content, and soil water. *Agron. J.* 79:957-965.

Rose, D. A. 1983. The description of the growth of root systems. *Plant and Soil* 75:405-415.

Scott Russell, R. and D. T. Clarkson. 1976. Ion transport in root systems. p. 401-411. In N. Sunderland (ed.) *Perspectives in Experimental Biology.* Pergamon Press, Oxford.

Smart, J. S. 1978. The analysis of drainage network composition. *Earth Surface Processes* 3:129-170.

Snapp, S. S. and C. Shennan. 1994. Salinity effects on root growth and senescence in tomato and the consequenses for severity of Phytophthora root rot infection. *J. Amer. Soc. Hort. Sci.* 119:458-463.

Strahler, A. N. 1952. Hypsometric (area altititude) analysis of erosional topology. *Geol. Soc. Amer. Bull.* 63:1117.

Sugihara, G. and R. M. May. 1990. Applications of Fractals in Ecology. *TREE* 5:79-86.

Tatsumi, J., A. Yamauchi and Y. Kono. 1989. Fractal analysis of plant root systems. *Anal. Bot.* 64:499-503.

Van Pelt, J. and R. W. H. Verwer. 1983. The exact probabilities of branching patterns under terminal and segmental growth hypotheses. *Bull. Math. Biol.* 45:269-185.

Van Pelt, J. and R. W. H. Verwer. 1984. Cut trees in the topological analysis of branching patterns. *Bull. Math. Biol.* 46:283-294.

Vogt, K. A. and J. Bloomfield. 1991. Tree root turnover and senescence. In A.E.Y. Waisel and U. Kafkafi (eds.) *Plant Roots, The Hidden Half.* Marcel Dekker, New York.

Waisel, Y. and A. Eshel. 1991. Multiform behavior of various constituents of one root system. p. 39-52. In A.E.Y. Waisel and U. Kafkafi (eds.) *Plant Roots, The Hidden Half.* Marcel Dekker, New York.

Weibell, E. R. 1963. *Morphometry of the Human Lung*. Academic Press, New York.

Werner, C. and J. S. Smart. 1973. Some new methods of topologic classification of channel networks. *Geographical Analysis* 5:271-295.

Whitney, G. G. 1976. The bifurcation ratio as an indicator of adaptive strategy in woody plant species. *Bulletin of the Torrey Botanical Club* 103:67-72.

Quantification of Soil Microtopography and Surface Roughness

C.-h. Huang

Contents

I Introduction

Surface relief, or topography, represents the geometric composition of the surface. Surface topography encompasses a wide range of scales as the areal extent changes from centimeters to thousands of kilometers. For landscapes with areas in the order of 10 to 1000 km, the surface relief is characterized by hills and valleys formed by geologic processes in a time span of hundreds to millions of years. Features of the landscape are quantified by topographic contour maps with a typical contour spacing of 1.5 m (5 ft). The landscape topography remains relatively unchanged except in areas of active volcanic and tectonic activities. One major impact of the landscape scale topography is on weather, especially on rainfall and wind patterns, due to its influences in energy balance and air circulation.

Soil microtopography is referred to as the millimeter to centimeter scale topographic variations within a meter scale area. Unlike the macro-scale surface topography, which remains relatively unchanged, soil microtopography sometimes changes rapidly. In agricultural fields, the soil surface microtopography is altered by tillage operations, livestock trampling, consolidation, and erosion and deposition from rain and wind. Cycles of wetting/drying and freezing/thawing can also change soil microtopography.

Immediately after a tillage operation, soil microtopography consists of tillage marks, or oriented roughness, and randomly oriented, differently sized roughness elements from clods to individual grains (Allmaras *et al.*, 1966; Zobeck and

ISBN 1-56670-105-8

Onstad, 1987). The oriented roughness is characteristic to a specific tillage tool, therefore, it is relatively easy to quantify the periodic pattern by a simple geometric model. The challenge is to quantify the spatial distribution of these randomly oriented, differently sized elements on the surface.

Why is it necessary to quantify soil microtopography? The answer is not apparent until one understands the interaction between the length scale of a physical process on the surface and the scale of roughness elements. It is well known that the same surface that appears smooth from a distance becomes rough at close inspection. Therefore, the degree of smoothness or roughness is only relative and somewhat meaningless unless it is associated with a specific length scale. The need to quantify soil microtopography is further illustrated by the following example.

The processes of soil detachment by raindrop impact and transport by surface runoff have characteristic scales in the millimeter range because sizes of raindrops and detached soil materials and the depth of the runoff water are all in millimeter scales. Therefore, soil microtopography in the millimeter scale range becomes important because it is where the detachment and transport processes are occurring. Other surface processes affected by soil microtopography are: infiltration, depressional storage, wind erosion, gas exchange, evaporation and heat flux. Surface microtopography not only influences these surface processes; many of these surface processes, *e.g.* erosion and deposition, also cause changes in soil microtopography. Therefore, changes of microtopography are also indicative of the extent of these processes. The understanding of physical processes that cause changes in microtopography can be greatly enhanced through the quantification of surface microtopography.

The scope of this chapter is to provide an overview of some recent developments in techniques of measuring soil microtopography and the use of scale-dependent roughness models, especially the fractal concept, to quantify the microtopographic data set.

II Microtopography Measurement Techniques

Quantification of soil surface roughness consists of two steps: (1) collection of surface elevation data, and (2) analysis of elevation data sets. Although the effects of surface microtopography on surface processes have long been recognized, there are relatively small numbers of published papers focusing on this subject. Romkens and Wang (1987) have attributed the lack of publications on soil surface roughness to "the complex and seemingly random nature of surface roughness and the difficulty of its mathematical description". We believe that the lack of progress in soil microtopography quantification was attributed to laborious field techniques which produced digitized soil topography at low resolutions. Consequently, the development of analytic procedures was limited by the quality of the data.

The most widely used technique for measuring soil surface topography is the pin technique, in which a single probe or a set of multiple pins are lowered onto the surface and the pin position is registered either electronically or photographically and later digitized (Burwell *et al.*, 1963; Podmore and Huggins, 1981; Radke *et al.*, 1981; Romkens *et al.*, 1986). Surface microtopography can also be digitized photogrammetrically by analyzing a stereo pair of the surface image (Welch *et al.*, 1984). These two techniques produced digitized elevations in grid points ranging from 5 to 50 mm apart. Therefore, their main utility was to quantify large scale ($> 10cm$) surface features such as tillage marks or oriented roughness and geometry of concentrated flow channels. To quantify microtopography resulting from centimeter to millimeter scale aggregates, the surface needs to be digitized at grid spacings on the order of 1 mm or less.

The development of noncontact optical transducers in the late 80's broke the barrier in acquiring a detailed microtopographic data set. In general, these optical transducers use a light source, most commonly a low power HeNe (Helium-Neon) laser beam, to project an illuminated spot onto the surface and an optoelectronic image sensing system to detect the position of the reflected beam. Variations among different transducers are due to differences in light source, optical geometry, detector, and signal processing circuit or algorithm (Rice *et al.*, 1988; Romkens *et al.*, 1988; Khorashahi *et al.*, 1987; Huang *et al.*, 1988; Destain *et al.*, 1989; Bertuzzi *et al.*, 1990a).

Most opto-electronic systems mentioned above are used in a 1-dimensional (1-D) surface profiler. A 2-D digitizing system capable of producing multiple surface profiles at close spacings is needed for the quantification of the microtopography. Based on the optical transducer of Huang *et al.* (1988), Huang and Bradford (1990) developed a portable 2-D scanner which is capable of digitizing soil microtopography in 0.5 mm grids with an elevational resolution on the order of 0.1 to 0.3 mm. Briefly, the laser scanner uses a 2 mW HeNe laser as the light source and a 512-element photodiode array mounted inside a regular 35-mm camera as the optical detector for the reflected laser beam. The camera is mounted at a small angle (5 to 25 degree) off the laser beam and the line of the diode array is aligned with the laser beam. The position of the laser spot on the diode array, focused through the camera lens, is related to surface elevation by triangulation. The range of elevation measurement is adjustable and is dependent on laser-camera angle and distance, focal length of the lens, and total length of the array sensing region. The adjustment of elevation measurement range is easily done by either changing the laser-camera angle or switching to a different focal length lens. With the 512-element diode array, the elevational resolution is approximately 0.1% of the adjusted range. The laser-camera assembly is mounted on a carriage which can be positioned precisely by stepping motors. The laser scanner now uses a linear slide and cable chain drive mechanism, replacing the lead screw drive reported earlier (Huang and Bradford, 1990) to move the carriage. With the new traversing mechanism, the scanner is now capable of digitizing a meter-long transect in 6 seconds with elevation data taken at every 0.5 mm grid points.

In addition to the 1 m by 1 m laser scanner, we have also fabricated a large 3 m by 1 m scanner. Both the 1-m and 3-m scanners are used routinely in rainfall simulation studies. Recently, the 3-m scanner was also used to quantify microtopographies of fractured limestone specimens and grazed rangelands.

III Roughness Indices and Function

Soil roughness is a measure of variations in surface microtopography. Probably the most commonly employed method for reporting soil surface roughness is the random roughness, or root mean square (rms) roughness index. It is equivalent to or related to the standard deviation of elevations from a mean surface (Kuipers, 1957; Allmaras et al, 1966; Currence and Lovely, 1970). There are several proposed filtering and transformation procedures to adjust for the effects of plot slope and tillage marks before the calculation of standard deviation is carried out (Currence and Lovely, 1970). Interestingly, there was also a proposal to remove both the top and bottom 10% of the measurements to smooth out effects of erratic readings (Allmaras *et al.*, 1966), a step that seems to contradict to the intended nature of the measurement. Although the concept of standard deviation is easily comprehensible, the lack of a standard procedure to pre-process the digitized data set for the effects of slope and oriented roughness makes it difficult to compare the rms roughness index from different studies (Burtuzzi *et al.*, 1990b).

Romkens and Wang (1986) introduced a dimensionless parameter, MIF, which is the product of microrelief index (MI) and peak frequency (F). The value of MI is equivalent to the mean absolute deviation of measured elevation from a reference plane, and F is the number of elevation peaks per unit length of a transect. This parameter was used to examine different tillage effects (Romkens and Wang, 1986) and to quantify roughness changes due to rain (Romkens and Wang, 1987). Lehrsch *et al.* (1987, 1988) further analyzed how soil properties contribute to variation of MIF parameter and identified MIF as the most sensitive roughness parameter among a set of eight different measures for describing surface roughness and its change after simulated rain.

Both rms roughness and MIF indices provide an overall surface roughness characteristic at the areal extent of the topographic data set without any information on the spatial structure of the roughness elements on the surface. The main drawback is that many different surface topographic patterns can yield an identical rms roughness or MIF index. Both indices are also biased by the presence of large roughness elements. Despite the shortcomings of these indices, they are simple and useful measures of roughness when the characteristic length scale of physical process is much greater than the largest roughness elements on the surface and the roughness effects are mainly contributed from the largest elements. A typical usage of the roughness indices is in the fluid dynamics energy balance calculation to account for frictional losses when the depth of the flow is much greater than the roughness height and most of the energy is dissipated by the largest roughness elements.

Since roughness is a statistical measure of topographic variations, it is apparent that, for a small areal extent, the variation in surface elevation tends to be small. As the areal extent is increased, the probability to encounter larger surface features which produce higher elevational differences is also increased. Therefore, soil roughness increases as the areal extent is increased. The measure of topographic variation should be associated with the areal extent, or in other words, roughness should be expressed generically as a scale-dependent function, not merely an index. Once the roughness is expressed as a function of the spatial scale, it can be fitted to generalized models. Laser-scanned surface microtopography contains sufficient spread in spatial scales to enable the calculation of a roughness function.

IV Roughness Measures and Models

Different measures have been used to represent the scale-dependent roughness function. These are:

- elevation difference, $\Delta Z(h)$:

$$\Delta Z(h) = E\left(\left|Z_{(\mathbf{x})} - Z_{(\mathbf{x}+h)}\right|\right), \tag{1}$$

- semivariance, $\gamma(h)$:

$$\gamma(h) = 0.5E\left(Z_{(\mathbf{x})} - Z_{(\mathbf{x}+h)}\right)^2, \tag{2}$$

- autocovariance, $R(h)$ and autocorrelation, $\rho(h)$:

$$R(h) = E\left(Z_{(\mathbf{x})} - Z_{(\mathbf{x}+h)}\right), \tag{3}$$

$$\rho(h) = R(h)/R(0), \text{ and} \tag{4}$$

- power spectral density, $S(f)$:

$$S(f) = \int R(r) \cos 2\pi \; f \; r \; dr \tag{5}$$

where $Z_{(\mathbf{x})}$ and $Z_{(\mathbf{x}+h)}$ are elevations at positions a horizontal distance h apart, $E()$ denotes the expectation operator and f is the spatial frequency. Here we have assumed the surface microtopography as being *statistically homogeneous*, which implies that statistical properties do not depend on the position, $\mathbf{x}$, and only depend on the spatial separation, h.

Using the elevation difference measure, Linden and van Doren (1986) showed a linear relationship between $1/\Delta Z$ and $1/h$.They defined two indices for their model: the limiting difference (LD) and limiting slope (LS). The LD parameter is defined as being the value of ΔZ when h approaches infinity, thus it is a measure of elevation difference at large spatial intervals. The LS index is the value of $\Delta Z/h$, or slope, when h approaches zero. Their model is one of the first ones to express soil roughness as a function of separation length scale. The definition of LS index contradicts the non-differentiability criteria of the fractal concept (cf Baveye and Boast, 1997).

Traditional usage of the correlation and spectral analyses in processing 1-D time series data is to identify the major scale of periodic variation. Extensive theoretical developments on the expansion of 1-D random process models used in time series analyses to 2-D surface data analyses were presented by Longuet-Higgins (1957) and Nayak (1971). The use of correlation and spectral analyses to quantify soil microtopography started in the early 70's (Currence and Lovely, 1970; and Merva *et al.*, 1970). At that time, the objective of using these analytic procedures was to identify the scale of periodic topographic variation or the oriented roughness. Dexter (1977) used both the autocorrelation and spectral density functions to examine roughness changes due to rain for different tillage operations and concluded that these functions, when compared to a variance term alone, did not yield additional information about rain effects. Podmore and Huggins (1980) used spectral analysis techniques to examine surfaces of a laboratory flume and concluded that their test surfaces were similar to random roughness surfaces with no characteristic frequency.

The term "random roughness" has been widely used to describe soil microtopography. The usage can be misleading because surface roughness may appear random at one scale and in an orderly fashion in another scale. The word "random" is used to imply the non-orderly fashion of the spatial distribution of clods, aggregates and grains on the surface. It is not to be interpreted as the surface topographic data being randomly distributed and lacking spatial correlation.

The use of the semivariance function, or the variogram, as the roughness measure gives a direct relationship between an elevation difference term and the separation length scale. The semivariance function does not require the existence of a global variance for a proper usage of the covariance function.

Here we present three types of scale-dependent models which can be used to quantify the roughness function. These models are (1) *random walk*, (2) *fractal* and (3) *Markov-Gaussian*. Examples of semivariance functions for these three models are presented in Figure 1.

The *random walk* or *Brownian motion* (Bm) model is characterized by the increment function

$$\Delta Z(h) \propto h^{0.5} \tag{6}$$

where $\Delta Z(h)$ is the absolute elevation difference, between points a distance h apart. The variogram, a plot of $\gamma(h)$ *versus* h, is thus a linear function of spatial separation, *e.g.*,

$$\gamma(h) \propto h \tag{7}$$

The variogram, plotted in log-log scales, is a straight line with slope s = 1. The spectral density function, also plotted in a log-log fashion, is a straight line of s = -2. Podmore and Huggins (1980) found that slopes of spectral functions from their test surfaces had values around -2 and concluded that their test surface had random elevation. Note the differences between a *random walk* and a *white noise* type "random" process. A *completely random* or *white noise* type

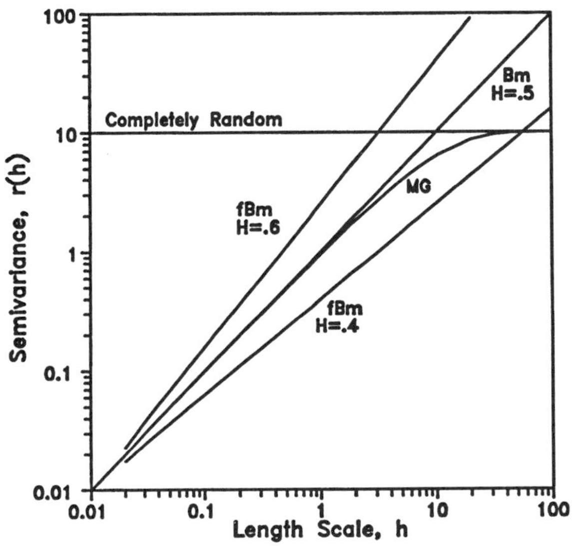

Figure 1: Semivariance functions, or variograms, for completely random, Brownian motion (Bm), fractal or fractional Brownian motion (fBm) and Markov-Gaussian (MG) models.

elevational variation has no spatial correlation. The variogram is depicted by a horizontal line and the constant semivariance is equal to the variance of the data set. Theoretically, these two processes are related and the Brownian motion is the integral of the white noise.

The *fractal* model for soil roughness quantification is an expansion of the Brownian random walk model by Mandelbrot and van Ness (1968). It is called *fractional Brownian motion* (fBm) when the exponent of the increment function of a Bm model is allowed to vary from its generally accepted value of 0.5 to a range from 0 to 1. The elevation difference of a fBm model is given by

$$\Delta Z(h) \propto h^H, \qquad 1 > H > 0 \tag{8}$$

with the following semivariance function:

$$\gamma(h) \propto h^{2H} \tag{9}$$

Plotted in log-log scales, the variogram of a fBm model is a straight line with $2 > s > 0$. The slope of the power spectral density function for the fBm model, plotted in log-log scales, equals $-2H - 1$. For surface topography, the H parameter is related to the fractal dimension, D, by $D = 3 - H$. The random walk or Brownian motion model can be considered as a special case of the fBm model.

In addition to the commonly known fractal dimension, D, which can be obtained from the slope of the straight line portion of a variogram or power spectrum plotted in log-log scales, the fractal roughness model requires a second parameter to define the relative position of the straight line. Huang and Bradford (1992) defined the crossover length scale, l, for the semivariance function of the fractal model:

$$\gamma(h) = l^{2-2H}\ h^{2H} \tag{10}$$

The term crossover is used because $\gamma(h) = h^2$ when $h = l$. The definition of the l parameter was modified from the crossover length scale used by Wong (1987) for pore geometry and by Brown (1987) for rock surface topography. They defined the crossover length based on a power-law-type variance structure, *e.g.*,

$$\sigma^2(\lambda) = b^{2-m}\lambda^m \tag{11}$$

where $\sigma^2(\lambda)$ is the variance from local regions with size scale λ, and b is the crossover length. The b parameter was also called topothesy, κ, by Sayles and Thomas (1978) for the case when $m = 1$ (*i.e.*, the Bm process) and in a more general fashion by Berry and Hannay (1978).

The *Markov-Gaussian* (MG) model has an exponential-type correlation structure. The semivariance of the MG model is given by

$$\gamma(h) = \sigma^2(1 - e^{-h/L}) \tag{12}$$

where σ^2 is the variance and L the correlation length scale. The MG process is characterized by two length scales, σ and L. The log-log variogram of the MG model (Figure 1) showed a varying slope between 1 and 0. At small lag spacings, $h << L$, the variogram can be approximated by a straight line with $s = 1$, similar to the Bm model. At large lag spacings, the variogram tapers off toward a horizontal line, or in other words, the semivariance is no longer dependent on the length scale.

V Applications

In this section, we show how scale-dependent models, especially the fractal concept, are applied to quantify roughness measures calculated from digitized microtopography.

Using microtopographic data collected by the laser scanner, Huang and Bradford (1992) showed the semivariance roughness functions for fallow plots exposed to natural rainfall, and field and laboratory erosion plots under simulated rainfall. They found that all variograms, when plotted in log-log scales, contain either regions of straight lines or a curvilinear portion wedged between straight lines. They used a combination of fractal (fBm) and MG models at different scales to quantify the semivariance roughness function. The fractal regime was used to describe the sloping straight line portions of the variogram whereas the MG model was used to describe the region when the slope decreases gradually toward 0 when an apparent local homogeneous scale was reached. The following discussion will be focused on the fractal characteristics of the semivariance

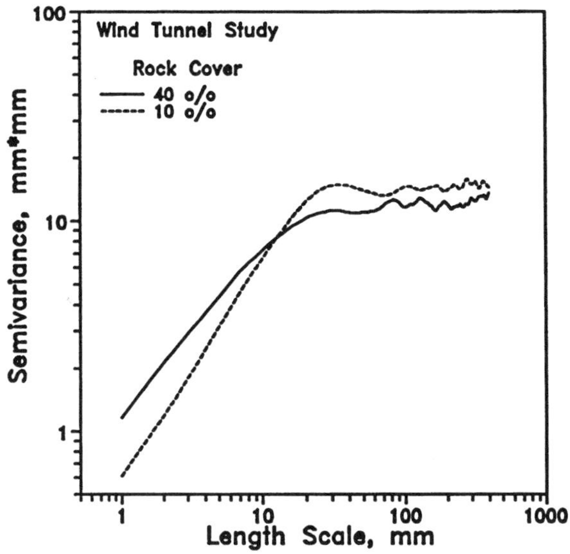

Figure 2: Semivariance roughness functions from soil trays having 10% and 40% surface cover of rock fragments.

roughness function. A detailed discussion of combining fractal and MG models to quantify soil roughness is given in a previous paper (Huang and Bradford, 1992).

We now use two additional examples to illustrate the utility of the fractal model to quantify surface microtopography. Two soil trays were prepared for a wind tunnel experiment related to the effects of rock fragment on soil detachment by wind shear. The trays were 0.5 m wide and 1 m long. Rock fragments, roughly 4 cm in size, were buried halfway on the soil surface to create approximately 10% and 40% cover, respectively. These two surfaces were laser scanned in a 1-mm grid along the long dimension and in a 5-mm grid between scan lines. Semivariance roughness functions from these two trays are plotted in Figure 2. Both curves showed a straight line portion to approximately 10-mm lag spacing. Fractal parameters are $D = 2.47$, $l = 0.56$ mm for 10% rock cover and $D = 2.62$, $l = 1.26$ mm for 40% cover. The standard deviation or the rms roughness index is similar for both surfaces, *i.e.*, 3.90 mm for 10% cover *versus* 3.76 mm for 40%, indicating its inability to quantify microtopographic features that are distributed differently on the surface.

In surface roughness quantification, the fractal dimension can be considered as a relative measure of the distribution of different-sized elements on the surface. A steeper slope on the variogram indicates a higher contrast on the surface size distribution. When the rock fragment is increased, the relative proportions of different elevations become more gradual, therefore, the slope of the vari-

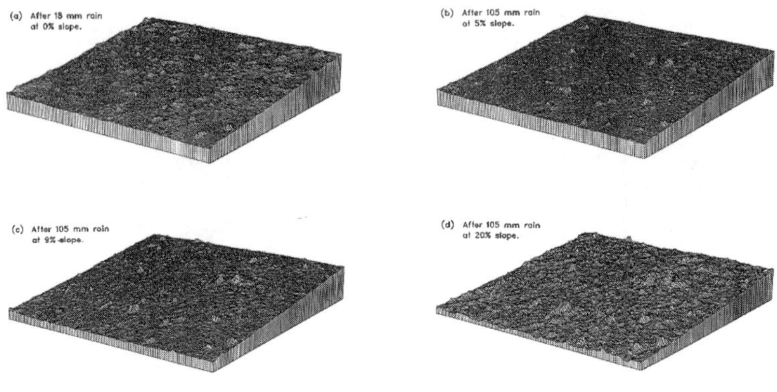

Figure 3: Three-dimensional plots of a 0.4 m by 0.4 m area from the soil pan (a) after 18 mm rain at 0% slope, (b) after 105 mm rain at 5% slope, (c) after 105 mm rain at 9% slope, and (d) after 105 mm rain at 20% slope.

ogram becomes lower. The crossover length puts the actual size scale into the proportional size distribution on the surface. Without the crossover length, it is like reading a map without a reference scale. Much emphasis has been placed in deriving the fractal dimension from digitized data and little is given to the crossover length. The importance of the crossover length scale in quantifying changes of soil microtopography is demonstrated in the next example.

The second example shows changes in soil roughness and associated fractal parameters from a laboratory soil pan exposed to simulated rainfall events. This experiment is designed specifically to show the effects of slope gradient on the dominant surface processes which in turn change the surface microtopography.

The soil pan, 1.2 m by 1.2 m in area and 0.12 m deep, was subjected to 10 simulated rain events. Immediately after placing soil materials into the pan, a simulated rainstorm of 15-minute duration at 70 mm/h intensity was applied with the soil pan at its level position. The initial rain was used to consolidate the loose aggregates to avoid irregularities from preparing the soil pan and to bring the surface soil to a uniformly sealed and moist condition. After the initial rain, the soil pan was set to 5% slope and rained three times, on a daily interval, for 30 minutes at 70 mm/h. After three events at 5% slope, the slope was increased to 9% for another three 30-minute rains. The procedure was repeated for 20% slope. After each rain event, the surface microtopography was digitized by the laser scanner at 0.5-mm grids along the scan line with 2-mm spacing between scan lines. This experiment was replicated twice.

Figure 3 shows the 3-dimensional (3-D) plots from a 0.4 m by 0.4 m area of the test surface after the first and three other rainfall events at 5%, 9% and 20% slope. These plots were drawn with an identical scale factor for all 3 axes to provide a realistic view of the soil surface and of its topographic changes with time. Corresponding semivariance functions are given in Figure

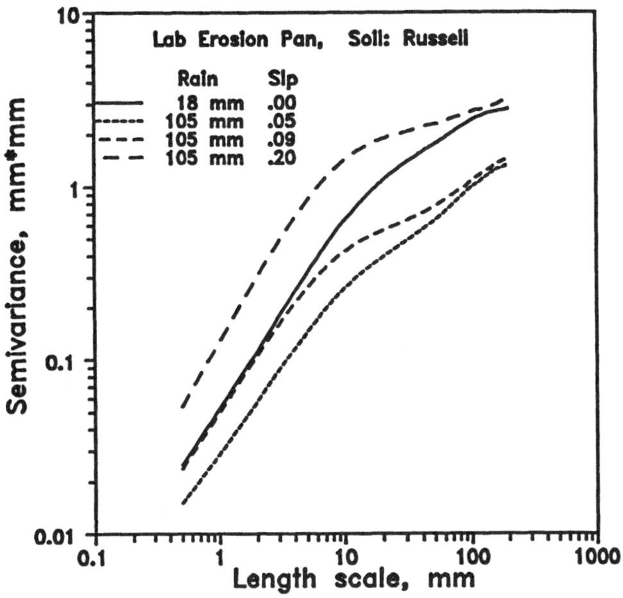

Figure 4: Changes in roughness functions from a laboratory soil pan exposed to a sequence of 10 simulated rainfall events. Data after the first (on 0% slope), fourth (5%), seventh (9%) and the tenth (on 20% slope) rainfall events are plotted.

4. All variograms displayed the general trend of a straight line portion up to approximately 10 mm lag spacing, and then a gradual taper at higher lags. The straight line portion of the variogram is quantified by the fractal model.

The soil roughness is decreased after raining on the 5% slope surface (Figures 3a and 3b). At this slope steepness, the transport capacity of the runoff is limited and the surface is dominated by seal formation from raindrop impact and deposition. After 105 mm of rainfall, the soil roughness reached its lowest levels as shown by the 3-D plot (Figures 3b and 4).

The roughness is increased after setting the pan to 9% slope when erosion started to overtake the deposition process. Some of the previously depositied materials are removed due to flow concentration, causing an increased roughness at small scales (Figure 3c). At this slope gradient, erosion is not severe enough to cause a significant change in large scale roughness. At 20% slope, erosion becomes a dominant process. Severely rilling caused an increased roughness at all scales (Figure 3d). This example shows that soil roughness decreases with an increasing amount of rainfall only when erosion is not a dominant process. Erosion causes the roughness to increase.

Both fractal parameters, the fractal dimension and crossover length, are calculated from the straight line portion of the variogram and plotted in Figures. 6 and 7. The 3-D plots given in Figure 3 and the semivariance functions in

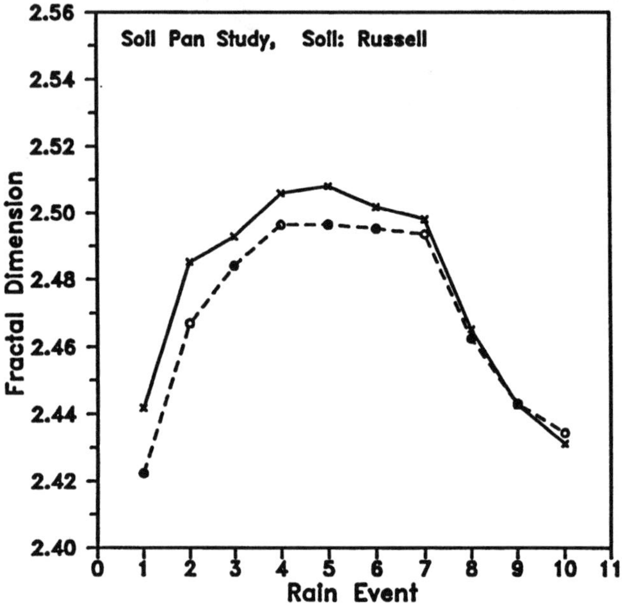

Figure 5: Fractal dimensions from digitized microtopography data after each rainfall event during a sequence of 10 rain events. Data from two replicate soil pans are plotted.

Figure 4 were corresponded to soil topographies after rainfall events 1, 4, 7, and 10. The fractal dimension displayed an increasing trend after raining at 5% slope, a gradual decrease at 9% slope and a sharp decrease at 20% slope (Figure 5). As explained earlier, a higher fractal dimension is associated with a lower slope in the log-log variogram, thus the distribution of surface heights is more gradual. The variability in fractal dimension among the set of 10 microtopographic data is relatively small, showing a maximum change of 0.08 units. On the other hand, values of crossover length showed a much greater change in relative scale, varying from 0.025 mm to 0.1 mm (Figure 6). Although values of the semivariance function were least after the fourth rain event, the crossover length for this particular case wasn't the lowest because it is also related to the fractal dimension. These two parameters should be used together to quantify soil microtopography.

A study in which soil microtopography under field, natural rainfall situations was monitored also showed that the crossover length is more sensitive to changes in microtopography due to rainfall than the fractal dimension (Eltz, 1993).

The fractal characteristics are found in a limited scale from the semivariance roughness function of the microtopography data. This means that the roughness or topographic variation does not scale up proportionally as the areal scale is increased. This is common in agricultural fields because the maximum size of

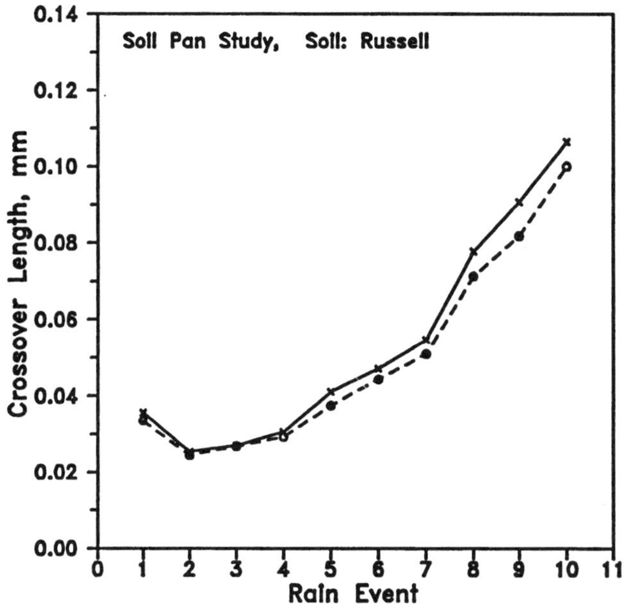

Figure 6: Crossover lengths from digitized microtopography data after each rainfall event during a sequence of 10 rain events. Data from both replicates are plotted.

the clod is controlled by the tillage tool. A flat surface or smooth landscape implies that it contains a scale range where the semivariance roughness function does not increase within the range of scales. The fractal process at a limited scale range is called *pseudofractal* by Orford and Whalley (1983; Whalley and Orford, 1989). We envision the roughness function extended to landscape scales to appear like a terraced hillslope. The sloping banks are pseudofractal regions and the plateaus are regions where localized homogeneity is reached. This is in accordance with the multiscale concept discussed by Burrough (1983a, 1983b).

VI Summary

From analyzing laser-scanned surface roughness data sets, we showed that surface roughness is a function of spatial scale and any proposed model for its quantification should reflect the scale-dependent characteristics. The semivariance, or the variogram, was used to plot the roughness function, and a combination of *fractal* (fBm) and *Markov-Gaussian* (MG) processes was proposed to characterize roughness at different scales (Huang and Bradford, 1992).

Both processes each required two separate parameters: fractal dimension, D, and crossover length, l, for the fractal process, and local variance scale, σ^2, and correlation length, L, for the MG process. The fractal process used to quantify the experimental variogram is applied to a limited range of scales, a process also referred to as pseudofractal. In this chapter, we emphasize the importance of the l parameter, in addition to the commonly known D, for a fractal-type roughness model in order to differentiate different degrees of soil roughness. Examples are given to demonstrate differences in both D and l values related to different surface microtopographies.

Many surface boundary processes are controlled by the microtopography. A detailed surface microtopography provides information on the spatial distribution of surface processes. For soil erosion, these processes are: raindrop impact effects, surface ponding pattern, flow meandering and shear, sediment deposition, and wind exposure and sheltering. Changes of roughness after erosive events further reflected the soil behavior against erosive forces. A statistical model of soil roughness facilitated the scaling of these surface processes.

References

Baveye, P. and C.W. Boast. 1997. Fractal geometry, fragmentation processes and the physics of scale-invariance: an introduction, (Chapter 1, this volume).

Berry, M.V. and J.H. Hannay. 1978. Topography of random surfaces. *Nature* 273:573.

Brown, S.R. 1987. A note on the description of surface roughness using fractal dimension. *Geophys. Res. Lett.* 14(11):1095-1098.

Burrough, P.A. 1983a. Multiscale source of spatial variation in soil. I. The application of fractal concepts to nested levels of soil variation. *J. Soil Sci.* 34:577-597.

Burrough, P.A. 1983b. Multiscale source of spatial variation in soil. II. A non-Brownian fractal model and its application in soil survey. *J. Soil Sci.* 34:599-620.

Bertuzzi, P., J.M. Caussignac, P. Stengel, G. Morel, J.Y. Lorendeau and G. Pelloux. 1990a. An automated, noncontact laser profile meter for measuring soil roughness in situ. *Soil Sci.* 149(3):169-178.

Bertuzzi, P., G. Rauws and D. Courault. 1990b. Testing roughness indices to estimate soil surface roughness changes due to simulated rainfall. *Soil Till. Res.* 17:87-99.

Burwell, R.E., R.R. Allmaras and M. Amemiya. 1963. A field measurement of total porosity and surface microrelief of soils. *Soil Sci. Soc. Am. Proc.* 27:697-700.

Currence, H.D. and W. G. Lovely. 1970. The analysis of soil surface roughness. *Trans. ASAE* 13:710-714.

Destain, M.F., G. Descornet, C. Roisin and M. Frankinet. 1989. Investigation of degradation by means of an opto-electronic microreliefmeter. *Soil Till. Res.* 13:299-315.

Dexter, A. R. 1977. Effect of rainfall on the surface micro-relief of tilled soil.

J. Terramechanics 14(1):11-22.

Eltz, F.L.F. 1993. *Surface Roughness Changes as Affected by Tillage and Rainfall Erosivity.* Unpublished Ph.D. dissertation, Purdue Univ., West Lafayette, Indiana.

Huang, C., I. Whita, E.G. Thwaite and A. Bendeli. 1988 A non-contact laser system for measuring soil surface topography. *Soil Sci. Soc. Am. J.* 52:350-355.

Huang, C. and J.M. Bradford. 1990. Portable laser scanner for measuring soil surface roughness. *Soil Sci. Soc. Am. J.* 54:1402-1406.

Huang, C. and J.M. Bradford. 1992. Application of a laser scanner to quantify soil microtopography. *Soil. Sci. Soc. Am. J.* 56(1):14-21.

Khorashahi, J., R.K. Byler and T.A. Dillaha. 1987. An opto-electronic soil profile meter. *Computers and Electronics in Agriculture* 2:145-155.

Kuipers, H. 1957. A reliefmeter for soil cultivation studies. *Neth. J. Agri. Sci.* 5:255-267.

Lehrsch, G.A., F.D. Whisler and M.J.M. Romkens. 1987. Soil surface roughness as influenced by selected soil physical properties. *Soil Till. Res.* 10:197-212.

Lehrsch, G.A., F.D. Whisler and M.J.M. Romkens. 1988. Selection of a soil parameter describing soil surface roughness. *Soil Sci. Soc. Am. J.* 52:1439-1445.

Linden, D.R. and D.M. Van Doren, Jr. 1986. Parameters for characterizing tillage-induced soil surface roughness. *Soil Sci. Soc. Am. J.* 50:1560-1565.

Longuet-Higgins, M. S. 1957. Statistical properties of an isotropic random surface. *Phil. Trans. Royal Soc.* A250:157-174.

Mandelbrot, B.B. and J.W. Van Ness. 1968. Fractional Brownian motions, fractional noises and applications. *SIAM Review* 10(4):422-437.

Merva, G.E., R.D. Brazee, G.O. Schwab and R.B. Curry. 1970. Theoretical considerations of watershed surface description. *Trans. ASAE* 13:462-465.

Nayak, P.R. 1971. Random process models of rough surfaces. *Trans ASME J. Lubrication Tech.* 93:398-407.

Orford, J.D. and W.B. Whalley. 1983. The use of the fractal dimension to quantify the irregular-shaped particles. *Sedimentology* 30(5):655-668.

Podmore, T.H. and L.F. Huggins. 1980. Surface roughness effects on overland flow. *Trans. ASAE* 23:1434-1439, 1445.

Radke, J.K., M.A. Otterby, R.A. Young and C.A. Onstad. 1981. A microprocessor automated rillmeter. *Trans. ASAE.* 24:401-404.

Rice, C., B.N. Wilson and M. Appleman. 1988. Soil topography measurements using image processing techniques. *Computers and Electronics in Agriculture* 3:97-107.

Romkens, M.J.M., S. Singarayar and C.J. Gantzer. 1986. An automated non-contact surface profile meter. *Soil Tillage Res.* 6:193-202.

Romkens, M.J.M. and J.Y. Wang. 1986. Effect of tillage on surface roughness. *Trans. ASAE* 29:429-433.

Romkens, M.J.M. and J.Y. Wang. 1987. Soil roughness changes from rainfall. *Trans. ASAE* 30:101-107.

Romkens, M.J.M., J.Y. Wang and R.W. Darden. 1988. A laser microreliefmeter. *Trans. ASAE* 31:408-413.

Sayles, R.S. and T.R. Thomas. 1978. Surface topography as a nonstationary random process. *Nature* 271:431-434.

Welch, R., T.R. Jordan and A.W. Thomas. 1984. A photogrammetric technique for measuring soil erosion. *J. Soil Water Conserv.* 39:191-194.

Whalley, W.B. and J.D. Orford. 1989. The use of fractals and pseudofractals in the analysis of two-dimensional outlines: review and future exploration. *Comput. Geosci.* 15(2):185-197.

Wong, P. 1987. Fractal surfaces in porous media. p. 304-318. In J. Banavar, J. Kopilk and K. Winkler (eds.) *Physics and chemistry of porous media* II. AIP Conf. Proc. Vol. 154. Amer. Inst. Phys., New York, New York.

Fractal Models of Fragmented and Aggregated Soils

M. Rieu and E. Perrier

Contents

I Introduction

Soils are extremely heterogeneous porous media made of a solid phase embedding a pore space. These two-phase systems can be considered in two different ways. First, soils can be viewed as a coherent medium where the packing of solid elements forms large aggregates and generates lacunar parts or pores. The constitutive elements can be loose or hierarchically structured. Properties like

ISBN 1-56670-105-8

solid density or porosity of this lacunar medium characterize the whole spatial organization. From another standpoint, a soil is a fragmented medium where a network of more or less connected pores surrounds the solid elements. Its complete fragmentation produces solid elements whose number-size distribution can be characterized experimentally.

Because of the intrinsic heterogeneity of soils, many measurements related to the soil fabric depend either on the size of the soil cores or the resolution length of the measuring device. Fractal geometry appears to be a useful tool for describing and analyzing such heterogeneous structures. On one hand, it is well known that several soil characteristics scale as power laws of the length scale, and on the other hand, simple self-similar fractal models exhibit measures scaling as the observed ones. Hence the power laws may be reinterpreted, and their exponents explained, in terms of fractal geometry.

Section II of this chapter deals with deterministic fractal geometric objects that exhibit power-law behaviors similar to those encountered in real soils. An attempt is made to clarify the difference between two main types of fractal modeling concepts. This constitute the theoretical framework of our presentation of fragmented and aggregated soil fractal models. We present in section III different models of fragmentation and aggregation processes that result in scaling laws and could explain observed data. Without necessarily going into much detail in the analysis of physics of these processes, it is nevertheless important to account for both the fragmented and aggregated states to develop realistic structure models. Various attempts at developing such models are described and compared.

Section IV is a non-exhaustive review of available experimental data on fragmented or aggregated soils and of the methods currently in use to assess the degree to which given soils exhibit fractal behavior. In each case, it is pointed out which underlying fractal structure the observed power laws may reveal. Fractal geometric models could provide a representation of soil systems that could be useful in order to predict specific properties such as hydraulic characteristics.

The last section of this chapter (section V) returns to the issue of modeling and is devoted to a brief discussion of the predicting ability of fractal models as far as hydraulic properties are concerned.

II Theoretical Background for a Comparative Analysis of Fractal Soil Models

Physical characterization of soils is carried out by measuring the distribution either of the solid elements, or of the pores. This typically results in mass-size distributions of the solid elements or volume-size distributions of the pores, which can be more or less easily converted into number-size distributions of individual constituting elements (*e.g.*, Gardner, 1956; Kemper and Rosenau, 1986). Generally, the soil structure itself is not measured but a global approximation is given by solid density or porosity values (*e.g.*, Monnier et al, 1973;

Chepil, 1950; Wittmus and Mazurak,1958).

Many soils have been characterized by means of fractal geometry. When the cumulative number-size distribution of a collection of objects follows a power-law distribution in the form

$$[N_r > l] \propto l^{-D} \tag{1}$$

where $[N_r > l]$ is the number of individual objects of linear size greater than l, and the symbol $\propto$ means "proportional to", the collection may be called a fractal of dimension D (Mandelbrot, 1983). Several number-size distributions of aggregates, or primary solid particles, or pores, have been so described by fractal dimensions. The organization or reciprocal location of the voids and solid elements in an aggregated soil may be represented by geometric models that exhibit number-size power laws similar to those observed in real soils. These models provide idealized representations of the soil structure, enabling relative properties of the void and solid phases to be calculated. For example, a fractal analysis of the aggregated state in soil may lead to respective void or solid densities, scaling as power laws of the length scale as follows:

$$\rho(L) \propto L^{D-E} \tag{2}$$

where ρ is the density of either the void or the solid phase in a portion of the soil model of linear size L.

All the fractal characterizations of fragmented and aggregated soils implicitly refer to geometric models. Despite a common background, these theoretical models (part III), as well as the corresponding practical measurements (part IV), do not always refer to the same modeling concepts. The common background is a deterministic lacunar fractal geometric object whose widely used properties are recalled and derived only once in this chapter (following section II.A). The different modeling concepts are presented in section II.B and are the basic framework of all subsequent discussions.

III Basic Properties of Deterministic Lacunar Fractal Objects

III.A Deterministic Lacunar Fractal Objects

Let us recall that the construction of a deterministic fractal object of fractal dimension D, embedded in a Euclidean space of dimension E, is based: First, on an initiator of Euclidean dimension E and linear size L (Figure 1a), which can be divided into n equal smaller replicates of linear size rL, paving the whole object so that $n(rL)^E = L^E$. The similarity ratio r is defined by

$$nr^E = 1 \tag{3}$$

Second, on a generator that replaces the n small replicates of the initiator by N identical parts (Figure 1b), which are then replaced by the generator scaled

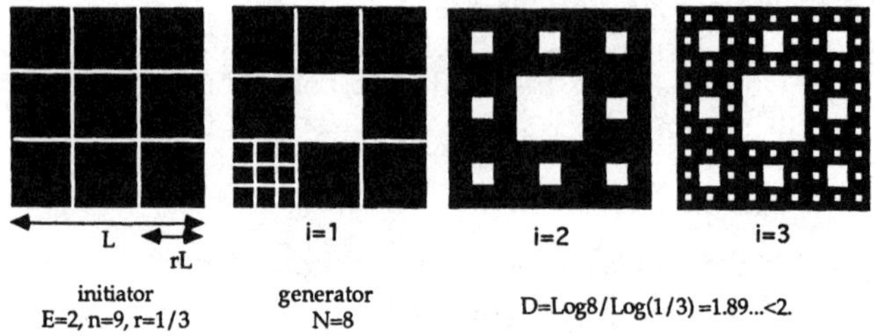

Figure 1: Construction of the Sierpinski carpet (first three iterations). The object is initiated on a square and N parts out of $n = r^{-E}$ are retained at each iteration.

by a factor r (Figure 1c), and so forth at subsequent levels i (Figure 1d) in a recursive manner. The fractal dimension D is given by:

$$D = \frac{\log N}{\log \frac{1}{r}} \tag{4a}$$

or equivalently

$$Nr^D = 1 \tag{4b}$$

A lacunar fractal object is obtained by creating at each stage a number N of replicates which is smaller than n (a process coined "curdling" by Mandelbrot, 1983): $N \leq n$ thus $D \leq E$. The Cantor set ($E = 1$), the Sierpinski Carpet ($E = 2$, first iterations shown on Figure 1) and the Menger Sponge ($E = 3$) are well-known examples.

Lacunar models are particularly well suited to model soils because they can represent both the solid and the void phase. Consequently, all the fractal models used to represent fragmented or aggregated porous media will be presented in the following via reference or comparison to lacunar fractal objects. For the sake of clarity, most of the illustrations will be given in two dimensions and some examples are given on Figure 2.

The properties of deterministic fractal lacunar objects are established thereafter. Let us distinguish, first, the properties of the self-similar fractal set F (illustrated by the black area on Figures 1 and 2). This fractal set is made of either connected or diconnected parts but is considered as a single geometric object. Secondly, the properties of the fractal collection of gaps G (white parts in Figures 1 and 2) should be considered. These gaps may be connected or disconnected but they are looked at as discrete individual elements.

III.A.a Properties of Self-Similar Fractal Sets

Top-Down Measurements When the fractal set F is bounded by an upper cut-off of scale denoted $l_{\max}$, it can be measured at increasing resolution: using E-tiles of decreasing linear size l_i and E-volume l_i^E, the number $N_r(l_i)$ of tiles needed to pave F can be calculated for each discrete value of l_i, and the

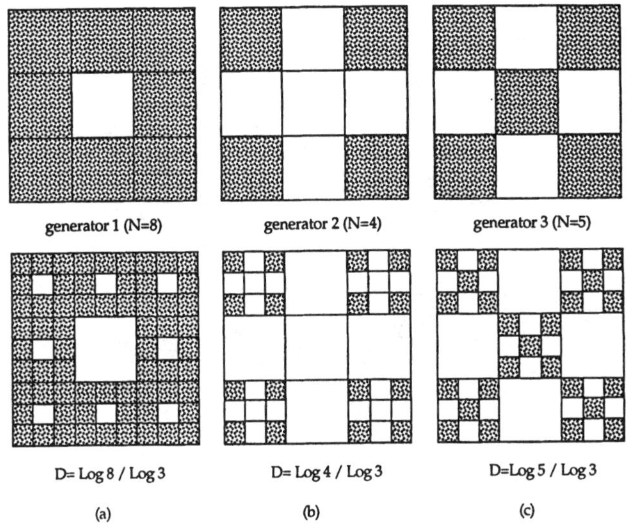

Figure 2: Deterministic lacunar fractal models built from a square initiator ($E = 2$, $n = 9$, $r = 1/3$) and different generators.

volume $V_E(l_i)$ of the set F can be deduced as well as the density $\rho_E(l_i)$ of F relatively to the whole volume $l_{\max}^E$.

The results are then extended to any continuous value of length-scale l and lead to general formulas expressing the measured quantities as power laws of the length scale. This general derivation can be understood in the particular example shown on Figure 3 , where $E = 2$, the tiles are squares of decreasing side length l, and $V_E(l)$ is simply the area of F measured at the resolution given by l.

Number-Size Property The successive linear sizes of the tiles are $l_1 = rl_{\max}$, $l_2 = rl_1 = r^2 l_{\max}$, ...$l_i = r^i l_{\max}$ (with $l_i < l_i - 1$), and the number of tiles used to cover F at each step is $N_r(l_1) = N$, $N_r(l_2) = N^2$, ...$N_r(l_i) = N^i$. Using the equalities $(l_i)^{-D} = (r^i l_{\max})^{-D} = (l_{\max})^{-D}(r^{-D})^i$ and introducing Equation (4b) in the form $N = r^{-D}$, it follows that $(l_i)^{-D} = (l_{\max})^{-D} N^i = (l_{\max})^{-D} N_r(l_i)$, thus:

$$N_r(l_i) = (l_i/l_{\max})^{-D} \tag{5a}$$

or simply

$$N_r(l) \propto (l)^{-D} \tag{5b}$$

Volume-Size Property The volume of F measured at each step is equal to the number of tiles multiplied by the volume of each tile, *i.e.*, $V_E(l_i) = N_r(l_i) l_i^E$. Introducing Equation (5a), one obtains:

$$V_E(l_i) = (l_i/l_{\max})^{-D} l_i^E = (l_{\max})^D (l_i)^{E-D} \tag{6a}$$

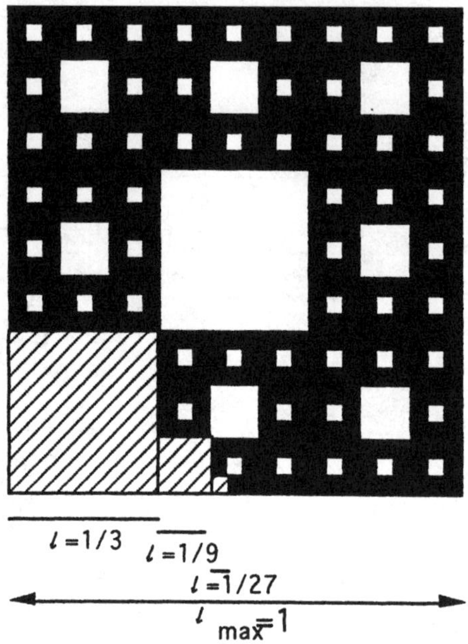

Figure 3: Measurement of the area of a Sierpinski carpet with normalized upper size $l_{\max} = 1$ through pavement with squared tiles with decreasing linear size l.

or simply

$$V_E(l) \propto (l)^{E-D} \tag{6b}$$

Density-Size Property The density of F equals the volume of F divided by the volume of the whole object of size $l_{\max}$, $\rho_E(l_i) = V_E(l_i)/(l_{\max})^E$. Introducing Equation (6a), one obtains:

$$\rho_E(l_i) = (l_{\max})^{D-E}(l_i)^{E-D} = (l_i/l_{\max})^{E-D} \tag{7a}$$

or simply

$$\rho_E(l) \propto (l)^{E-D} \tag{7b}$$

The density of the fractal set F measured at increasing resolution is proportional to the measure unit.

Bottom-Up Measurements If there exists a smallest typical size denoted $L_{\min}$ limiting the extension of the fractal set, one can carry out measurements of portions of the fractal set included in progressively larger and larger volumes.

Let us consider a fractal set with increasing linear size L and measure it by paving with tiles of constant linear size $L_{\min}$ and constant elementary volume $L_{\min}^E$ (*cf.* Figure 4) .

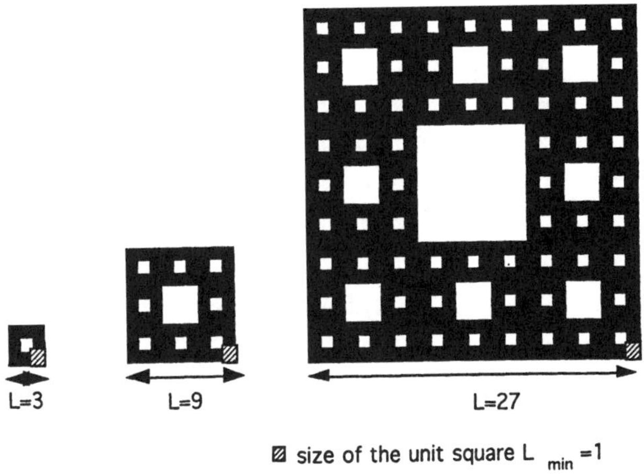

Figure 4: Measurement of the area of Sierpinski carpets with increasing side length L through pavement with constant unit squares of normalized size $L_{\min} = 1$.

Number-Size Property The successive linear sizes of the measured objects are $L_1 = r^{-1}L_{\min}$, $L_2 = r^{-1}L_1 = r^{-2}L_{\min}$, $...L_i = r^{-i}L_{\min}$ (with $L_i > L_i - 1$). The number of tiles of size $L_{\min}$ used to cover a fractal set of size L is $N_r(L_1) = N$, $N_r(L_2) = N^2$, $....N_r(L_i) = N^i$. Using the equalities $(L_i)^D = (r^{-i}L_{\min})^D = (L_{\min})^D(r^{-D})^i$ and introducing Equation (4b) in the form $N = r^{-D}$, one finds that $(L_i)^D = (L_{\min})^D N^i = (L_{\min})^D N_r(L_i)$, thus:

$$N_r(L_i) = (L_i/L_{\min})^D \tag{8a}$$

or simply

$$N_r(L) \propto (L)^D \tag{8b}$$

Volume-Size Property The volume of F included in an object of size L equals the number of tiles multiplied by the volume of each tile, $V_E(L_i) = N_r(L_i)L_{\min}^E$. Introducing Equation (8a), one obtains:

$$V_E(L_i) = (L_i/L_{\min})^D L_{\min}^E = (L_{\min})^{E-D}(L_i)^D \tag{9a}$$

or simply

$$V_E(L) \propto (L)^D \tag{9b}$$

Density-Size Property The density $\rho_E(L_i) \equiv V_E(L_i)/(L_i)^E$, the volume $V_E(L)$ of F divided by the total volume occupied by the whole object of size L. Introduction of Equation (9a) in this expression gives:

$$\rho_E(L_i) = (L_{\min})^{E-D}(L_i)^{D-E} = (L_i/L_{\min})^{D-E} \tag{10a}$$

or simply

$$\rho_E(L) \propto (L)^{D-E} \tag{10b}$$

The density of any part of the fractal set F is proportional to its linear size L.

If a fractal object presents both upper and lower cut-offs of scale, it is possible to use compact and simple expressions for all the preceding results, using the dimensionless variable l/L (*e.g.*, Vicseck,1989). In this context, analogues of Equations (5b) and (8b), on one hand, and of Equations (7b) and (10b), on the other, are given respectively by:

$$N^r(l/L) \propto (l/L)^{-D} \tag{11}$$

$$\rho_E(l/L) \propto (l/L)^{E-D} \tag{12}$$

These expressions can be used either keeping the object size L constant and decreasing the measure unit l (case 1) or keeping l constant and increasing L (case 2). They must be interpreted with caution. Case 1 considers the measure of a given object at increasing resolution whereas case 2 considers the measure of objects of varying size at the same resolution.

III.A.b Properties of Fractal Collections of Gaps

The collection G of lacunar parts, or gaps (the white parts on previous figures), forms the complementary part of the fractal set F. In the mathematical fractal obtained in the limit of an infinite number of iterations, the volume of F tends towards 0, the gaps fill all the initiator and their total volume is finite; the total volume of G scales as L^E and G is not a fractal set. Nevertheless, the gap size distribution or number/diameter relationship is a power-law distribution that can be considered as a fractal distribution of a set of discrete objects of varying size. Let us derive the most widely used properties of such distributions. We consider here only a fractal set bounded by an upper limit of scale $l_{\max}$, which refers in this section to the maximum gap size (see Figure 5).

Number-Size Property The successive linear sizes l_i and numbers $N_r(l_i)$ of gaps scale are as follows:

$$l_1 = l_{\max}, \quad l_2 = rl_1 = rl_{\max}, \quad \ldots \quad l_i = r^{i-1}l_{\max}(l_i < l_i - 1) \tag{13}$$

On the other hand, $N_r(l_1) = n - N$, $N_r(l_2) = N(n-N)$, ...$N_r(l_i) = (n - N)N^{i-1}$, thus $(l_i)^{-D} = (r^{i-1}l_{\max})^{-D} = (l_{\max})^{-D}(r^{-D})^{i-1}$. Introducing Equation (4b) in the form $r^{-D} = N$, one gets: $(l_i)^{-D} = (l_{\max})^{-D}N^{i-1} = (l_{\max})^{-D}\frac{N_r(l_i)}{n-N}$, and the number-size relationship can be written:

$$N_r(l_i) = (n-N)(l_{\max})^D(l_i)^{-D} \tag{14a}$$

or simply

$$N_r(l) \propto (l)^{-D} \tag{14b}$$

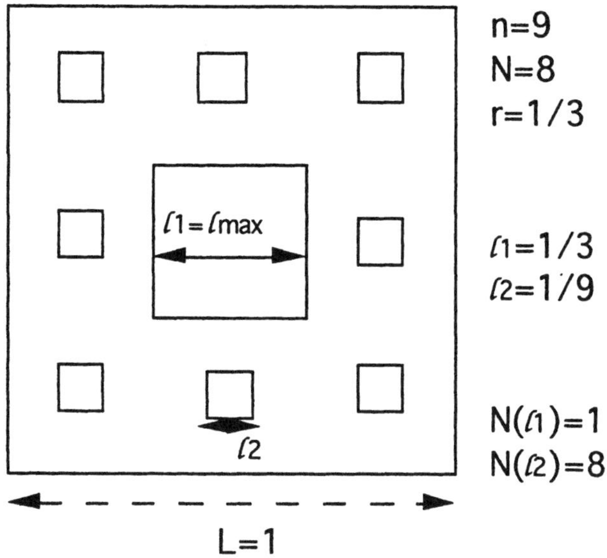

Figure 5: A fractal collection of gaps. The number $N(l_i)$ of gaps follows a power law of the linear size gap l_i.

Cumulative Number-Size Property The cumulative number $[N_r > l_i]$ keeps the same scaling law as $N_r(l_i)$ if the number i of iterations tends towards infinity or can be considered high enough to approximate infinity. It follows from Equation (14a) that $[N_r > l_i] = \sum_{j=1}^{i-1} N_r(l_j) \propto \sum_{j=1}^{i-1}(l_j)^{-D}$. Introducing Equation (13), one obtains: $[N_r > l_i] \propto \sum_{j=1}^{i-1}(r^{-D})^{j-1}$. This last expression is the summation of the successive terms in a geometric series[1], so, $[N_r > l_i] \propto \frac{1-(r^{-D})^{(i-1)}}{1-(r^{-D})} \propto (1-(r^{-D})^{(i-1)})$. From a straightforward computation, one can also write $[N_r > l_i] \propto (r^{-D})^{(i-1)}((r^D)^{(i-1)} - 1)$. Let us note that, if $i \to \infty$, then $(r^D)^{(i-1)} \to 0$ (due to the condition $r^D < 1$), hence $[N_r > l_i] \propto (r^{i-1})^{-D}$. From Equation (13), $(r^{i-1})^{-D} \propto l_i^{-D}$, and one finds that :

$$[N_r > l_i] \propto l_i^{-D} \quad \text{for } i \text{ high enough} \tag{15a}$$

or equivalently,

$$[N_r > l] \propto l^{-D} \quad \text{for } l \text{ small enough} \tag{15b}$$

Volume-Size Property The total volume of the gaps of size l_i and elementary volume $(l_i)^E$ is $V_E(l_i) = N_r(l_i)(l_i)^E = (n-N)(l_{\max})^D(l_i)^{E-D}$. The corresponding gap density is obtained when dividing the volume $V_E(l_i)$ of G by the volume of the whole object, and the latter volume can be calculated

[1] $q^a + q^{a+1} + q^{a+2} + ... + q^{b-1} + q^b = \frac{q^a - q^{b+1}}{1-q}$.

as the sum of n sub-volumes of linear size $l_{\max}$. Thus, $\rho_E(l_i) = \frac{V_E(l_i)}{n l_{\max}^E} = (1 - N/n)(l_{\max})^{D-E}(l_i)^{E-D} = (1 - N/n)(l_i/l_{\max})^{E-D}$.

Partial Density-Size Property The partial density of gaps of size lesser than a given size l_i can be calculated. Denoting i_{max} the maximum number of iterations taken into account, one obtains: $[\rho_E \leq l_i] = \sum_{j=i}^{i_{max}} \rho_E(l_j) = \sum_{j=i}^{i_{max}} (1 - N/n)(l_i/l_{\max})^{E-D}$. Introducing Equation (13), one obtains $[\rho_E \leq l_i] = \sum_{j=i}^{i_{max}} (1 - N/n)(l_{\max})^{D-E}(r^{j-1} l_{\max})^{E-D}$, that is, $[\rho_E \leq l_i] = (1 - N/n) \sum_{j=i}^{i_{max}} (r^{E-D})^{j-1}$. Performing a summation of the geometric series, one finds that: $[\rho_E \leq l_i]/(1 - N/n) = \frac{(r^{E-D})^{(i-1)} - (r^{E-D})^{(imax)}}{1-(r^{E-D})}$. Knowing from Equations (3) and (4b) that $1 - N/n = 1 - r^{-D}/r^{-E} = 1 - (r^{E-D})$, one obtains: $[\rho_E \leq l_i] = (r^{E-D})^{(i-1)} - (r^{E-D})^{(imax)}$. Let us note that, if $i_{max} \to \infty$, $(r^{E-D})^{(imax)} \to 0$ (because $r^{E-D} < 1$) so $[\rho_E \leq l_i] \to (r^{i-1})^{E-D}$. From Equation (13), $(r^{i-1})^{-D} = (l_i/l_{\max})^{-D}$, and one finds that:

$$[\rho_E \leq l_i] \to (l_i/l_{\max})^{E-D} \quad \text{, for } l_{\min} \to 0 \tag{16a}$$

or equivalently,

$$[\rho_E \leq l] \to (l/l_{\max})^{E-D} \quad \text{, for } l_{\min} \to 0 \tag{16b}$$

The partial density of gaps of linear size greater than l_i can be calculated straightforwardly. The same principles as above are used and, without details: $[\rho_E > l_i] = \sum_{j=1}^{i-1} \rho_E(l_i) = (1 - N/n) \sum_{j=1}^{i-1} (r^{E-D})^{j-1} = (1 - N/n) \frac{1-(r^{E-D})^{(i-1)}}{1-r^{E-D}} = 1 - (r^{E-D})^{(i-1)}$. Then, $[\rho_E > l_i] = 1 - (l_i/l_{\max})^{E-D}$ and one finds that:

$$[\rho_E > l_i] = 1 - (l_i/l_{\max})^{E-D} \tag{17a}$$

or equivalently,

$$[\rho_E > l] = 1 - (l/l_{\max})^{E-D} \tag{17b}$$

To summarize the above mathematical developments, a geometric model of a fractal lacunar medium of dimension D exhibits number-size power laws. These power laws express either the distribution of the sub-objects revealed by the measure of F at increasing resolution (Equation (5b)) or the distribution of gaps of decreasing size (Equation (14b)). In both cases, the same power laws can be derived for the cumulative distribution of sub-objects or gaps (see the derivation of Equation (15b) from 14b) and the power-law exponent is $(-D)$. The spatial arrangement of the constitutive elements inside a whole fractal object enables the derivation of new properties: the volume and the density of F and G are also expressed as power laws of the length scale and the power-law exponent involves the fractal dimension D.

III.B Fractal Models of Distributions of Solid Fragments or Pores and of Two-Phase Aggregated Soils

When modeling a soil fabric, two main approaches are implicitly used. In

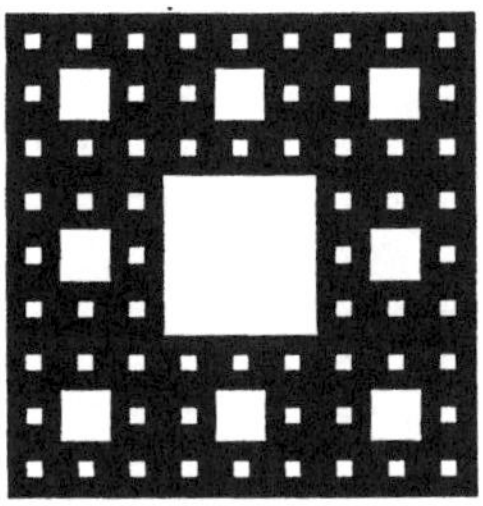

(a) Model type A

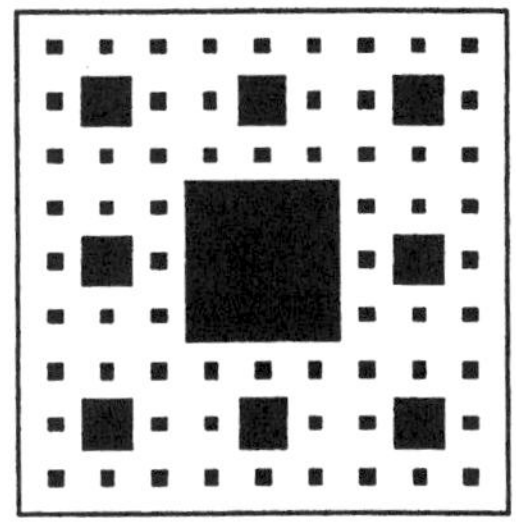

(b) Model type B

Figure 6: Two different models of a fractal porous medium. The lacunar parts (in white) represent the pores in column (a) while the same lacunar parts (in black) represent the solids in column (b). The second model behaves as a photographic "negative" of the first one.

the first approach (model type A), the solid phase appears coherent but can be divided into small elements that have almost the same size (as the tiles of size $L_{\min}$ in Figure 4), whereas pores of different sizes are represented by the lacunar parts (Figure 6a). In the second approach (model type B), loose, solid elements of different sizes, represented by the lacunar parts, are surrounded by a pore space that can be divided into small elements of identical size (Figure 6b). Both models are geometrically similar and result in related yet different scaling laws whose particularities have not always been clearly identified in the literature.

i When the fractal model is the unbounded mathematical object obtained with infinite iterations, the volume of F vanishes and the fractal object is only made of gaps. Hence, the first approach models only the pore space, while the second models only the solid phase. The mathematical model is only a geometric illustration of a distribution and not a model of a porous structure considered as a two-phase geometric object.

ii When a lower cut-off of scale is introduced, the same but truncated mathematical model can represent a geometrically reasonable space partition of a natural porous medium: the model considers actually two phases. Depending on the modeling concept, the fractal set F or the collection of gaps G represents either the void space or the solid phase. It must be noted that in both cases the interface of the voids and solids is the same and can model a fractal surface with adequate spatial configuration in a 3-dimensional space (*e.g.*, Friesen and Mikula, 1987).

The basic formulas that have been established on a general lacunar model can be used directly in both of the situations just outlined (*cf.* Table 1). When the volumes V_E and densities ρ_E refer to the solid space, they are proportional, respectively, to solid masses M or solid densities σ (assuming an idealized

Basic formulas	**Model type A** **(mass fractal)** $F \Leftrightarrow$ solid phase $G \Leftrightarrow$ pore phase	**Model type B** **(pore fractal)** $F \Leftrightarrow$ pore phase $G \Leftrightarrow$ solid phase
Two-phase models with cut-offs of scale $l_{\min}$ and $l_{\max}$		
	Properties of the fractal set F	
	fractal solid phase, mass M	fractal pore space, volume P
(9b)	$M(L) \propto L^D$ (18)	$P(L) \propto L^D$ (19)
	solid density σ	porosity $\Phi(L)$, total porosity F
(10b)	$\sigma(L) \propto L^{D-E}$ (20)	$\Phi(L) \propto L^{D-E}$ (21)
(7a) or (10a)		$\Phi = (l_{min}/l_{max})^{E-D}$ (22)
	number of sub-objects of size l centered on the solid phase	number of sub-objects of size l centered on the pore space
(5b) and	$N_r(l) \propto l^{-D}$ (23)	$N_r(l) \propto l^{-D}$ (24)
if $l \ll l_{\max}$	$[N_r > l] \propto l^{-D}$ (25)	$[N_r > l] \propto l^{-D}$ (26)
	Properties of the fractal collection of gaps G	
	pore number-size	solid elements number-size
(14b) $\Rightarrow$	$N_r(l) \propto l^{-D}$ (27)	$N_r(l) \propto l^{-D}$ (28)
(15b) if $l \ll l_{max}$	$[N_r > l] \propto l^{-D}$ (29)	$[N_r > l] \propto l^{-D}$ (30)
	partial porosity, pores of size $>l$	partial density, solids of size $>l$
(17b)	$[\Phi > l] = 1-(l/l_{max})^{E-D}$ (31)	$[\sigma > l] \propto 1-(l/l_{max})^{E-D}$ (32)
	l_{max} maximum pore size	l_{max} maximum size for the solid elements
Mathematical fractal obtained in the limit of an infinite number of iterations ($l_{\min} \rightarrow 0$) to represent only a fractal distribution of solid fragments or pores		
(15b) if $l \ll l_{max}$	pore number-size	solid elements number-size
	$[N_r > l] \propto l^{-D}$ (27)	$[N_r > l] \propto l^{-D}$ (34)
	partial porosity, pores of size $\leq l$	partial mass, solids of size $\leq l$
(16b)	$([\Phi \leq l]/\Phi) \rightarrow (l/l_{max})^{E-D}$ (35)	$[M \leq l] \propto (l/l_{max})^{E-D}$ (36)

Table 1: Main properties of the models of type A and B.

uniform density for the basic material) (Equations (18), (20) and (32)), whereas when the volumes and densities refer to the pore space, they represent, respectively, pore volumes P and porosities F (Equations (19), (21), (22) and (31)).

In the model of type A, where F represents the solid phase (Figure 6a) and is bounded by two cut-offs of scale, the mass of the solid phase scales as a power law of the object size: we call this model a mass fractal even in the limit case where the lower bound $l_{\min}$ tends towards 0 and only pores are represented!

Note that, in this limit case, the partial void density ρ, which has been defined as a partial volume of gaps divided by the total volume of the object, is here equal to a partial volume of pores divided by the total volume made only of pores: it is not stricto-sensu equal to a partial porosity $[\Phi \leq l]$, but equal to $[\Phi \leq l]/\Phi$ (Equation (35)).

In the model of type B, where F represents the void phase (Figure 6a) and is bounded by two cut-offs of scale, the volume of the pore space scales as a power law of the object size: we call this model a pore fractal even in the limit case where the lower bound $l_{\min}$ tends towards 0 and only solids are represented! In this limit case, the partial solid density r is a partial volume of solids divided by the total volume made only of solids, it is simply proportional to a partial mass M of solids (Equation (36)).

Many expressions used to characterize fractal fragmented or aggregated soils may be so derived and classified according to the modeling concepts outlined in Table 1. Possible misunderstandings when comparing different approaches used in fragmentation or aggregation modeling may be traced back to the existence of two different interpretations (model A versus model B) of the same lacunar fractal model (*e.g.*, the Sierpinki carpet or the Menger Sponge) when representing soils. For example, Turcotte (1992) presented several results in his chapter entitled "Fragmentation", in particular how Equation (34) results from scale invariant fragmentation processes (there, a model of type B was used, see following section), and also how the scale invariant porosity represented by Menger sponge's gap explains Equation (31) (there, a model of type A was used). He also mentioned the results from Katz and Thompson (1985) pointing to the existence of a fractal porosity in sandstones. The equation used by the latter authors to describe a fractal porous medium is equivalent to Equation (22), and was justified only by the following explanation: "In a volume element of size $l_{\max}$, the "volume" of the pore-filling surface is given in units of $l_{\min}$ by$A(l_{\max}/l_{\min})^D$, where $l_{\min}$ is the lower limit of the self similar region, $l_{\max}$ is the upper limit, and A is a constant of order 1. Hence the porosity is simply $\Phi = A(l_{\max}/l_{\min})^{3-D}$". This sentence has been judged by some (*e.g.*, Feder, 1988, p. 243) not to be self-explanatory, and in any case cannot apply to the porosity modeled by Figure 6a. Another example is given by Tyler and Wheatcraft (1990), who derived Equation (35) using a model of type A, then later (Tyler and Wheatcraft, 1992) established Equation (36) using a model of type B (see section IV). The difference between these two approaches deserves to be underlined. Let us give a last example from Rieu and Sposito (1991b), where a model of type A has been used to represent fragmented soils and where Equation (31) is one of the basic equations established by these authors (see section V).

Therefore, utmost care must be exercised in interpreting experimental results on fractal soils. Our discussion will focus on the type of fractal behavior that may be expected in fragmented and aggregated soils.

IV Fragmentation and Aggregation Modeling

IV.A Fragmentation Modeling

IV.A.a Fragmentation Model 1

The fragmentation model described by Turcotte (1986) has been the starting point of numerous developments carried out in soil science. A cube of linear size L is first split into $\mathfrak{N} = 8$ sub-cubes of linear dimension $L/2$. Then each of these 8 sub-cubes is split again into 8 sub-cubes of linear size $L/22$, and so forth in subsequent fragmentation stages (*cf.* Baveye and Boast, 1997). The fractal model introduces an incomplete fragmentation probability: the probability p that a cube of size $L/2^i$ will fragment into 8 cubes of size $L/2^{i+1}$ is taken to be constant at each iteration i in order to preserve scale invariance. Turcotte (1986,1992) shows that the number $[N_r > l]$ of fragments with size greater than $l = L/2^i$ (when ^{l} is small enough) is proportional to l^{-D}, where $D = \log(8p)/\log(2)$ and $1/8 < p < 1$, $0 < D < 3$. The existence and the value of a real scale-invariant probability of fragmentation p are modeled by different assumptions on rock fragility using an approach based on the renormalization group theory (in Allègre *et al.*, 1982). The value of p (and that of the exponent D) depends on the spatial distribution of the sound and fragile cells in a given rock which determines preexisting zones of weakness at all scales.

This example can be generalized (Turcotte, 1992); a fragment of size l is divided into $\mathfrak{N}$ subfragments of size l/r, with$r = \mathfrak{N}^{-1/E}$. The number $\mathfrak{N}$ of subfragments can assume any value, the probability p may have any value between $1/\mathfrak{N}$ and 1, and D can be calculated as follows:

$$D = E\frac{\log \mathfrak{N}p}{\log \mathfrak{N}} \quad \text{with} \quad 0 < D < E, \quad \text{or equivalently,} \quad r = \mathfrak{N}^{-1/E} \tag{37}$$

Clearly, Turcotte's model defined by the parameters ($\mathfrak{N}$,p) can be associated with a lacunar model (n, N, r) with $n = \mathfrak{N}$, $N = \mathfrak{N}p$, and $r = \mathfrak{N}^{-1/E}$. And the definition of D given by Equation (4a) ($D = \frac{\log N}{\log(1/r)}$) is equivalent to that given by Equation (37) ($D = E\frac{\log \mathfrak{N}p}{\log \mathfrak{N}}$). The lacunar model is of type B, for infinite iterations, and leads only to a distribution of gaps (colored black in Figure 7 as in Figure 6a) to model solid fragments. The general equations apply, and Turcotte's result on the number of fragments is a special case of Equation (34) ($[N_r > l] \propto l^{-D}$) established for the number of gaps[2].

In Turcotte's approach, the initial material is represented by a bulk Euclidean square, which splits into several solid fragments, and the porosity is neglected. This kind of fractal distribution of fragments may result from the crushing of a piece of quartz or of any compact rock. Here D does not reflect any fractal characteristic of the fragmented material but is solely a measure of fragmentation.

---

[2] As in Turcotte's derivation, this result can be expressed in terms of cumulative numbers only if the number of iterations in the model is high enough to justify such an approximation.

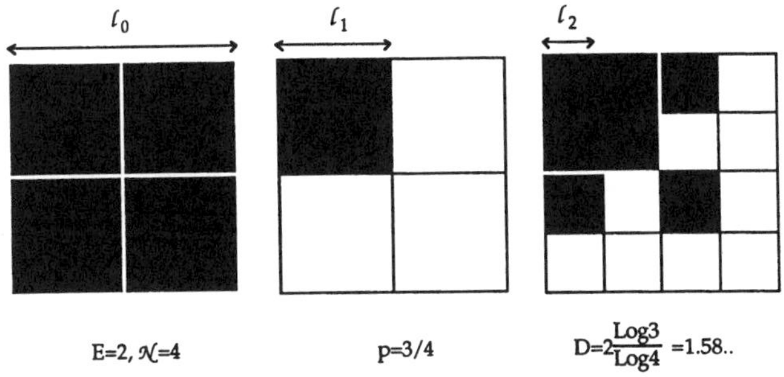

Figure 7: Illustration of a two-dimensional version of Turcotte's (1986) model ($\mathfrak{N} = 4$, $p = 3/4$) that can be viewed as lacunar model defined by ($n = 4$, $N = 3$, $r = 1/2$).

Yet the fragmentation dimension D could reveal a fractal character of the fragmented material in some special cases. For example, Ghilardi *et al.* (1993), referring to both Turcotte (1986) and Katz and Thompson (1985), showed that a fractal porous medium of dimension D, represented by a model of type B (Figure 6b) bounded by cut-offs of scale, is made of solid particles whose number-size distribution again follows Equation (34). The complete dissociation of such a porous medium should release particles (solid fragments) with a fragmentation dimension equal to that of the pore space volume. It can also be noted that Tyler and Wheatcraft (1989, 1992) used the same model to represent a fractal distribution of the particles in a soil, but they did not consider, in the referred papers, the properties of the pore space.

IV.A.b Fragmentation Model 2

Rieu and Sposito (1991b) proposed a general theoretical framework for a fractal representation of a soil as a naturally fragmented porous material. The organization of such a medium is described on successive levels of structural fragmentation. At each level any given volume of material is distributed into both void fractions and porous sub-volumes. Each sub-volume is similar to the whole. Figure 8 illustrates a representative volume of a soil by means of an initial cube of size d_o. The cube is divided into $\mathfrak{N}$ parts, which are reduced by a ratio r to produce $\mathfrak{N}$ sub-volumes of size d_1 and a void space of size f_o. At each iteration, each sub-volume of linear size d_i is split into $\mathfrak{N}$ self-similar sub-volumes of linear size d_{i+1}, using a constant similarity ratio $r = d_{i+1}/d_i < 1/\mathfrak{N}^3$, and creating pores of size f_i scaling with the same similarity ratio r. The last level m of fragmentation in the hierarchy of embedded structural units is constituted of solid particles all of size d_m. The fractal dimension of this self-similar structure is

$$D = Log\mathfrak{N}/Log(1/r) \quad 0 < D < E, \quad \text{or equivalently,} \quad \mathfrak{N} = r^{-D} \tag{38}$$

This fragmentation model can be compared to Turcotte's (1986) by noting

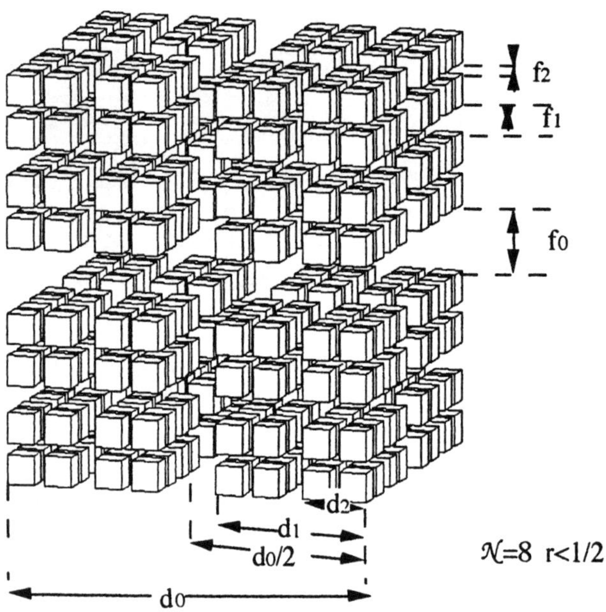

Figure 8: Illustration of an idealized mode for a fragmented porous medium (Rieu and Sposito, 1991b)

that the fragmentation process illustrated in Figure 8 creates the same number $\mathfrak{N}$ of sub-volumes as in a complete, Euclidean fragmentation, but that the fractal nature of the fragmented material is due to the reduction of the sub-volume sizes by a ratio $r < \mathfrak{N}^{-1/E}$, a reduction which effectively creates a porous space.

The sub-volumes resulting from Rieu and Sposito's (1991b) fragmentation may be considered as soil fragments having a coherent organization in an undisturbed structure. Rieu and Sposito (1991c) showed that the model enables a reasonably realistic representation of embedded soil fragments whose number $N_r(l)$ increases with the fragment size l according to a power-law relationship, $N_r(l) \propto l^{-D}$. The same sub-volumes can be considered as embedded soil aggregates whose density-size properties are presented in section IV.B. The pore-size distribution and porosity properties are presented in section V.A. These properties are also power laws of the length scale, whose exponent involves the fractal dimension D. The latter dimension D has become a characteristic of the whole soil organization. Rieu and Sposito's model can be viewed as a fractal lacunar model: the initial definition by means of the two parameters (N, r) is conceptually equivalent to that of a lacunar model (n, N, r)[3] with $n = r^{-E}$ and $N = r^{-D} = \mathfrak{N}$. The example given in Figure 9 with $\mathfrak{N} = 4$ is equivalent to the example given in Figure 2b with $n = 9$. This model is of type A, and it is clear

[3] More generally, in the equivalent definition as a lacunar model, n and N must be high enough to remain integers while taking account of the small size of pores in a realistic approach.

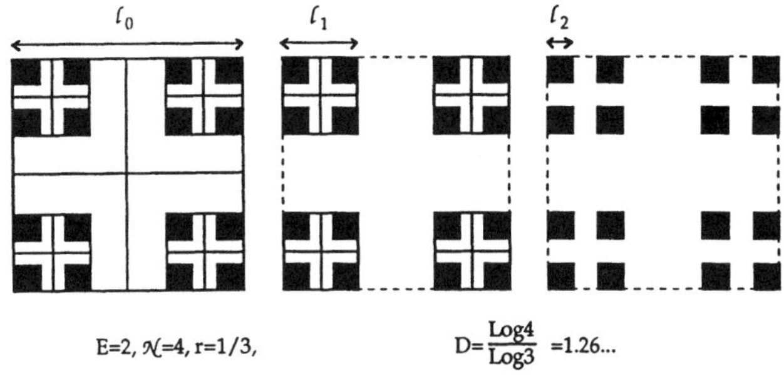

Figure 9: Two-dimensional illustration of Rieu and Sposito's (1991b) fragmentation model ($\mathfrak{N} = 4$, $r = 1/3$) , which can be viewed as a lacunar model defined by ($n = 9$, $N = 4$, $r = 1/3$) (cf. Figure 2.b).

that infinite iterations would lead to a vanishing solid space and a representation of the pore space only. Therefore, D can be identified to a mass fractal dimension. Indeed, the general results of section II apply: we know from Equation (23) that the number $N_r(l)$ of sub-objects, here the number of sub-aggregates, scales as $N_r(l) \propto l^{-D}$ and that the cumulative number given by Equation (25) follows a similar power law, provided that the number of fragmentation levels is high enough. It appears therefore that this model represents again a fractal number-size distribution of fragments, but the fragments considered here refer to embedded constituents of an aggregated soil, and not to the solid particles obtained at the last level of fragmentation.

Crawford (1993) remarked that experimental number-size analyses do not usually take into account this type of hierarchical clustering and instead destroy the relative distribution of the components. One might therefore ask what happens when the organization is partially disaggregated by applying an artificial fragmentation process. Suppose that an undisturbed fractal soil core is well represented by Rieu and Sposito's model (*cf.* Figure 9) and by a mass fractal dimension D (*cf.* Equation (38)), and assume that the method of aggregates separation is similar to the fragmentation process described in Turcotte's model (*cf.* Figure 7) and associated to a fragmentation dimension D' (*cf.* Equation (37)). If a constant probability of disaggregation p is assumed at each fragmentation level as in Turcotte's analysis, a proportion $(1-p)$ of aggregates remains intact, whereas a proportion p releases smaller aggregates, and so forth, at each level (*cf.* Figure 10). It is possible to show that the resulting number-size distribution of intact aggregates still follows a power-law relationship. The exponent of this power law can be associated to a fractal dimension D'' given by (*cf.* Appendix):

$$D'' = \frac{DD'}{E} \tag{39}$$

The dimension D'' is a combination of the mass fractal dimension D of the aggregated medium and a fragmentation dimension D' characterizing a scale-

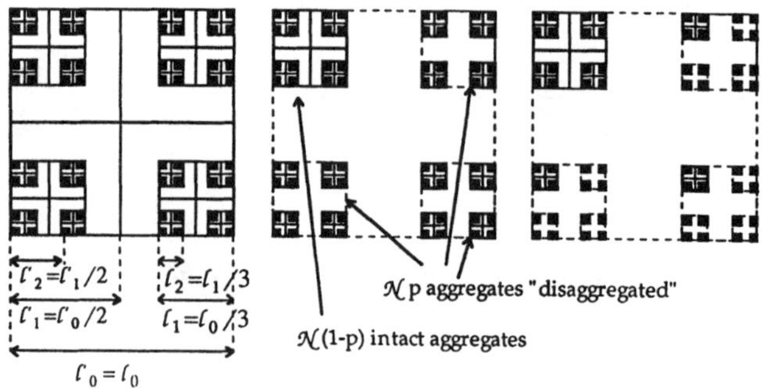

Figure 10: Scale-invariant fragmentation (or disaggregation) of the aggregated porous medium represented in Figure 9.

invariant fragmentation process. D'' is always less than D and less than D' (as $D < E$ and $D' < E$). Assuming that such a model is valid, one obtains a fractal dimension D'' which is not solely a measure of the fragmentation process associated with a probability of failure of the porous medium under a fragmentation process, but also a measure of the intrinsic, fragmented constitution of an aggregated porous medium. The combination of these two effects leads to a fractal dimension D'' less than that, D and D', associated with either one of the two interacting fragmentation processes. This could explain the relation between mass fractal dimension and fractal dimension of a number-size aggregate distribution, whose existence was suggested by Perfect and Kay (1991).

Rieu and Sposito (1991b) had already distinguished two fractal dimensions $\mathfrak{D}_r$ and $\mathfrak{D}$ in a fragmented porous medium. $\mathfrak{D}_r$ referred to the structured state and was associated with other properties of the medium (see sections IV.B and V.A). $\mathfrak{D}_r$ is equivalent here to the mass dimension denoted by D. $\mathfrak{D}$ characterized the aggregates resulting from the complete fragmentation of the structure and was always smaller than $\mathfrak{D}_r$. An analysis of the spatial representation that could account both for the difference between $\mathfrak{D}$ and $\mathfrak{D}_r$ and for a realistic stability of the hierarchy of aggregates in the model was achieved by introducing bridges linking some aggregates, thereby filling a given proportion of the pore space at each level. These bridges were made of a porous matter similar to the whole and may be considered as the proportion of sub-aggregates which are disaggregated in the artificial fragmentation process. The fractal dimension $\mathfrak{D}$, therefore, could have the same meaning as D''.

IV.A.c Fractal Characterization of Fragmented Soils and Fragment Number-Size Power Laws

The presented fragmentation models briefly described above can account for fragment distributions scaling as a power law of the fragment size, provided the fragmentation process is scale-invariant. Modeling attempts in cases where the fragmentation process is assumed not to be scale invariant are beyond the scope of the present chapter; the resulting distribution often turns out to be significantly more complicated to analyze (Crawford *et al.*, 1993; Perfect *et al.*, 1993).

Nevertheless, the interpretation of the power-law exponent as a fractal dimension D is not always made in the context of a particular fractal model of the whole structure, made of both void and solid components. Some models are explicitly distribution models: a distribution of fragments in Turcotte (1986), a distribution of solid particles in Tyler and Wheatcraft (1989, 1992). Crawford *et al.* (1993) pointed out that no relation can be estabished with assurance between a distribution of fragments and alleged fractal characteristics of the parent structure once this parent structure has been destroyed by several "agitations" in order to individualize and count the fragments. The parent material might be a homogeneous solid represented by a bulk Euclidean volume. It might also be a porous material as modeled by a lacunar model of type B, where D would be both the fractal dimension of the particle distribution and the fractal dimension of the void volume in the parent material (Ghilardi *et al.*, 1993). The parent material might also be a naturally fragmented porous medium as modeled by a lacunar model of type A, where D would be a combination of the mass fractal dimension of the solid space in the parent structure and of a pure fragmentation dimension due to a scale invariant disaggregation of the originally coherent material (Rieu and Sposito, 1991b). Crawford *et al.* (1993) introduced other fragmentation models. In a first example, the parent material is a compact solid aggregate having a fractal boundary of dimension D; a complete scale-invariant fragmentation process leads to a fractal collection of fragments with the same[4] fragmentation dimension D. In the subsequent examples depicted by Crawford *et al.* (1993), either the fragmented material has a non-fractal rough surface, or the fragmentation process is incomplete, and one still obtains a fractal distribution of fragments without any direct link to the parent material. It should be noted that the model of parent structure exhibited by these authors is essentially conceptual, because the porous structure is made of one big solid aggregate surrounded by a void space. This model is not a lacunar model where the pores and solids are intertwined to provide a realistic image of a porous medium.

To summarize, it appears that a scale-invariant fragmentation process leads to a fractal distribution of fragments, but that such a distribution cannot indicate the fractal or non-fractal nature of the fragmented parent material.

IV.B Aggregation Modeling

IV.B.a A Widespread Physical Mechanism

Aggregation is defined by Jullien and Botet (1986) or Gouyet (1992) as an irreversible physical phenomenon causing the clustering of elementary microparticles or microaggregates into macroscopic structures called aggregates. Aggregates are encountered in many areas of physics or chemistry; they characterize the state of several types of colloids, where basic particles in suspension in

[4] Perfect and Kay (1991) suggested that D could be simultaneously a measure of fragmentation and irregularity.

a liquid irreversibly "stick" together under the action of short- or long-range attractive forces (*e.g.*, gold or silica colloids). Similar phenomena occur in aerosols (microparticles initially dispersed in a gas), but also in the formation of polymeric chains, in metallic deposits observed in electrolysis experiments, in particles packings induced by sedimentation or filtration. The specific ramified or tree-like shapes of such aggregates have been frequently reported in the past. Similar shapes are observed in widely different situations (clouds formation, dielectric breakdown, viscous fingering or percolation clusters in porous media) and the differences or similarities between the observed patterns have been well described. The quantitative description of the aggregate geometry is comparatively recent and is due to the introduction of fractal concepts, whereas aggregation kinetics is better understood as a result of numerous numerical simulations of growth processes that have been carried out in recent years.

IV.B.b Numerical Simulations of Growth Processes

Many studies have been devoted to the search of simple rules of dynamic aggregation that could lead to observed aggregation patterns (*cf.* Vicsek, 1989, with regard to fractal growth phenomen or Jullien and Botet, 1986, on aggregation and fractal aggregates).

Numerical experiments have been done on computers to "build" aggregates. A widely known simulation process is called a particle-cluster aggregation: the algorithms begin with one initial particle located on a given site of a regular lattice; new particles, generally identical to the first one, are iteratively added and "stuck" to the previous ones to form a cluster with increasing size. The choice of the site receiving the new particle determines the shape of the aggregate; it can be any empty neighboring site of the growing aggregate. Below are some examples of simulation processes.

In the simplest version of the well-known Eden model (a model first used to represent cellular growth and the evolution of tumors), the site accepting the new particle is chosen randomly among all the aggregate neighboring sites. This procedure results in compact aggregates, like that shown in Figure 11a. The aggregation process may be limited by diffusion (diffusion-limited aggregation). The Brownian diffusion of particles before aggregating is represented by a random walk on the lattice. When a wandering particle meets the cluster, it immediately sticks to it (Figure 11b). A ballistic aggregation according to a preferential direction (*e.g.*, induced by the gravity field in sedimentation) can be represented by letting particles move straight down and stick to the growing aggregate. Many extensions of these examples have been explored, by introducing clustering probabilities, modifying the particle-cluster aggregation (where moving particles stick to a fixed aggregate), or by considering cluster-cluster aggregation (where moving aggregates join and stick together to form larger aggregates). The aggregates obtained by simulation resemble closely those obtained in real experiments; they often have a fractal structure. In the Eden model, only the aggregate boundary is fractal; the volume is compact and non-fractal. In most numerical experiments, however, the aggregates exhibit a fractal volume characterized by what is often called a cluster dimension. The

Figure 11: Numerical simulations of aggregation processes. (a) Eden model: a fractal boundary enclosing a non-fractal mass (5,000 particles, reprinted with permission from Vicsek, 1989, p. 184); (b) diffusion-limited aggregation: the Witten-Sander mode, a mass fractal with $D = 1.71$ (50,000 particles, simulated by E. Perrier).

cluster dimension (Feder, 1988, p. 32) D is defined by the relation $N(L) \propto L^D$, where $N(L)$ is the number of particles in a volume of linear size L. This equation is equivalent to the already-defined mass dimension $M(L) \propto L^D$ (Equation (18)), assuming that the clustered particles are nearly identical, and it can be reformulated in terms of a density property $\sigma(R) \propto R^{D-E}$ (Equation (20)). Feder's sentence "A fractal cluster has the property that the density decreases as the cluster size increases in a way described by the exponent in the number-radius relation $N(L) \propto L^D$" refers to fractal aggregates whose solid density decreases as the aggregate size increases in a way described by the fractal mass dimension D in Equation (20).

IV.B.c Fractal Characterization of Aggregated Soils and Aggregate Mass-Size Power Laws

With respect to the lacunar fractal models of aggregated soils, it is obvious that only a model of type A can represent fractal aggregates, if we adopt the above Feders's definition of a fractal clusters or aggregates. Rieu and Sposito's illustration (Figure 9) is one geometric realization, among others. Its advantage is to provide a realistic picture of hierarchical aggregation in many soils. More realism is achieved (Perrier *et al.*, 1992, Perrier, 1994) in statistical extensions (an example is shown on Figure 14) that keep the same global properties as the deterministic initial model.

The mass-size property has been established mathematically by Rieu and Sposito (1991b) on a general basis and in terms of density. It has been shown that $\sigma_i/\sigma_0 = (d_i/d_0)^{D-3}$, where d_i represents the linear size of aggregates of fragmentation level i. The solid matter density decreases, as larger and larger component aggregates are considered, because larger and larger pores become enclosed.

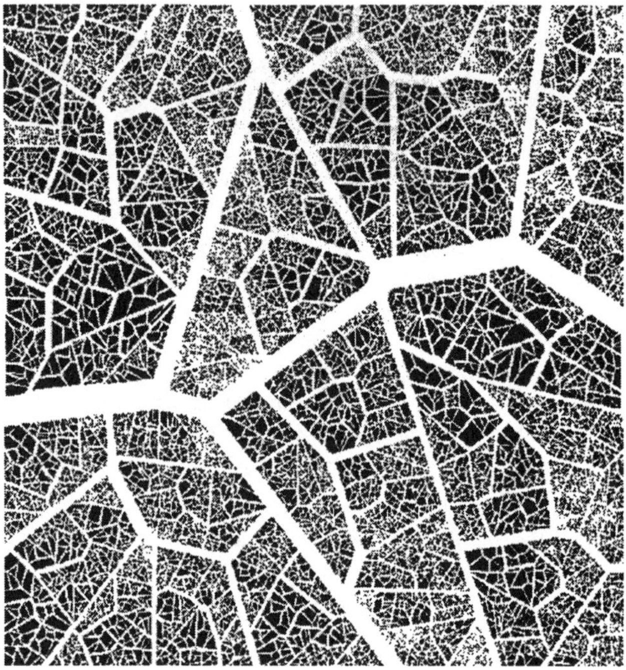

Figure 12: Statistical model of a hierarchically aggregated soil (Perrier, 1994). The partition into N squares has been replaced by a partition into N irregular polygons.

Other spatial configurations of lacunar models of type A could be used (*cf.* Figure 2), for they all share the mass-size property expressed by Equations (18) and (20). The main differences are not due to the value of the fractal dimension characterizing the way the aggregated soil fills the Euclidian space it is embedded in; they are due to the choice of the generator, which determines the whole configuration of the cluster, its connectivity or its ramification. The configuration proposed by Rieu and Sposito (1991b) is very different from that of a Menger Sponge used by Turcotte (1992) to establish a fractal density-size power law, even if they belong to the same class of models (lacunar models of type A), and the resulting pattern of aggregates is also dissimilar.

Crawford *et al.* (1993) modeled soil aggregates as having a fractal boundary and a non-fractal mass, as in Eden's model. However, such compact aggregates constituted of tightly packed particles, without any holes, strongly differ from those encountered in soils where the measured aggregate density has been shown to vary with the aggregate size (Crawford *et al.*, 1991). Anyway, this model of aggregation can represent other types of aggregated soils, and aggregates with a fractal boundary can also be called fractal aggregates!

Ghilardi *et al.* (1993) carried out numerical simulations of a ballistic aggregation of particles. The particles were chosen from a fractal distribution resulting from the fragmentation of a bulk solid, and then launched downwards to simulate a sedimentation process. The resulting porous structure was modeled by the authors using a model of type B and was called a loose granular aggregate. Ghilardi et al's (1993) work refers to an extended definition of the word aggregate in the sense of any coherent spatial organization of particles in a porous medium.

IV.C Comparison and Discussion

Most analyses of fragmentation processes conclude that fragments follow a fractal number-size distribution of fragments whereas in terms of aggregation processes, including other materials than soils, a fractal mass-size distribution seems often to be the rule. For example, Van Damme and Ben Ohoud (1990) wrote: "There are strong arguments for saying that, contrary to aggregation of colloidal particles which is known to lead to mass or surface (multi)fractal structures in a wide range of conditions, fragmentation of colloidal packings reveals a scaling behavior which applies to the open porosity".

Under natural or artificial events, however, an aggregated medium can fragment, and fragments can aggregate. Fragmentation and aggregation phenomena are manifested simultaneously in many soil structure organizations. The approach of Rieu and Sposito (1991b) can be considered as an attempt to represent both aggregated and fragmented behaviors of soils; the mass-size power law may express the aggregated nature of the soil model (mass fractal dimension $\mathfrak{D}_r$) whereas the number-size power law may micmic the naturally fragmented nature of the same model (fragmentation dimension $\mathfrak{D}$).

In a first analysis based on fractal lacunar models, one could say that a model of type B is a good model to represent a fractal number-size distribution of primary particles with a wide range of particle sizes. A model of type A, meanwhile, is well fitted by a fractal mass-size distribution or a fractal number-size distribution of aggregates made of nearly identical particles. It is clear, however, that these models are too schematic and that they are not the only ones. One might conceive of other geometric models that present one or another of the experimentally observed scaling properties without exhibiting any kind of self-similarity or other simple fractal characteristics. Mandelbrot (1983, p. 124) wrote that one may encounter scaling laws such as $M(L) \propto L^Q$ (Equation (16) in Table 1) but "*such a formula does not by itself guarantee that Q is a fractal dimension*". Crawford *et al.* (1993) showed that so-called fractal number-size distributions can be generated by non-fractal structure models. One may be tempted to interpret the exponent in such powers laws as a fractal dimension of an underlying structure. If the experimental observations (section III) of power laws compatible with any presented model cannot "validate" it, they can lead to reject some models in some cases or show the necessity to improve their ability to represent a real soil. If a fractal model is not used only to illustrate an observed power law, but also to find a realistic yet simple model of a real structure, secondary properties of the models must be analyzed and tested to verify that the model is adequate to describe a complex reality.

V Fragmented and Aggregated Soils: Measures of Solid Phase Organization

V.A Fractal Particle-Size Distributions and Aggregate-Size Distributions in Fragmented Soils

Many authors referring to Turcotte (1986) and to fragmentation processes have measured cumulative number-size relations satisfying Equation (34) (Table 1). Turcotte himself (1992) considered a variety of geological materials, including soils, but mainly studied the distribution of fragments obtained when rocks are naturally or artificially "broken" (weathering, explosions, impacts,...). He found that different empirical equations that have been used to characterize the distribution of fragments are equivalent to Equation (34) and that several sets of experimental data (on interstellar grains, disaggregated gneiss or granite, broken coal, sandy clay,... etc.) are described quite well by this equation. In all cases, the quality of the fitting and the fractal dimension were estimated from $\log[N_r > l]$ versus $\log l$ adjustments.

A clear distinction must be made between soil studies dealing with particle size distributions (PSD) and those involving aggregate size distributions (ASD).

V.A.a Fractal Particle-Size Distributions (PSD)

Tyler and Wheatcraft (1989) analyzed particle-size distributions obtained by mechanical analysis of soils with varied textures. To determine the number $N_r(l)$ of particles with a given size l, these authors defined a mean particle size in each class as the arithmetic mean between two sieve sizes. They then calculated the particle number by dividing the total mass in each class by the mean mass of a spherical particle of size l, assuming a constant particle density. According to log / log adjustments, the soils showed clear fractal behavior but the estimated fractal dimensions often exceeded 3 (*cf.* Figure 13a), a physically impossible value for a geometric porous fractal embedded in a 3-dimensional Euclidean space.

Three years later, Tyler and Wheatcraft (1992) discussed the practical limitations of the cumulative number-based approach. Thay argued that D would be over-estimated using this approach because of artifacts due to the determination of approximate particle numbers from mass measurements. Their second approach uses the cumulative mass-size relation previously derived, $[M \leq l] \propto l^{3-D}$ (Equation (36), table 1), where $[M \leq l]$ refers to the mass of particles with linear size smaller than l, and D is the fractal dimension of the

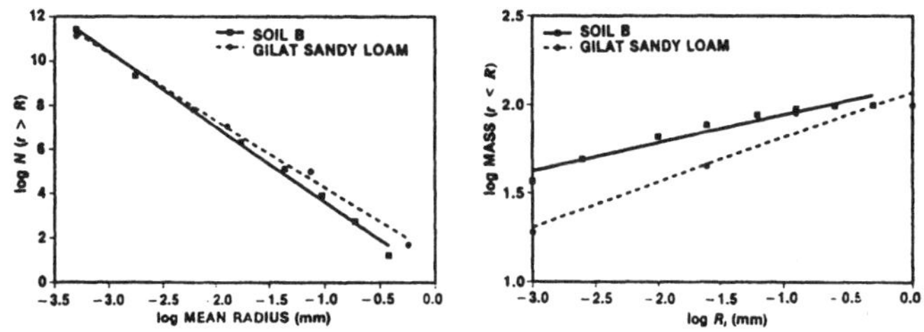

(a) Log/Log plot of the number of particles of mean radius r greater than a given mean radius R (soil B, D≅3.4, Gilat sandy loam, D≅3.0)

(b) Log/Log plot of the partial mass of all the particles of mean radius r smaller than a characteristic radius R (soil B, D≅2.8, Gilat sandy loam, D≅2.7)

Figure 13: Particle-size distributions satisfying a fractal number-size power law. (a) Direct fit of Equation (34) and number calculated from available mass data (Tyler and Wheatcraft, 1989); (b) fit of Equation (36) equivalent to Equation (34) and direct use of mass data (Tyler and Wheatcraft, 1992).

particle collection. This new method is equivalent to the previous one if the elementary particle density is assumed to be constant. It resulted in a smaller number of soils displaying strict fractal scaling being less numerous; the fractal domain was often restricted to a limited portion of the PSD, and the estimates of the fractal dimension D were within the allowed range (< 3) (*cf.* Figure 13b), with mean values of 2.4 for sandy soils to near 3 for clayey soils.

In a similar manner, Wu *et al.* (1993) studied the PSDs of four different soils over a wide range of textures, from 5 cm to 20 nm radius, using several measurement techniques (sedimentation, sieving, light-scattering methods, etc.). The results converged and corresponded closely to the power law of Equation (34), in spite of the fact that they exhibited two crossover regions. Surprisingly, the exponents were identical for all the studied soils, and the authors concluded to a possibly universal fractal dimension 2.8 ± 0.1 in the 50 nm - 100 mm range.

When the power laws of Equations (34) or (36) are applied in this fashion to model fractal particle-size distributions, a number of assumptions are implicitly made. It has been shown in section I that Equation (34) is expected to be verified in the case of cumulative data if the number of fragmentation levels is high enough; whereas Equation (36) is expected to be valid in lacunar models of type B only when they are unbounded by a lower cut-off of scale. Consequently, if a model of type B is considered to represent the whole biphasic soil structure, Equations (34) and (36) are only approximations, and using them may strongly bias the estimation of D.

V.A.b Fractal Aggregate-Size Distributions (ASD)

The power law of Equation (34) ($[N_r > l] \propto l^{-D}$) has been used by other authors to describe aggregate-size distributions, with $[N_r > l]$ referring not to particles but to the cumulative number of loose, porous aggregates with mean linear size greater than l. However, the definition of soil aggregates and the identification of ASDs is a delicate matter.

Perfect and Kay (1991) measured aggregate distributions on silt loam soil cores collected after different cropping treatments. Two main experimental techniques were compared to determine the ASDs before and after energy input (wet sieving), and several levels of energy were also compared. The values of D estimated from rather good log / log adjustments were found to be sensitive both to the cropping history and energy input. The dimension D was interpreted as a measure of fragmentation and its value increased with the degree of fragmentation. The D estimates varied with the method of aggregate dispersion and often exceeded 3! The numbers of aggregates were calculated assuming a constant dry density, an assumption that is not consistent with the results that we shall present in section V.B.

In Rasiah *et al.* (1993), the definition adopted for the fractal distribution of aggregates is again that of Equation (34). However, a new mass-based characterization is proposed for fractal ASDs. Rejecting Turcotte's assumption of a constant probability p of fragmentation and defining different probabilities of fragmentation at each length scale (or probabilities of failure at each sieving stage), Rasiah *et al.* (1993) reformulated Equation (34) in terms of successive mass ratios. This reformulation avoided the need to introduce assumptions on scale invariance in aggregate density and shape. The measured estimates of D appeared to provide good discriminators of different soil structural properties and cropping treatments. The authors performed statistical regressions between these estimations of D, those obtained using Equation (36) (the mass-based approach of Tyler and Wheatcraft, 1992), those obtained with direct application of Equation (34) on numbers computed from mass data, and those obtained by direct fitting of Equation (34) and by careful number-size determination (manually-counted aggregates). The latter estimates were chosen as reference values. Rasiah *et al.* (1993) concluded that the estimates were all statistically equivalent, except those obtained with Equation (36), and that the scale-invariant density assumption did not introduce any significant error in the estimation of D, which was again greater than 3.

In a third paper, Perfect *et al.* (1993) introduced a distribution function to model the scale-dependent probability of failure and to generate a multifractal extension of Turcotte's model. A method was developed for calculating probabilities of failure from tensile strength data, and the number of aggregates between two sieve sizes was determined by visual counting. The practical application was an analysis of the deviations from linearity in the log / log plot of the cumulative number-size distribution and the determination of a spectrum of fractal dimensions. Usefulness and validity appear to be doubtful since each "fractal" dimension was associated with a linear interpolation between any two points of the log / log plot.

In all the articles that we reviewed above, no assumption is made about the underlying fractal model of the biphasic soil structure. As aggregates are concerned, this model could be a fractal structure of type A (and the cumulative number-size equation would be denoted Equation (25), *cf.* Table 1). In this case, the choice of a constant density is in contradiction with the main result on this

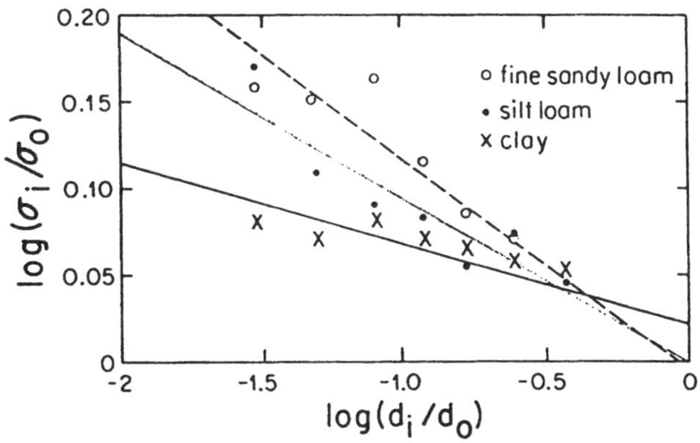

fine sandy loam, D≅2.88, silt loam, D≅2.91, clay, D≅2.95

Figure 14: A density-size power law in aggregated soils: tests of Equation (20) with the data of Chepil (1950), from Rieu and Sposito (1991c).

type of structure, that is, a density varying with the length scale.

Rieu and Sposito (1991c) took into account this density effect when converting mass data into number data. The fitting of experimental data (where both aggregate mass and density measurements were available in each size class) by Equation (25) was good but showed two different fractal regions in some soils. The estimated fractal dimensions were always smaller than 3.

V.B Fractal Scaling of Bulk Density in Aggregated Soils

Variations in the aggregate density or porosity with aggregate size have often been reported in the past (*e.g.*, Monnier *et al.*, 1973). The fact that available data could reveal fractal structures has provoked renewed interest in their analysis.

Rieu and Sposito (1991c), in the same soils as in the previous section, determined the mass-size relation using density measurements. Log/ log adjustments of Equation (20) showed a clear fractal behavior on undisturbed soil cores. Different sets of data were analyzed in this manner. The fit and estimates of D are presented for Chepil's (1950) data in Figure 14.

This method of determination of D is based on an estimation of the mass distribution inside structured soil aggregates and provides estimates of a bulk fractal dimension $\mathfrak{D}_r$, which was termed the "mass dimension" in section IV.A.b. As expected from the theoretical model, estimates of D ($\mathfrak{D}_r$) were always found to be greater than those ($\mathfrak{D}$) based on number-size analysis on the collection of aggregates resulting from the complete fragmentation of the porous structure. For example, $\mathfrak{D} = 2.95$ instead of $\mathfrak{D}_r = 2.84$ in the second fractal domain of the Sharpsburg soil (Rieu and Sposito, 1991c).

Young and Crawford (1991) directly measured the mass M of aggregates of varying size L. Surface aggregates from two sandy loam soils were collected with minimum disturbance, then dried and weighted. Their diameter was measured by visual inspection, a procedure which seems accurate for the most spherical aggregates and more approximate for highly irregular aggregates. Some corrections were also made for stone content. Then Equation (18) ($M(L) \propto L^D$) was tested using a linear regression of a plot of $\log M$ versus $\log L$. The authors provided persuasive arguments in support of a fractal nature of soil structure and for a mass fractal dimension characterizing the degree of soil heterogeneity under different types of cultivation.

Despite the analogy with the previous section, where mass and size were also measured on aggregate distributions, the methods and the implicit models (here, model of type A instead of model B) are different. Here, the mass M of any individual aggregate of size L is measured to characterize its internal structure whereas, earlier, the total mass of all the aggregates of size smaller than l was measured.

Bartoli *et al.* (1991) carried out a number of experiments to test the validity of Equation (18) in silty and sandy soils. The method adopted by these authors was based on image analysis of soil sections using photographs made at different resolutions. M(L) was calculated as the number of "solid" pixels included in squares of increasing side length L. All the soils studied were found to be mass fractals with a fractal dimension in a two-dimensional space significantly smaller than 2. The fractal dimension was extrapolated to 3-dimensional space by adding 1 to this value. On the other hand, when the soil porosity was measured on the thin sections, they found the pore space to be non-fractal because of poor agreement with Equation (19) and because the two-dimensional estimates of a pore fractal dimension were close to 2.

Among all the theoretical models presently available, only a lacunar model of type A can explain the variation of the soil mass as a power law of the length scale (*cf.* Equation (18)). A lacunar model of type B would, on the contrary, explain the variation of the pore volume as a power law of the length scale (*cf.* Equation (19)). In all the approaches presented in this section, a geometric fractal model is not always explicit, but the mass-size fractal power law (*cf.* Equation (18)) has been successfully tested.

VI Modeling the Pore-Space Properties in Fragmented and Aggregated Soils: Applications to Hydraulic Properties

VI.A Fractal Scaling of the Pore Space and Models of the Retention Curve in a Fractal Soil

Like the solid phase, the pore space of soils can be represented by fractal models of type A or B. On the one hand, the pore space can be viewed as surrounding loose solid fragments. It can be called fractal stricto-sensu if Equation (19) is satisfied as in model B (bounded by cut-offs of scale). All the pores can have nearly the same size, but they form a fractal geometric set. On the other hand, the pore space can also be viewed as a distribution of pores of varying size embedded in the solid aggregated phase. The pore-size distribution is called fractal when it presents the properties of the gap distribution in a lacunar fractal model. This corresponds to a type of soil structure which may be described by model A.

The hydraulic properties of soils are intimately reated to geometry of the pore space and their description is therefore affected by the choice of a particular fractal model.

Assuming a simple capillary model, the water retention curve can be associated to a pore-size distribution: the value water content θ at pressure h is assumed to be that of the porosity $[\Phi \leq l]$ due to pores of size smaller than $l = \alpha/h$ (Laplace law). On the basis of a fractal pore-size distribution of type A, two different modeling approaches have been used.

i) Tyler and Wheatcraft (1990) represented a fractal pore-size distribution using models of type A in the form of variants of the Sierpinski carpet (Figure 6a). They assumed no lower cut-off of scale, and rederived Equation (35), which is valid for any model of type A when $l_{\min}$ tends towards 0. From Equation (35), that is $[\Phi \leq l] = \Phi(l/l_{\max})^{E-D}$ (with $l_{\max}$ maximum pore size), direct use of Laplace's law leads to $\theta(h) = [\Phi \leq \alpha/h] = \Phi(h/h_{\min})^{D-E}$ (with $h_{\min}$ inversely proportional to $l_{\max}$). The analytic expression proposed by the authors to model $\theta(h)$ in a fractal pore space is

$$\theta(h) = \theta_{\max}(\frac{h}{h_{\min}})^{D-E} \tag{40}$$

This expression is equivalent to the simplified expression of Brooks and Corey (1964) that is widely used to model water retention data. The fractal dimension D could explain the empirical exponent of this power-law expression of $\theta(h)$. It must be stressed that the corresponding fractal model is only a distribution model with no representation of the solid phase.

ii) In a two-phase geometric model of type A, there exists a minimum cut-off of scale, $l_{\min} \neq 0$, as assumed by Rieu and Sposito (1991b). Here the value of $[\Phi \leq l]$ can be calculated from Equation (31) (that is, $[\Phi > l] = 1 - (l/l_{\max})^{E-D}$), which is equivalent to $[\Phi \leq l] = \Phi_{\max} - [\Phi > l] = \Phi_{\max} - 1 + (l/l_{\max})^{E-D}$. With the same assumptions as in the previous case (direct use of Laplace's law), one obtains $\theta(h) = [\Phi \leq \alpha/h] = \Phi_{\max} - 1 + (h/h_{\min})^{D-E}$. And the analytic expression proposed by the latter authors to model $\theta(h)$ in a fractal pore space is

$$\theta(h) = \theta_{\max} - 1 + (h/h_{\min})^{D-E} \tag{41}$$

Simulated retention data on the statistical realization shown in Figure 14 agree quite well with predictions based on Equation (41) (Perrier *et al.*, 1992). This equation is expected therefore to apply to irregular media globally scaling as the deterministic model.

It appears that different modeling concepts result in different mathematical relations (Equations (40) and (41)). In both cases, nevertheless, a fractal dimension appears to be a fundamental structural property that could enable the prediction of the pressure-water content relationship. We do not present here the modeling attempts with regard to the hydraulic conductivity, a second main hydraulic property of soils. The necessity to go beyond mere scaling exponent and to take into account the connectivity of the pore space is well recognized in the literature. A full account of this connectivity requires relying on geometric models of the spatial organization of both the solid- and pore space (Rieu and Sposito, 1991b, Perrier, 1994), and on the development of experimental techniques to measure connectivity dimensions (Gouyet, 1992).

VI.B Experimental Data

Thompson *et al.* (1987) measured a fractal pore space of type B in sandstones, using image analysis and density/density correlation functions[5]. Their results dealt mainly with the prediction of saturated hydraulic conductivity. Most applications dealing with unsaturated hydraulic properties, in contrast, are based on fractal lacunar models of type A. Many authors consider the inverse determination of a fractal dimension D from retention data (*e.g.*, Rieu and Sposito, 1991a,c, Brakensiek and Rawls, 1992, Toledo *et al.*, 1991, Tyler and Wheatcraft, 1990). Such indirect estimations of D depend on the validity of either Equation (40) or Equation (41). Perrier (1994) discusses the differences between the estimates obtained with Equations (40) and (41) and the consequences of neglecting hysteresis effects. Reversely, a fractal dimension determined from structural data might be used to predict the retention curve. For example, Agnese *et al.* (1994), using Rieu and Sposito's approach (1991b,c), found a very good agreement between the mass fractal dimension $\mathfrak{D}_r$ estimated from aggregate density measurements in a clayey soil and the exponent of the water retention curve measured on the same soil.

VII Conclusions

Because it allows geometric ideal representations of heterogeneous porous media, fractals provide an interesting way to improve the quantitative characterization of soils.

In a first stage, experimental number-size distributions of solid elements or pores can be reinterpreted in terms of fractal power laws, where the exponent is explained by the fractal dimension of the distribution. This operation does not involve explicitly any geometrical model.

In a second stage, geometric models of the soil construct can be built to represent the geometrical arrangement of the solid phase or the pore space. These models still exhibit experimental number-size power laws, but they allow properties of the soil construct such as solid density or porosity to be taken explicitly

[5] Working on the pore correlation function $C(l)$ calculated on all points l-distant and looking for the proportionality $C(l) \propto l^{D-E}$, is conceptually identical to working on the pore density in a volume of size L and looking for the proportionality $\Phi(L) \propto L^{D-E}$; but it provides a statistical average on many points (Jullien and Botet, 1986, p. 32).

into account. With the introduction of upper and lower cut-offs of scale, a deterministic, lacunar, fractal model appears well suited to represent soils viewed as aggregated or as fragmented media. Based on the same geometric pattern, two modeling concepts can be used:

- The model of type A represents a fractal distribution of pores of different sizes embedded in an aggregated medium, resulting in a mass fractal. This medium can be fragmented into smaller aggregates of different sizes.
- The model of type B represents a fractal distribution of solid elements or particles surrounded by a pore space (a cluster of pores), resulting in a pore fractal. This medium can be fragmented into solid elements or particles of uniform density.

Models of type A and type B having an identical geometric pattern allow the derivation of similar power laws that denote different properties, depending on the concept underlying the model. In both cases, a scale-invariant fragmentation process generates fragments whose number varies as power laws of the length scale. But the similarity of Equations (23) and (28) must not conceal the fact that model A deals with a fractal aggregate-size distribution and model B with a particle-size distribution. Thus, the number-size power laws often encountered in experimental data on fragmented soils may be confusing. To recall here only one example, model type A leads to a total porosity of the medium given by $\Phi = [\Phi > l_{\text{min}}] = 1 - (l_{\text{min}}/l_{\text{max}})^{E-D}$, according to Equation (31), whereas in model B, the total porosity is given by $\Phi = (l_{\text{min}}/l_{\text{max}})^{E-D}$, by virtue of Equation (22) and the same discrepancy may be observed concerning the solid density. Hence it is necessary to make clear the modeling concept used to interpret the experimental data.

It is noteworthy that, when these models are used as "true" fractals resulting from infinite iterations, they come back to mere representations of fractal distributions (pores in model A and solid elements in model B). From this point of view, they enlighten the limits of representations where a fractal distribution of natural objects is represented by a mathematical model without cut-offs of scale.

The use of fractal geometry can go beyond simple re-interpretation of power laws. Anyway, finding a geometric model of aggregated and fragmented soils by means of a simple fractal analysis of power-law exponents is a challenge. Two main classes of models have been proposed in the literature. Each of them may be valid, but on different soils. Complementary experimental data (pore space, solid phase, fragmented or aggregated states) must be obtained on the same soils to explore the degree of agreement between models and reality with regard to all properties of interest. Among these properties, the pore space accessibility in soils and the cohesiveness of the solid phase must be taken into account to enable the modeling of hydraulic properties. This will lead to further research and a more detailed comparison of different models in the same class with respect to their connectivity properties.

Appendix: Derivation of Equation (39)

The fractal dimension $D = \frac{\log \mathfrak{N}}{\log(1/r)}$ (with $r = \mathfrak{N}^{-1/D}$, Equation (38)) is the mass fractal dimension corresponding to the originally fragmented and aggregated structure, and associated to successive subaggregates of size l_i at each iteration i. The successive sizes l_i are given by iterated reductions of ratio r: $l_i \propto r^i \propto (N^{-1/D})^i$. On the example shown in Figure 9, $N = 4$, $r = 1/3$ and $D = \frac{\log 4}{\log 3}$.

A fragmentation dimension $D' = E\frac{\log N_p}{\log N}$ ($r' = N^{-1/E}$, Equation (37)) can be defined as in Turcotte's model and is associated to fragments of size and number denoted here respectively l'_i and $N'_r(l'_i)$, neglecting the pores. The successive sizes l'_i are given by iterated reductions of ratio r': $l'_i \propto r'^i \propto (N^{-1/E})^i$. On the example shown in Figure 7, $N = 4$, $r' = 1/2$, $p = 3/4$ and $D = 2\frac{\log 3}{\log 4}$.

An artificial fragmentation process is applied to the aggregated structure as can be imagined by superimposing Figure 7 on Figure 9. Each fragment of size l'_i that was not re-fragmented in smaller sub-fragments at higher iteration levels in Figure 7 is associated now to a reduced sub-aggregate of size l_i that remains intact in the artificial fragmentation (see Figure 10). The number $N''_r(l_i)$ of intact aggregates of size l_i is equal to the number $N'_r(l'_i)$ of fragments of size l'_i. Counting the fragments as in fragmentation model 1 leads to $N'_r(l'_i) \propto l_i'^{-D'}$. Knowing that $N''_r(l_i) = N'_r(l'_i)$ and that $l'_i \propto (l_i)^{D/E}$ [as $l_i \propto (N^{-1/D})^i$, and $l'_i \propto (N^{-1/E})^i$], one obtains that $N''_r(l_i) \propto l_i'^{-D'} \propto (l_i)^{-DD'/E}$, *i.e.*, that the number $N''_r(l_i)$ of intact aggregates of size l_i is proportional to $(l_i)^{-DD'/E}$. The exponent of this power law can be associated to a fractal dimension D'' given by: $D'' = \frac{DD'}{E}$ as required. On the example shown in Figure 10, $D'' = \frac{DD'}{E} = (\frac{\log 4}{\log 3})(2\frac{\log 3}{\log 4})/2 = 1$.

References

Allègre, C.J., J.L. Le Mouel and A. Provost. 1982. Scaling rules in rock fracture and possible implications for earthquake prediction. *Nature* 297:47-49.

Agnese, C., G. Crescimanno and M. Iovino. 1994. On the possibility of predicting the hydrological characteristics of soil from the fractal structure of porous media. *Annales Geophysicae*, Supp. II to Vol. 12:482.

Bartoli, F., R. Philippy, M. Doirisse, S. Niquet and M. Dubuit. 1991. Silty and sandy soil structure and self-similarity: the fractal approach. *J. Soil Sci.* 42:167-185.

Brakensiek, D.L. and W.J. Rawls. 1992. Comment on "Fractal processes in soil water retention" by S.W. Tyler and S.W. Wheatcraft. *Water Resour. Res.* 28:601-602.

Brooks, R.H. and A.T. Corey. 1964. *Hydraulic Properties of Porous Media.* Hydrol. Pap. 3. Colorado State Univ., Fort Collins.

Chepil, W.S. 1950. Methods of estimating apparent density of discrete soil grains and aggregates. *Soil Sci.* 70:351-362.

Crawford, J.W., B.D. Sleeman and I.M.Young. 1993. On the relation between number-size distributions and the fractal dimensions of aggregates. *J. Soil Science* 44:555-565.

Friesen, W.I. and R.J. Mikula. 1987. Fractal dimensions of coal particles. *J. Colloid Interface Sci.* 120(1):263-271.

Ghilardi, P., A.K. Kai and G. Menduni. 1993. Self-similar heterogeneity in granular porous media at the representative elementary volume scale. *Water Resour. Res.* 29(4):1205-1214.

Gouyet, J.F. 1992. *Physique et Structures Fractales.* Masson, Paris.

Jullien, R. and R. Botet. 1986. *Aggregation and Fractal Aggregates.*World Scientific, Singapore.

Katz, A.J. and A.H. Thompson. 1985. Fractal sandstones pores: implication for conductivity and pore formation. *Phys. Rev. Lett.* 54(12):1325-1328.

Mandelbrot, B.B. 1983. *The Fractal Geometry of Nature.* W.H. Freeman, San Francisco.

Monnier, G., P. Stengel and J.C. Fis. 1973. Une méthode de mesure de la densité apparente de petits agglomérats terreux. Application à l'analyse des systèmes de porosité des sols. *Ann. Agr.* 24:533-545.

Perfect, E. and B.D. Kay. 1991. Fractal theory applied to soil aggregation. *Soil Sci. Soc. Am. J.* 55:1552-1558.

Perfect, E., B.D. Kay and V. Rasiah. 1993. Multifractal model for soil aggregate fragmentation. *Soil Sci. Soc. Am. J.* 57:896-900.

Perrier, E., C. Mullon, M. Rieu and G. de Marsily. 1992. An "object-oriented" computer construction of deforming fractal soil structures. Determination of their water properties. p. 255-273. In *Porous or fractured unsaturated media: transports and behaviour*, Colloquium proceedings, EPFL, Lausanne, Switzerland.

Perrier, E.. 1994. *Structures Géométriques et Fonctionnement Hydrique des Sols. Simulations Exploratoires.* Ph.D. dissertation. Université Paris VI. Etudes et Thèses, Orstom, Paris.

Pfeifer, P. and D. Avnir. 1985. Chemistry in noninteger dimensions between two and three. I. Fractal theory of heterogeneous surfaces. *J. Chem. Phys.* 79(7):3558-3564.

Rasiah, V., B.D. Kay and E. Perfect. 1993. New mass-based model for estimating fractal dimensions of soil aggregates. *Soil Sci. Soc. Am. J.* 57:891-895.

Rieu, M. and G. Sposito. 1991a. Relation pression capillaire-teneur en eau dans les milieux poreux fragmentés et identification du caractère fractal de la structure des sols. *C.R. Acad. Sci. Paris* 312 Série II:1483-1489.

Rieu, M. and G. Sposito. 1991b. Fractal fragmentation, soil porosity, and soil water properties: I. Theory. *Soil Sci. Soc. Am. J.* 55:1231-1238.

Rieu, M. and G. Sposito. 1991c. Fractal fragmentation, soil porosity, and soil water properties: II. Applications. *Soil Sci. Soc. Am. J.* 55:1239-1244.

Thompson, A.H., A.J. Katz and C.E. Krohn. 1987. The microgeometry and transport properties of sedimentary rock. *Advances in Physics* 36:625-694.

Toledo, P.G, R.A Novy, H.T. Davis and L.E. Scriven. 1990. Hydraulic conductivity of porous media at low water content. *Soil Sci. Soc. Am. J.* 54:673-679.

Turcotte, D.L. 1986. Fractals and fragmentation. *J. Geophys. Res.* 91(B2):1921-1926.

Turcotte, D.L. 1992. *Fractals and Chaos in Geology and Geophysics.* Cambridge University Press, Cambridge, U.K.

Tyler, S.W. and S.W. Wheatcraft. 1989. Application of fractal mathematics to soil water retention estimation. *Soil Sci. Soc. Am. J.* 53:987-996.

Tyler, S.W. and S.W. Wheatcraft. 1990. Fractal processes in soil water retention. *Water Resour. Res.* 26:1045-1054.

Tyler, S.W. and S.W. Wheatcraft. 1992. Fractal scaling of soil particle-size distributions: analysis and limitations. *Soil Sci. Soc. Am. J.* 56:362-369.

Van Damme, H. and M. Ben Ohoud. 1990. From flow to fracture and fragmentation in colloidal media. 2: Local order and fragmentation geometry. p. 105-116. In J.C. Charmet, S. Roux and E. Guyon (eds.) *Disorder and Fracture.* NATO ASI series, Plenum Press, New York.

Vicsek, T.. 1989. *Fractal Growth Phenomena.* World Scientific, Singapore.

Wittmus, H.D. and A.P. Mazurak. 1958. Physical and chemical properties of soil aggregates in a Brunizem soil. *Soil Sci. Soc. Am. Proc.* 22:1-5.

Wu, Q., M. Borkovec and H. Sticher. 1993. On particle-size distributions in soils. *Soil Sci. Soc. Am. J.* 57:883-890.

Young, I.M. and J.W. Crawford. 1991. The fractal structure of soil aggregates: its measurement and interpretation. *J. Soil Sci.* 42:187-192.

Fractals and Soil Structure

F. Bartoli, Ph. Dutartre, V. Gomendy, S. Niquet, M. Dubuit and H. Vivier

Contents

I Introduction

Transport processes in variably-saturated porous media can be described at the macroscopic scale using mathematical models such as the Darcy-Buckingham or

ISBN 1-56670-105-8

the Richards equations. They have been extensively used to predict and model the transport of water, chemical, clays, microorganisms and energy in soils. In a number of situations, particularly in very heterogeneous soils and aquifer materials, the use of these macroscopic models has faced serious difficulties in recent years.

It is believed that quantitative characterization of spatial geometric variability is required to make possible the modeling of transport processes in such heterogeneous porous media. This has motivated the recent introduction in soil science of a unified conceptual framework, fractal geometry, which at least in principles allows one to get a handle on the geometric complexity of soils and to characterize it fully with just a few numbers, the fractal dimensions.

The present chapter reports recent preliminary results of this fractal approach to the study of soil structure. Soil structures seem to have unique features that result from underlying soil aggregation processes. Our main intent is to identify some fractal properties of soil structure both (i) from published and unpublished recent results obtained by our group in Nancy and (ii) from the available literature on this topic.

The present chapter is organized as follows. In the next section, concepts of soil structure, approaches used to describe the complex geometry of heterogeneous porous media and notions of fractal geometry applied to these porous media are summarized. Then, the fractal approach to the study of each of the three soil structure elements - the soil solid space, the soil pore space and the soil solid-pore interface - is presented. We conclude with a discussion of the potential utility of the fractal description of soil structure.

II Concepts of Soil Structure

In North America, the concept of structure of a soil is often restricted to the size, shape and arrangement of its solid constituents (McKeague *et al.*, 1986). Soil structure refers to "the natural organization of soil particles into units" (Soil Survey Staff, 1981) or to "the aggregation of primary particles into compound particles, which are separated from adjoining aggregates by planes of weakness". (Day, 1983).

In Western Europe and Australia, the concept of soil structure is more general, including the three geometric elements of the porous medium: the solid space, the pore space and the solid-pore interface. Soil structure has been defined either as "the spatial arrangement of the elementary constituents and any aggregates there and of the cavities occuring in the soil" (Jongerius, 1959) or as "the physical constitution of a soil material as expressed by the size, shape and arrangement of the solid particles and voids, including both primary particles to form compound particles and the compound particles themselves" (Brewer, 1964) or simply as "the arrangement of particles and pores in soils" (Oades, 1993).

III Scale-Invariant and Probabilistic Descriptors of Heterogeneous Porous Media

When heterogeneity and interactions occur, such as in soils, it is necessary to describe the geometry of the structures studied with scale laws. Usually, the geometric property relative to one of the three structural elements of a porous medium (the solid space, the pore space and the solid-pore interface) is a symmetry, that is, an invariance under a class of geometric transformations (Adler, 1992). Two geometric symmetries can be distinguished: translational and dilational.

Translational symmetry is the classical concept following which many materials look much the same at different locations. In most cases, real materials are random, but their properties can be derived approximately from those of a finite sample. The whole medium is then reconstructed by the juxtaposition of identical samples or cells in space (spatially periodic structure: Adler, 1992).

In contrast, dilational symmetry or similarity is a new concept that was introduced independently both by Miller and Miller (1956) and by Mandelbrot (1975, 1982). Many materials look virtually identical under different magnifications (Figure 1). Classical theories and scaling theory (Miller and Miller, 1956), which address homogeneous porous materials, can be characterized by a constant porosity, independent of the size of the volume of observation, while fractal approach (Mandelbrot, 1975, 1982), which addresses heterogeneous porous materials (the word "fractal" comes from the Latin adjective "fractus", which means irregular) involves scale laws for structural properties, included porosity (deterministic or statistical self-similarity concept). The porosity (and other structural features such as tortuosity) of a fractal self-similar object depends upon the size of the volume of observation.

Compared to conventional porosity, the major advantage of this self-similar structural approach is to give a unified framework of soil structure by maintaining a continuity between observations obtained at various length scales.

Another difference between these two similar structural approaches is that scale invariants are conceptual within the scaling theory but experimental, either geometric or probabilistic, within the fractal approach. Nevertheless, the conceptual scale invariants of scaling theory have been recognized as important and useful parameters for water retention and transport modeling in soils (Youngs, 1990; Warrick, 1990; Daamen *et al.*, 1990). In contrast, the fractal similitude dimensions have mostly physical meanings (Mandelbrot, 1975, 1982; de Gennes *et al.*, 1985; Feder, 1988; Adler, 1992; Gouyet, 1992) but their use for water retention and transport modeling in soils still seems in its infancy (Ahl and Niemeyer, 1989; Tyler and Wheatcraft, 1990; Toledo *et al.*, 1990; Rieu and Sposito, 1991a,b; Hatano *et al.*, 1992).

Another way to consider heterogeneity is to assume that it is a random spatial process, governed by probability. Most real materials are random, but the most original feature of randomness is that it tends to create fractal structures; it is a remarkable property that random objects such as percolation networks at the percolation threshold are fractal (Adler, 1992). This feature largely explains

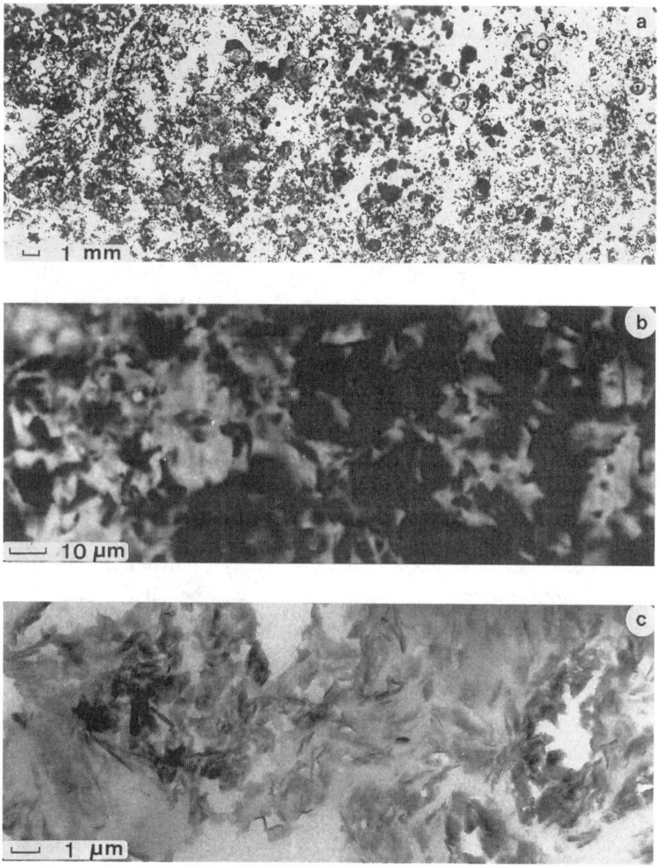

Figure 1: (a) Photograph of soil thin section, (b) backscattered S.E.M. of a small soil thin section and (c) T.E.M. soil ultra-thin section micrographs of the A1 (B) sandy acid brown soil studied by Bartoli *et al.* (1993) (average solid mass fractal dimension $D_m = 1.95$).

why the fractal concept has become an important approach for the description of the geometry of random structures produced by aggregation phenomena.

IV Notions of Fractal Geometry Applied to Heterogeneous Porous Media

Soils, and more generally porous materials, are characterized by their solid mass, m, pore mass, p, and solid-pore interface area, s. The term "mass" is used here loosely to denote the length, or the surface, or the volume of a given material, depending upon its topological dimension. In soils, the "mass" is mainly either the volume or the surface when analyses are carried out either in three or two dimensions, respectively.

As mentioned earlier, the first property of a fractal object is its ability to be described in terms of dilational self-similarity (heterogeneity is independent of scale). Following Van Damme and co-workers (Van Damme *et al.*, 1988; Van Damme and Ben Ohoud, 1990) and Pfeifer and Obert (1989), it is possible to look for simple fractal power relationships between the three bulk properties m, p and s of the material studied and the size R (square side, circle radius, aggregate or clod diameter or any yardstick at a given scale) of a region of space in which the material can be characterized. These power relationships are as follows:

$$m(R) \propto R^{D_m} \tag{1}$$

$$p(R) \propto R^{D_p} \tag{2}$$

$$s(R) \propto R^{D_s} \tag{3}$$

where D_m, D_p and D_s are the solid mass fractal, pore mass fractal and solid-pore interface fractal or surface fractal dimensions, respectively, and where the symbol $\propto$ denotes proportionality. The validity of a power law is not at all guaranteed *a priori*; power law (fractal) distribution functions may exist when mass or surface distributions have long tails that contrast with bell-shaped Gaussian functions.

When measurements are performed on a two-dimensional section such as a soil thin section, each of these D_m, D_p or D_s fractal dimensions can be obtained by adding 1 to the corresponding fractal dimension measured in the plane intersecting the porous medium (Mandelbrot, 1982; Gouyet, 1992; Adler, 1992).

If $D_m = D_p = D_s = 3$, the material is non-fractal. This is so for homogeneous regular three-dimensional porous aggregates such as silica gel (Avnir *et al.*, 1985) or kaolinite (Van Damme and Ben Ohoud, 1990).

In contrast, the dimension D of a fractal object is not an integer; for example, the surface fractal dimension of a fractal crumpled plane is between the Euclidean dimension of a plane (2 for a completely smooth surface) and that of a volume (3 for an infinitely crumpled plane). If solid mass and interface scale alike, the porous medium is called a solid mass fractal; if pore mass and interface scale alike, it is called a pore mass fractal; if only the interface is fractal, the porous medium is called a surface fractal (Van Damme *et al.*, 1988; Pfeifer and Obert, 1989; Van Damme and Ben Ohoud, 1990).

The previous fractal dimensions may not be sufficient to apprehend completely the heterogeneity of a fractal porous medium. For water transport modeling in soils, it may be interesting and necessary to characterize the connectivity properties of soil structure (spreading and connectivity dimensions, ramification, lacunarity) (Gouyet, 1992). For example, the spreading dimension d_e, which is an intrinsic connectivity property of fractal porous media, may be defined as follows (Gouyet, 1992). The number of locations $S(l)$ accessible from a reference initial site belonging to a given fractal porous medium is a function of the so-called chemical or connectivity distance l:

$$S(l) \propto l^{d_e} \tag{4}$$

This spreading dimension d_e therefore is controlled only by the connections between the elements of the fractal structure, and is correlated to the tortuosity. We have the following inequality between the propagation fractal dimension D and the spreading dimension d_e; $D \geq d_e$, with the equality obtained in the limit case where the euclidian distances are equal or proportional to the distances between points belonging to the fractal, as is the case for the Sierpinski sieves.

Because of statistical scale invariance of the structure by dilational self-similarity, the mean euclidian quadratic distance $R^2(l)$ between two sites separated by a connectivity distance l is as follows (Gouyet, 1992):

$$R^2(l) \propto l^{2/d_{\min}} \tag{5}$$

where $d_{\min}$, the connection dimension, is the trace fractal dimension of the shortest path:

$$l \propto R^{d_{\min}} \tag{6}$$

From Equations (4) and (6), one obtains:

$$d_e = D/dmin \tag{7}$$

The limit case where $d_e = D$ is only obtained if $d_{\min} = 1$, that is to say, if $R \propto l$.

V Solid Mass Space and Porosity

In soils, the "solid mass" is mainly either the volume or the surface when analyses are carried out either in three (soil aggregate or clod bulk density) or two dimensions (image analysis on thin and soil ultra-thin sections), respectively.

V.A Evaluation Based on Soil Aggregate and Clod Bulk Density

There have been converging results in the literature (*cf.* Figures 2-4; Wittmuss and Mazurak, 1958; Gumbs and Warkentin, 1976) indicating that the dry bulk density of soil aggregates or clods decreases (or their porosity increases) as a function of their mean diameter (0.1 to 2 mm and 0.4 to 15 cm diameter for aggregates and clods, respectively).

The first rational interpretation of these results was proposed by Currie (1966), who showed that the surface area of aggregates could have a substantial effect on their bulk density. He set up a model of spherical aggregates, considered the amount of pore space which was lost upon subdividing the aggregates (Figure 5), and derived an equation for the effective porosity. This equation partly fitted experimental data. The porosity exclusion principle (Currie, 1966;

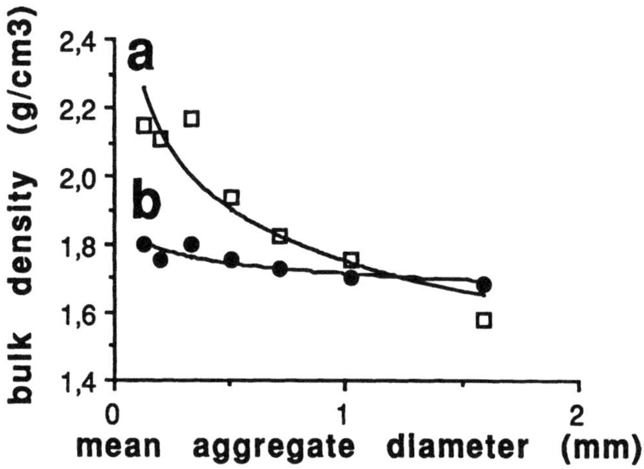

Figure 2: (a) Photogram of soil thin section, (b) backscattered S.E.M. of a small soil thin section and (c) T.E.M. soil ultra-thin section micrographs of the A1 (B) sandy acid brown soil studied by Bartoli *et al.* (1993) (average solid mass fractal dimension $D_m = 1.95$).

Dexter, 1988) of this aggregate hierarchy (Figure 5; Currie, 1966; Oades and Waters, 1992; Oades, 1993) mainly explains why smaller aggregates have the lowest porosity and the greatest contact between particles.

The aggregate hierarchical model of Currie is typically self-similar (solid mass deterministic mass fractal) and was therefore naturally reinterpreted using statistical fractal geometry by Rieu and Sposito (1991a,b), Young and Crawford (1991) and Bartoli *et al.* (1991).

The bulk density ρ_b of a porous medium is equal to the solid mass m divided by the enclosing aggregate volume V:

$$\rho_b = m/V \tag{8}$$

If the porous medium is a solid mass fractal, the solid mass m is a fractal power law function of the size R (Equation (1)). Furthermore, the enclosed aggregate volume is a function of R^3. Therefore, the bulk density ρ_b is related to the solid mass fractal dimension D_m by:

$$\rho_b(R) \propto R^{D_m}/R^3 \qquad \text{or} \qquad \rho_b(R) \propto R^{D_m - 3} \tag{9}$$

Similarly, from the classical equation $1 - m/V$, the scale law relationship between the size R and the porosity can be described as follows:

$$1 - \phi(R) \propto R^{D_m - 3} \tag{10}$$

Although some literature data (*cf.* Chepil, 1950; Nicou, 1974) show considerable scatter, they are always consistent in a statistical sense with Equations

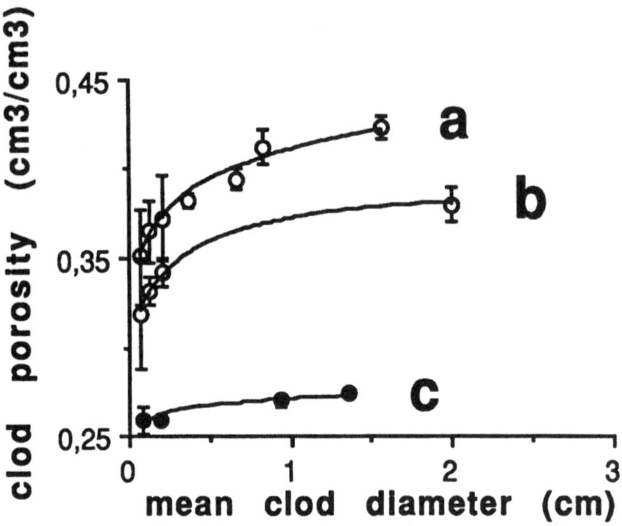

Figure 3: Porosity as a function of mean clod diameter for a silty loam (a), a silty clay loam (b) and a clay (c) soil (data from Monnier *et al.* 1973). The data were fitted using Equation (10) ($D_m = 2.94$ with $P < 0.001$, $D_m = 2.95$ with $P < 0.001$ and $D_m = 2.98$ with $P < 0.025$ for (a), (b) and (c), respectively).

(9) and (10) (*cf.* Figures 2-4; Rieu and Sposito, 1991a,b; Young and Crawford, 1991). Soils therefore are solid mass fractals within the $0.1 - 150$ mm scale range studied.

These preliminary results show that aggregates become significantly more homogeneous for clay soil aggregates than for silty ones (*cf.* Figures 2 and 3, within the $0.1 - 1.6$ and $1 - 20$ mm scale range fractal domain, respectively).

The mean clod porosity versus mean clod diameter data obtained by Monnier *et al.* (1973) in very meticulous experiments (replicated 10 times) are nicely fitted by Equation (10), within the $0.1 - 2$ cm scale range studied (Figure 3). At a given mean clod diameter, the mean dry clod porosity decreases as a function of the solid mass fractal dimension, from the silty soils to the clayey one, and the dry porosity variability, another soil structure heterogeneity parameter, also decreases as a function of the solid mass fractal dimension, from the silty soils to the clayey one. The dry porosity variability seems important within the 1-4 mm scale range but decreases thereafter. It should be attributed to a complex fragmentation process discussed elsewhere by Rieu and Sposito (1991a,b) and Crawford *et al.* (1993a).

Although few data exist on the effect of cultivation on aggregate or clod heterogeneity, it appears that the $0.2 - 2$ mm aggregates in the sandy loam soil studied by Young and Crawford (1991) became more homogeneous, as a result of cultivation, their solid mass fractal dimensions increasing fom 2.75 ± 0.10 to

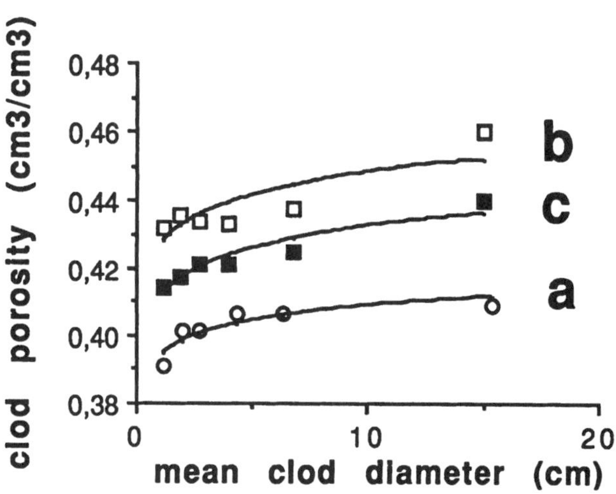

Figure 4: Porosity as a function of mean clod diameter for a tropical sandy topsoil before (a: control) and after cultivation (b: ploughing without human tillage and c: ploughing with human tillage) (data from Nicou 1974). The data were fitted using Equation (10) ($D_m = 2.98$ with $P < 0.025$, < 0.1 and < 0.01, for (a), (b) and (c), respectively).

2.95 ± 0.03. In contrast, the effect of cultivation on the heterogeneity of tropical sandy soil clods, within a larger scale range (1 – 15 cm), was not statistically significant, whereas the clod porosity significantly increased (Figure 4).

V.B Evaluation Based on the Fitted Squares Method on Images of Thin and Ultra-Thin Soil Sections

The computation of a mass fractal dimension (either a solid mass or a pore mass fractal dimension) is usually carried out using the so-called fitted squares or circles method (Kaye, 1989) on digitized binary images. The fitted squares method was recently introduced in soil science by Bartoli *et al.* (1991). Black (pore space) or white (solid space) 1×1 pixel squares were analyzed in each of these fitted squares of side length R and additions of either black (pore mass $p(R)$) or white (solid mass $m(R)$) square units were calculated. The validity of Equations (1) and (2) was checked by plotting $p(R)$ or $m(R)$ *versus* R in a log-log graph.

This fitted squares mass method is easy to use. However, it has the drawback of being affected by (i) the error on the slope of the $\log p(R)$ or $\log m(R)$ *versus* $\log R$ straight line, and (ii) the location of the center of the fitted squares. The 90% confidence interval on the slope of the $\log p(R)$ or $\log m(R)$ *versus* $\log R$ straight line decreased as a function of the solid mass fractal dimension; the more heterogeneous the solid mass, the more variable its corresponding solid

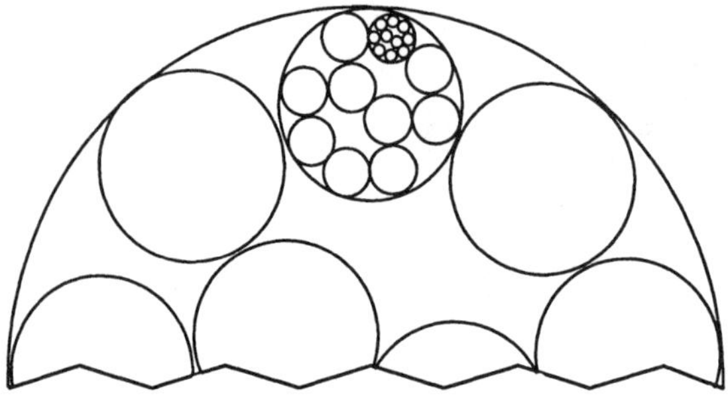

Figure 5: Illustration of Currie's (1966) concept of aggregate hierarchy.

mass fractal dimension (*cf.* Figure 6).

Computation of statistical averages of at least 4 or 5 measurements of the solid mass fractal dimension at different locations usually removed the drawback of the location of the center of the fitted squares, as it is shown in the two following examples.

An image of a thin gravel sandy soil section (Figure 7a) was divided into four subimages, which were analyzed within the $4-51.2$ mm scale range. Their planar solid mass fractal dimensions were 1.95 ± 0.02, 1.99 ± 0.04, 1.97 ± 0.02 and 1.96 ± 0.01, with an arithmetic average of the four fractal dimensions of 1.97, exactly the value obtained for the whole image. The mean porosity $\phi(R)$ *versus* square side R curve was well fitted by Equation (10) using this mean solid mass fractal dimension (Figure 7b).

Similarly, the Sierpinski sieve (Figure 7c) was randomly subdivided into four subimages. Their planar solid mass fractal dimensions were 1.89 ± 0.00, 1.93 ± 0.02, 1.90 ± 0.01 and 1.94 ± 0.08, with an arithmetic average of the four fractal dimensions of 1.91, close to the theoretical solid mass fractal dimension of the whole image (1.89). Their mean porosity $\phi(R)$ versus square side R curve was well fitted by Equation (10) using the theoretical solid mass fractal dimension, whereas a slight discrepancy occurred using the mean experimental solid mass fractal dimension (Figure 7d).

In order to miniminize this drawback of the location of the center of the fitted squares we also adopted another strategy by writing an image analysis package in which the center of the fitted squares must be a solid mass and not a pore mass. This method was used in the following sections.

We analyzed about 70 thin sandy soil sections (Figure 8a; Dutartre, 1993; Dutartre *et al.*, 1994) and 40 silty soil sections (Figure 8b) and always found that the observed microstructures were solid mass fractals within a $0.01 - 1.28$ mm scale range (backscattered electron SEM micrographs, unpublished data). At a definite square side (1.28 mm on Figure 8), porosity decreased as a function of the solid mass fractal dimension, following Equation (10): the least porous

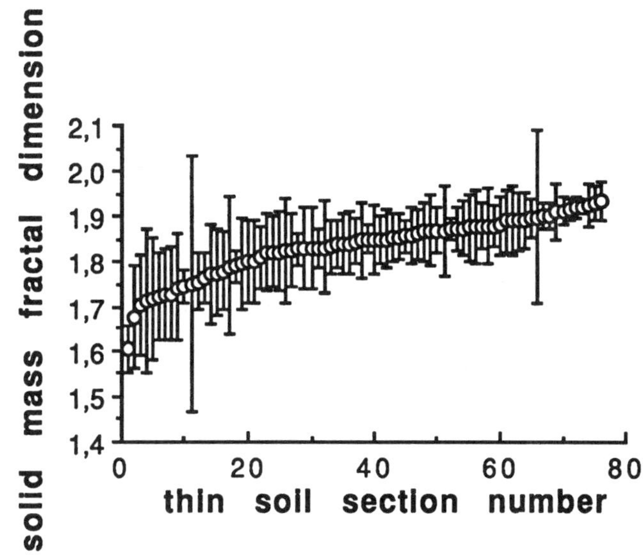

Figure 6: Ninety percent trust confidence intervals of a set of solid mass fractal dimensions D_m (evaluated from SEM micrographs of thin sections of tropical sandy topsoils) (Dutartre *et al.*, unpublished data 1993).

the structure, the least heterogeneous it was (Figure 8). The tropical sandy soil microstructures also appeared significantly more porous and heterogeneous than the temperate silty soil ones (Figure 8).

Within this 0.01 – 1.28 mm scale range, the decrease of porosity, from the upper A_p horizons to the deeper B_{tg} horizons of silty leached brown soils, was followed by an increase of soil microstructure homogeneity (Figure 9). These preliminary results therefore confirmed those showing that the 0.1 – 20 mm aggregates became significantly more homogeneous for clay soil aggregates than for silty ones (Figures 2 and 3).

Other unpublished results that we obtained demonstrated that the solid mass fractal dimension of silty topsoils, within this 0.01 – 1.28 mm scale range, seasonally increased as a function of the number of wetting-drying cycles in the topsoil (Figure 10). This decrease of structural heterogeneity was attributed to an irreversible aggregation process revealed by seasonal water retention curves and soil ultra-thin sections.

Each of the three temperate silty and sandy soil structures studied by Bartoli *et al.* (1991) was characterized by a statistically unique solid mass fractal dimension within the $10^{-9} - 10^{-1}$ m scale range studied by image analysis on thin, very-thin and soil ultra-thin sections, with different corresponding resolution powers.

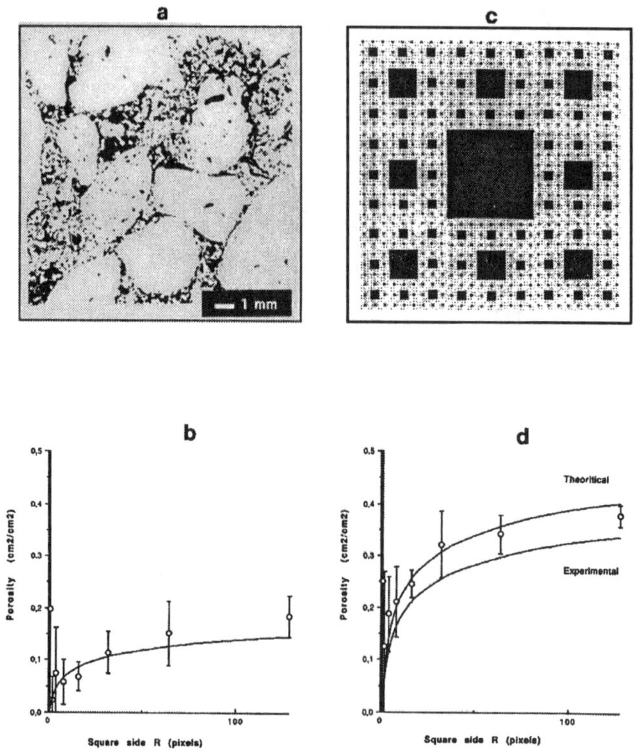

Figure 7: Mean areal porosity and its standard error ($n = 4$) as a function of square side of fitted squares for a photogram of a block of a tropical gravel sandy topsoil (a, b) and the Sierpinski sieve after 6 iterations (c, d). The fitted curves used mean D_m values - 1.97 and 1.91 for (b) and (d), respectively - and Equation (10). The Sierpinski sieve theoretical curve corresponds to $D_m = 1.89$ (Dutartre *et al.*, unpublished data 1993).

In contrast, results on a set of tropical sandy soils revealed that their structures were much more heterogeneous within the $0.01 - 1.28$ mm scale range (solid mass fractal dimension: $1.75 - 1.88$) than within the $4 - 51.2$ mm scale range (solid mass fractal dimension: $1.9 - 1.98$); in that case there were two fractal domains within a gigogne spatial structure (Serra, 1968). These multiscale sources of solid mass spatial variation may be due to the superposition of two independent effects: particle - heap (clay plasma - sandy quartz) aggregation limited by diffusion (Gouyet, 1992) for the $0.01 - 1.28$ mm scale range geometry and, possibly, macrostructuration by soil ploughing, soil drying and mesofaunal activity within the $4 - 51.2$ mm scale range.

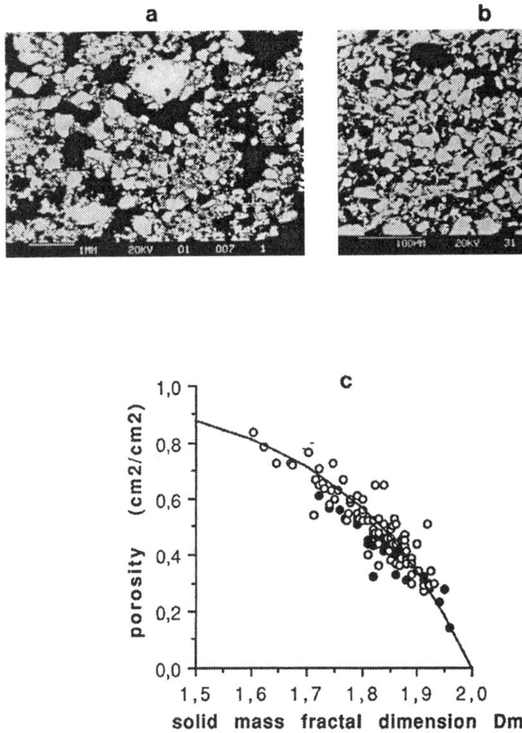

Figure 8: Porosity as a function of solid mass fractal dimension D_m at 1.28 mm square side scale (backscattered SEM micrographs of soil thin sections) for a large set of tropical sandy topsoils (a, open circles) and temperate silty topsoils (b, dark circles). The curves correspond to Equation (10) (Dutartre *et al.* 1994; Gomendy *et al.*, unpublished data 1994).

V.C Solid Mass Fractal Dimensions of Wetted Soil Structures

Rieu and Sposito (1991a,b) showed that water retention curves may be approximated reasonably well with the following analytical expression, derived from fractal fragmentation theory:

$$\theta(P) + 1 - \theta_s = (P_b/P)^{3-D_r} \tag{11}$$

In this expression, $\theta(P)$ and θ_s are the volumetric water content at matric potential P and at saturation, respectively, P_b is the air-entry or bubbling pressure in the initially water-saturated porous medium, and D_r is the bulk fractal dimension (or the solid mass fractal dimension D_m of the wetted soil structure, using the terminology of the present chapter).

The validity of Equation (11) was confirmed to be valid by Rieu and Sposito

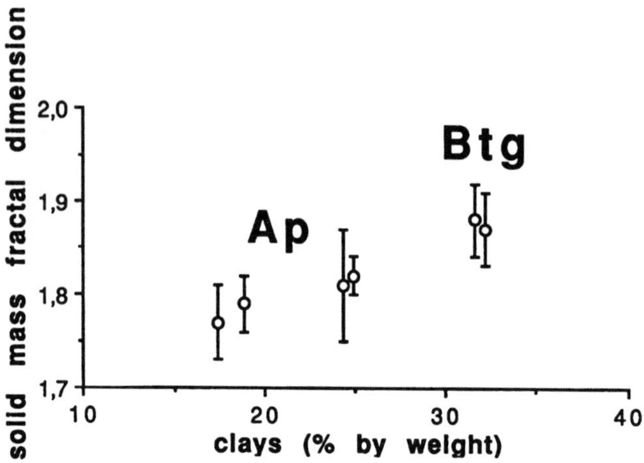

Figure 9: Solid mass fractal dimension D_m (evaluated from backscattered SEM micrographs of thin silty soil sections) as a function of the percentage by weight of clays for a set of leached brown silty soils (Ap and Btg horizons) sampled in September 1991 (Bartoli *et al.*, unpublished data 1994).

(1991b) and Perrier (1994) over the range of length scales probed by the values of matric potential for a set of sandy and silty soils. Consistently high D_m values ($D_m = 2.9 - 2.97$) were obtained, very close to the non-fractal limit (3.0).

VI Pore Mass Space and Connectivity

VI.A Pore Mass Fractal Dimensions Obtained with the Fitted-Squares Method

From all the published (Bartoli *et al.*, 1991) and, mainly, unpublished image analyses carried out by our group, it appeared that the pore mass fractal dimensions, in two dimensions, were not statistically different than 2.0 (non-pore fractal). Nevertheless, the statistical variability of the fractal dimension was often larger for the pore mass fractal dimension than for the corresponding solid mass fractal dimension (*e.g.*, Bartoli *et al.*, 1991) without, at the time of writing this paper, any rational explanation of that fact.

VI.B Pore Mass Fractal Dimensions Obtained with the Box-Counting Method

The so-called box-counting method is not center-dependent. It was recently

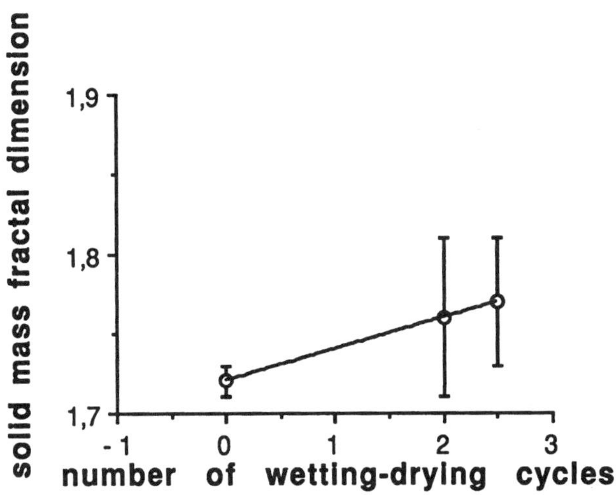

Figure 10: Solid mass fractal dimension D_m of a temperate silty topsoil (evaluated from backscattered SEM micrographs of soil thin sections) as a function of the number of *in situ* wetting-drying cycles (March-September 1991, Brie, France) (Bartoli *et al.*, unpublished data 1994).

introduced in soil science by Hatano *et al.* (1992) and Crawford *et al.* (1993b). The digitized binary image is covered with a grid consisting of square boxes of size r. The number of square boxes $N(r)$ which contain a pore (this can be a whole pore or a fraction of pore space) may follow a scaling power-law relationship:

$$N(r) \propto r^{-D_{po}} \tag{12}$$

where D_{po} is the pore space fractal dimension (D_s and D_N in the papers of Hatano *et al.* (1992) and of Crawford *et al.* (1993b), respectively).

Although Hatano *et al.* (1992) related this fractal dimension to the basic shape of the soil-pore interface, it is rather, in our opinion, a pore mass fractal dimension (Kaye, 1989; Vignes-Adler *et al.*, 1991; Adler, 1992; Muller and McCauley, 1992). In contrast, the number of square boxes needed to cover the lines of contact between black and white regions may be scale-dependent through a solid-pore interface fractal power law (Kaye, 1989; Muller and McCauley, 1992).

The experiments reported by Hatano *et al.* (1992) and Crawford *et al.* (1993b) appear to have been carefully carried out and the results, at least in one case, were analyzed using sophisticated statistical techniques (Crawford *et al.*, 1993b). Nevertheless, physical meaning of these results is uncertain. As a matter of fact, Vignes-Adler *et al.* (1991) found some systematic biases in the small and large scales of the square-box counting method. Furthermore, even though between these two limits the $\log N(r) - \log r$ slope is usually between -1 and -2, Vignes-Adler *et al.* (1991) demonstrated that, in most cases, this transition regime does not correspond to a fractal.

Hatano *et al.* (1992) studied the pore space of B horizons of andisols stained by methylene blue. This staining pattern was attributed to preferential water flow paths. These authors found that the pore space fractal dimension decreased with an increase in soil depth, ranging from 0.59 to 2.0. A surprising (in our opinion) D_{po} value of less than 1 was obtained for stains found in parts of cracks and in root channels in the deeper layers of three clay loam or silty soil columns, whereas a value greater than 1.7 was obtained for large stains found throughout the soil column (even with non-fractal homogeneity, *i.e.*, with $D_{po} = 2$, at 3.5 cm depth). The volcanic regosol had a relative small variance with an average D_{po} value of 1.48.

Crawford *et al.* (1993b) used the same square-box counting method with 200 randomly chosen origins. The heterogeneity of the pore network was greater within the angular blocky soil structure with cracking patterns ($D_{po} = 1.71$) than within the crumby one ($D_{po} = 1.94$).

By visual inspection, it seems that the crumby soil structure could be considered as a solid mass fractal structure, and not as a pore fractal. It is fascinating nonetheless to observe that the fractal dimension of the soil crack network found by Crawford *et al.* (1993b) ($D_{po} = 1.71$) was very similar to the one obtained on a controlled fractured monolayer, with similar cracking patterns, also obtained using the same box-counting method (Skeltorp and Meakin, 1988; Gouyet, 1992) ($D_{po} = 1.68 \pm 0.06$). In both cases, the underlying crack-growth process explored by these authors clearly suggests that the cracking patterns might be described in terms of the concepts of fractal geometry.

VI.C Pore Connectivity on Images of Soil Thin Sections: the Fracton or Spectral Dimension

Connectivity is an important concept when flow problems are considered. Two points of the pore space are connected if a continuous path inside the void space goes from one point to the other.

In soil science, Crawford *et al.* (1993b), using a Monte Carlo simulation, were the first to introduce the fracton or spectral dimension (Pfeifer and Obert, 1989) characterizing the connectivity of fractal porous media (see part IV of this paper and Gouyet, 1992). They have shown that the tortuosity of the angular blocky clay structure (cracking pattern) was greater than the tortuosity of the crumb structure.

VII Solid-Pore Interface

VII.A Determination of D_s from Pore-Volume Distributions

A method available to characterize indirectly the features of solid-pore interfaces is the use of a surface probe such as nitrogen (Avnir *et al.*, 1985; Van Damme and Ben Ohoud, 1990) or mercury (Friesen and Mikula, 1987; Bartoli *et al.*, 1991, 1992a,b, 1993). In these cases, the surface fractal dimension D_s is only a similitude fractal dimension because the surface probe mostly describes the surface accessible by the probe and not the real surface. Surface accessibility may also vary with the scale and the probe-type used. For example, a large pore to which access from outside the soil sample may only be through a narrow channel will be counted as a relatively small pore.

Compared to the traditional $N_2 - B.E.T.$ method to estimate the specific surface of soils, mercury porosimetry as an intraaggregate-structure probe has the disadvantage to follow soil-pore interfaces with poorer resolution (pore neck effects) but the advantage that a wider range of sizes may be probed. Compared to image analysis techniques, mercury suction and intrusion experiments have the significant advantage that they take very little time.

The surface pore structure can be characterized by a distribution, $V_p(R)$, defined as the cumulative volume of pores with constriction radii greater than R. If $V_p(R)$ follows a power-law distribution, the total surface area of pores $S(R)$ with constriction radii greater than R is given by:

$$S(R) \propto dV_p(R)/dR \tag{13}$$

If the solid-pore interface is fractal, it has features with a distribution of sizes such that its surface area depends on the length scale $l(R)$ according to:

$$S(R) \propto l(R)^{2-D_s} \tag{14}$$

where D_s is the solid-pore interface or surface fractal dimension (see part IV of this paper and Pfeifer and Obert, 1989).

For simplicity we may assume that the pores have an average length scale $l(R)$ proportional to the constriction pore radius R:

$$l(R) \propto R \tag{15}$$

and that they are connected in a hierarchical structure. Then, combining Equations (12), (13) and (14) one finds that:

$$dV_p/dR \propto -R^{(2-D_s)} \tag{16}$$

The fractal dimension of the surface can therefore be deduced from the slope of the linear log-log plot of dV_p/dR *versus* R. The interpretation of D_s as a fractal dimension restricts D_s to the range $2 < D_s < 3$. Equation (16) was also previously derived either from the Menger sponge self-similar deterministic fractal structure (Pfeifer, 1984; Friesen and Mikula, 1987) or from a topological approach (Fripiat, 1989).

In combination with the Laplace (or Washburn) equation ($P \propto R^{-1}$), Friesen and Mikula (1987) found:

$$dV_p/dP \propto -P^{(D_s-4)} \tag{17}$$

where P is the pressure applied within the mercury experiments.

One fractal domain was always distinguished from the log-log dV/dR *versus* R curves of the sets of soil aggregates we studied, confirming the validity of

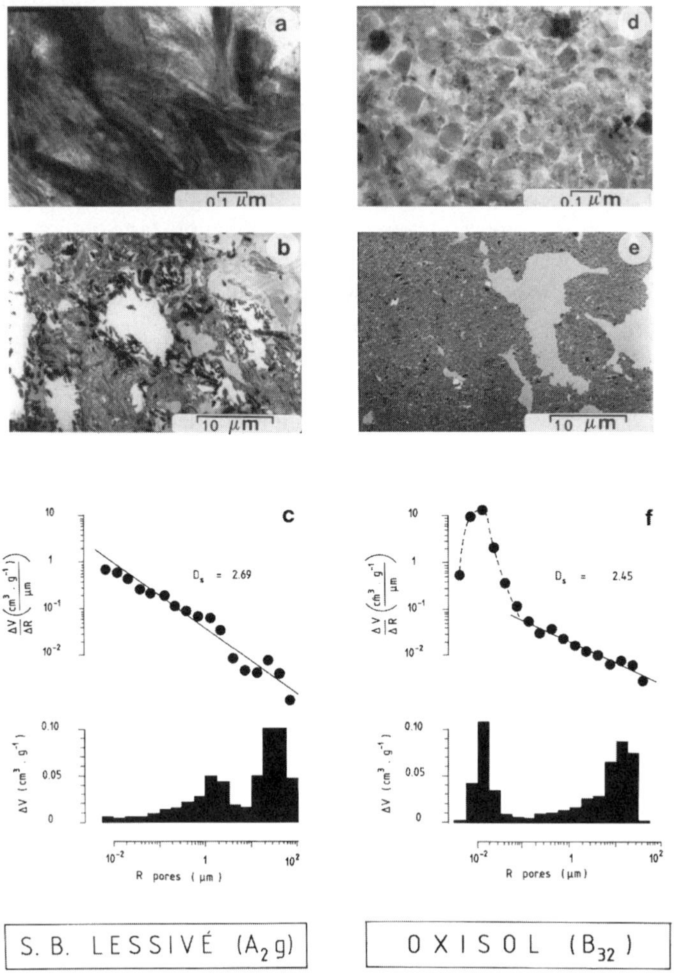

Figure 11: Surface fractal dimension D_s calculated from aggregate porosimetry using Equation (15) for a temperate silty topsoil (a, b: soil ultra-thin sections, c: mercury porosimetry) and an Oxisol B horizon rich in gibbsite (d, e: soil ultra-thin sections, f: mercury porosimetry).

Equation (16) (Figure 11c and f; Bartoli *et al.*, 1991, 1992a,b, 1993). In each soil aggregate studied, the upper limit/lower limit ratios of the aggregate fractal regime largely exceed $2^{1/D_s}$, which is the minimal condition to accept an experimental D_s value over a range of fractal self-similarity (Pfeifer and Obert, 1989). The existence of a negative relationship between D_s and the total fractal pore volume (Bartoli *et al.*, 1991, 1992a,b, 1993) confirms that D_s is an intrinsic measurement of the degree of volume filling of the different pores (Mandelbrot, 1982; Friesen and Mikula, 1987; Pfeifer and Obert, 1989).

For temperate silty and sandy soils, recently studied by Bartoli *et al.* (1991, 1992a, 1993), the focal point was the repetition of disorder at all length scales within the 0.004 – 100 μm scale range (Figure 11c), the repetition of structure within structure. The relative flexibility of the clays in these temperate soils (Figure 11a) may explain this large degree of disorder. Similarly, Wong and Howard (1986) suggested that the fractal properties of the pore surfaces of the sandstones and shales are due to the presence of fractal, flexible clay minerals on pore surfaces of these rocks; in contrast, the pore surfaces of limestones, free from clay, were smooth ($D_s = 2$) on length scales larger than about 5 nm.

In contrast, for tropical Oxisols, limits of the range of fractal self-similarity correspond to the limits of the range of inter-assemblage and inter-aggregate pores (Figure 11e and f; Bartoli *et al.* , 1992b). There exists a lower limiting characteristic length ranging from 7.8 to 250 nm, below which the aggregates cannot be described as fractals and a peak or a shoulder centered at 7-40 nm pore radius can be seen (Figure 11f; Bartoli *et al.*, 1992b). This may be attributed to the existence of card-house-type assembages of rigid kaolinite-gibbsite or kaolinite-goethite associations, as observed on TEM micrographs of soil ultra-thin sections (Figure 11d). The surface fractal dimensions D_s of these Oxisol aggregates increased as a function of organic carbon. The fractal approach suggests possible underlying aggregation mechanisms by adsorption of organic materials onto clay surfaces, forming inter-particle bridges (long-range inter-particle binding forces) (Bartoli *et al.*, 1992b).

Fractal geometry was also recently applied to the study of rhizospheric soils (Bartoli *et al.*, 1993; Amellal *et al.*, 1994). Sandy soil structure was characterized by variations in heterogeneities and fractal scaling laws along a 23 m transect under 3 fir trees (Bartoli *et al.*, 1993). The surface fractal dimension D_s was significantly smaller (larger degree of soil structure disorder) for the root zone soils than the non-root zone soils and appeared to be a better discriminator than porosity between root zone and non-root zone soils.

Wheat rhizosphere soil structures were also self-similar and it was demonstrated that inoculation of soil and wheat seeds with Pantoea dispersa decreased the structural disorder and the interface rugosity of the aggregates ($D_s = 2.47$ and 2.59 for the non-inoculated control and the inoculated root-adhering soil structures, respectively) (Amellal *et al.*, 1994).

VII.B Determination of D_s from Pore Areas and Perimeters

The so-called area-perimeter method (Lovejoy, 1982; Feder, 1988; Mandelbrot, 1982; Adler, 1992; Gouyet, 1992) was recently introduced in soil science by Hatano *et al.* (1992) and by Kampichler and Hauser (1993). This method is limited to similar non-connected objects such as the macropores studied by these authors. The surface fractal similitude dimension D_s of these objects can be found using the following power-law relationship between area, A, and perimeter, p:

$$A \propto p^{D_s/2} \tag{18}$$

In fact, this relationship is not a general one and must be used with care, as we shall now demonstrate. In the formalism introduced earlier, A is a mass, a pore mass in the case of both Hatano's and Kampichler's studies and p is the intersection of the solid-pore interface with the plane of thin section. From Equations (2) and (3) above, it is possible to write the following two scale power-law functions:

$$A(R) \propto R^{D_p} \tag{19}$$

$$p(R) \propto R^{D_s} \tag{20}$$

After reorganization of terms in these two Equations, one finds that:

$$A \propto p^{D_s/D_p} \tag{21}$$

When and only when the pore area mass is not fractal, which was checked neither by Hatano *et al.* (1992) nor by Kampichler and Hauser (1993), we have, within the two dimensions of the soil thin section:

$$A(R) \propto R^2 \tag{22}$$

$$p(R) \propto R^{D_s} \tag{23}$$

which finally lead to the area-perimeter Equation (18).

Hatano *et al.* (1992) have found that this surface fractal dimension D_s, estimated in a plane, was almost the same in each soil column studied, varying from 1.3 for a brown andisol to 1.64 for a volcanic regosol.

The nearly constant D_s value found by Kampichler and Hauser (1993) (within the range 1.26 − 1.39 for a set of different temperate soils) may correspond, in our opinion, to the fact that the studied macropores are pore mass fractals with a nearly constant D_s/D_p ratio.

VII.C Determination of D_s from Chord-Length Distributions

Intercept length distributions, either pore or solid intercepts, are well-known image analysis tools (Dehoff and Rhines, 1972; Stoyan *et al.*, 1987). They were introduced in soil science by Ringrose-Voase (1990). Intercept lengths evaluated by this author were exponentially distributed, as is expected within a Boolean model (Matheron, 1967; Stoyan *et al.*, 1987; Herman and Ohser, 1993).

For a fractal structure, because the total number of chord lengths multiplied by 4 is equal to the specific surface by volume unit S/V (Dehoff and Rhines, 1972), such a cumulative chord distribution should be scale invariant (Thompson *et al.*, 1987) and should follow a power-law distribution on a plot of the cumulative number $N(l)$ of chords of a particular length against the length l:

$$N(l) \propto l^{2-D_s}/l^3 = l^{-D_s-1} \tag{24}$$

where D_s is the solid-pore interface similitude fractal dimension (fingerprint of the structural disorder), in a two-dimensional space.

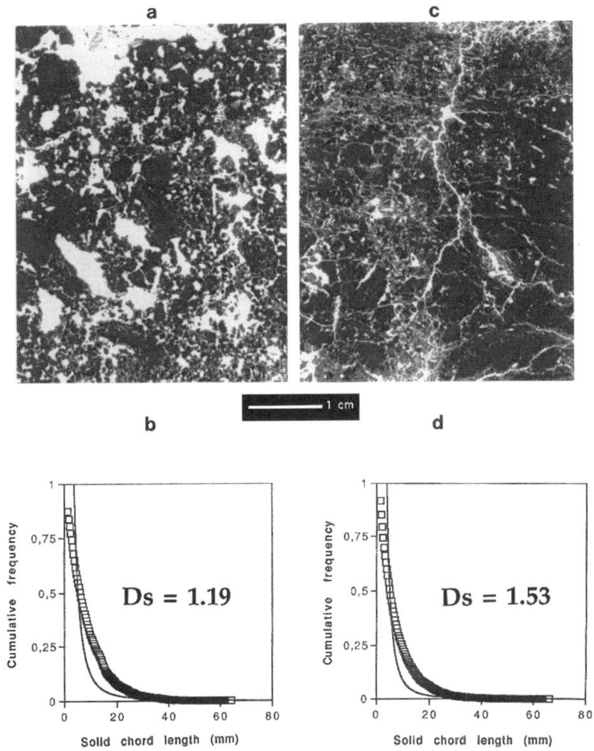

Figure 12: Cumulative solid intercept chord-length distributions, and their corresponding fitted power laws (b, d) calculated from a set of two photographs of blocks of the same silty topsoil sampled in August (a) and in September 1991 (c), at the end of the driest period. From these power law coefficients and Equation (24), $D_s = 1.19$ (b) and 1.53 (d), with $P < 0.001$ (Gomendy *et al.*, unpublished data 1994).

Preliminary unpublished results carried out in our group showed that Equation (24) is statistically well verified on a set of soil thin section images. The solid-pore interface (solid intercepts) within the same silty topsoil appeared much more heterogeneous in August (roundy isolated macropores with $D_s = 1.19$) than in September (interconnected crack patterns with $D_s = 1.53$), during the driest period (Figure 12).

Another focal point is that, at large distances, the cumulative solid and pore chord-length distributions evolve in a similar way, with the same surface fractal dimensions. Similar results were recently obtained on cements by Levitz and Tchoubar, (1992) who analyzed three types of disorder: the long-range Debye randomness, the correlated disorder and finally, such complex structures where length-scale invariance properties can be observed.

VII.D Surface Fractal Dimensions of Wetted Soil Structures

The surface fractal dimensions D_s referred to in the preceding sections were calculated from structural data obtained on dried soil. Water retention curve modeling may also be carried out, using the pore-solid interface fractal approach, recently introduced in soil science and hydrology by Ahl and Niemeyer (1989), Tyler and Wheatcraft (1990) and Toledo *et al.* (1990).

De Gennes (1985) examined two families of fractal extremes of pore-solid interface roughness: iterative pits (self-similar pits within pits) and iterative flocs (self-similar grains fused to grains). In each case, using the Laplace Equation, this author obtained the following power-law proportionality:

$$S \propto P^{D_s - 3} \tag{25}$$

or:

$$\theta(P)/\theta s \propto (Pb/P)^{3-D_s} \tag{26}$$

where $\theta(P)$ and θ_s are the volumetric water content at matric potential P and at saturation, respectively, $S = \theta(P)/\theta s$ is the water saturation ratio, P_b is the air-entry or bubbling pressure in the initially water-saturated porous medium, and D_s is the fractal dimension of the pore-solid interface empty of water.

Toledo *et al.* (1990) demonstrated that the deterministic fractal Menger sponge also agrees with de Gennes's proportionality and followed the conclusion of de Gennes: the fractal proportionality of Equation (26) applies to disordered porous media that have asperities distributed over a range of length scales (pendular water structures occupying asperities predominate). The validity of Equation (26) was effectively confirmed over the range of length scales probed by the values of matric potential imposed by Toledo *et al.* (1990) to several sandy soils.

Using a mathematical expression equivalent to Equation (17) above, Ahl and Niemeyer (1989) derived the following analytical relation:

$$V_p \propto P^{D_s - 3} \tag{27}$$

where V_p is the cumulative pore volume, P the matric potential and D_s is the surface fractal dimension. Equation (27) is equivalent, in our opinion, to Equation (26).

Ahl and Niemeyer (1989) confirmed that Equation (27) was valid for a set of three temperate soils and found that the values of D_s calculated from the water retention curves increased from the deeper, less porous horizons ($D_s = 2.1 - 2.46$) to the upper, more porous horizons ($D_s = 2.7 - 2.84$).

Both Ahl and Niemeyer (1989) and Toledo *et al.*, (1990) related the fractal soil water - matric potential relationship of Equation (26) to the mathematical one of Brooks and Corey (1964). In this context, Brooks and Corey's curve-fitting index λ is related to the pore-size distribution, $\lambda = 3 - D_s$.

Tyler and Wheatcraft (1990), combining the deterministic fractal Sierpinski carpet and the Laplace law, obtained the following analytical expression, within a two-dimensional space:

$$\theta(P)/\theta_s \propto (P_b/P)^{2-D} \tag{28}$$

which is equivalent to Equation (26), applicable within a three-dimensional space. Similarly, their fractal dimension D was also related to the Brooks and Corey λ index as follows: $\lambda = 2 - D$, in a two-dimensional space.

The fractal dimension D defined by Equation (28) is inversely related to soil texture, larger values of D corresponding to finer-textured soils (Tyler and Wheatcraft, 1990). Brakensiek and Rawls (1992) used a linear relationship between the Brooks and Corey index λ and the fractal dimension D in their analysis of the 5350 water retention curve data of Rawls *et al.* (1982). They confirmed and extended the previous preliminary results of Tyler and Wheatcraft (1990), finding a non-linear increase of the fractal dimension from 1.41 to 1.87 (2.41 to 2.87 in three-dimensional space) when the granulometry went from coarse (sandy) to fine (clayey).

In our opinion, this fractal dimension D is not a surface fractal dimension but a fractal dimension of the pore-size and pore-number distribution. Perrier (1994) recently demonstrated that, for both the deterministic fractal Menger sponge and the Sierpinski carpet, $D = D_s$. This explains why the conclusions reached by Tyler and Wheatcraft (1990) on the basis of Equation (28) coincide with those based on Equation (26) (de Gennes, 1985).

Finally, using Equation (16), Bartoli *et al.* (1992a) found that the surface fractal dimensions D_s calculated from water retention curves (wetted soil structures) were always higher than the corresponding surface fractal dimensions D_s calculated from mercury porosimetry data (dried soil structures).

VII.E Fractal Approach Applied to the Geostatistical Analysis of Images of Soil Thin Sections

Geostatistical analysis (Matheron, 1965; Warrick *et al.*, 1986), in particular the semivariogram, quantifies and summarizes the spatial variation of a soil character or property within a transect or a region. The observed values of the so-called regionalized variables are said to be realizations of a spatial random function (RF) that has a certain probability distribution and expresses both the random and structural aspects of these regionalized variables (Matheron, 1965).

Using fractional Brownian motion it is possible to further characterize such spatial structures by scale invariants: the fractal dimensions D_i (Mandelbrot and Van Ness, 1968; Mandelbrot, 1975, 1982; Burrough, 1981, 1983a,b, 1984, 1989; Amstrong, 1986; Feder, 1988; Huang and Bradford, 1992; Gouyet, 1992).

This geostatistical fractal approach was introduced in soil micromorphology by Dutartre (1993) and Dutartre *et al.* (1994) on a set of about 70 tropical sandy soil thin sections. For each image, we calculated autocorrelation functions and variograms on 2046 pixels (far the minimum 150 samples required to obtain reliable estimates of geostatistical parameters, as recently demonstrated by Webster and Oliver, 1992), perpendicular horizontal and vertical lines whose intersect is the center of either the grey-level or the binary image.

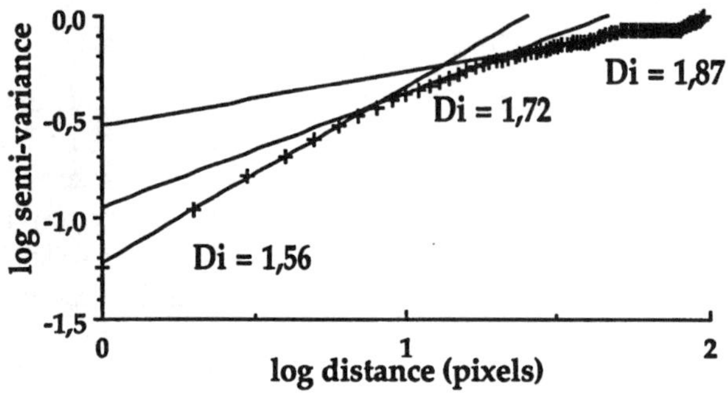

Figure 13: Log-log variogram of a horizontal transect passing through the geometrical center of a SEM micrograph of a thin section of a tropical sandy topsoil; the largest distance corresponds to 1.28 mm. The fractal dimensions D_i were calculated from the slopes of the linear parts of the graph (slope $= 4 - 2D_i$).

Each log-log variogram (slope $= 4 - 2D_i$) within a $0.01 - 1.28$ mm scale range (backscattered electron SEM micrographs, unpublished data) had a gigogne (Serra, 1968) spatial multifractal structure (Figure 13), similar to the non-Brownian fractal nested model described by Burrough (1983b). This spatial multifractal structuration occurred within a $0.1 - 0.3$ mm domain corresponding to the elementary association of some sandy quartz grains more or less coated and bridged with some clay plasma.

At very short-range distances, the fractal dimension D_i was nearly equal to 1.5, which corresponds to a pure completely random Brownian motion: the covariance is always zero and the fractional noise is identical to the Gaussian normal distribution of uncorrelated, independent variables (Mandelbrot and Wallis, 1969; Mandelbrot, 1977, 1982; Feder, 1988; Gouyet, 1992). The two other fitted fractal domains were characterized by fractal dimensions near 1.7 and 1.9 (Figure 13). These fractal dimension values higher than 1.5 imply that the increments along the solid-pore series tend to be negatively correlated with each other: positive increments tend to be followed by negative increments giving rise to short-range variations (Mandelbrot and Wallis, 1969; Mandelbrot, 1977, 1982; Burrough, 1983a; Feder, 1988; Gouyet, 1992). From the sill of the log-log variogram and thereafter, the fractal dimension D_i was 2, as expected from structures lacking spatial correlation (Gouyet, 1992) and demonstrated by Burrough (1983b) for soil clay variograms.

Within the $4 - 51.2$ mm scale range, the sill and the corresponding multifractal domain were larger than commonly in soil geostatistics (for example a 3200 m transect, sampled every 10 m, reanalyzed by Burrough, 1983a,b). In these cases, simple explanations such as changing rock and soil types can account for some of the results. The non-stationarity of the parameters and processes studied may also play a role (Burrough, 1983a,b; Warrick *et al.*, 1986).

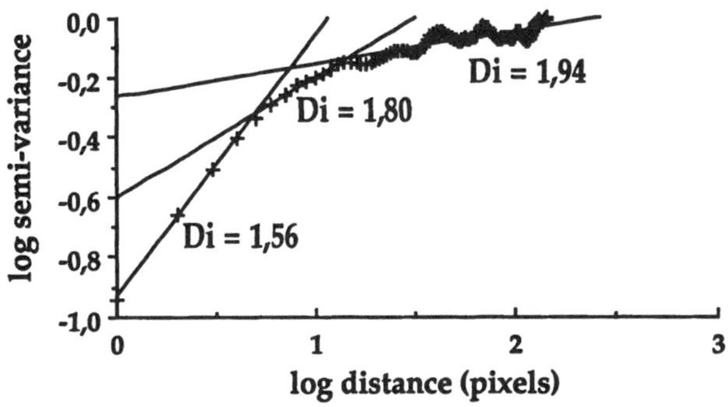

Figure 14: Log-log variogram of a horizontal transect passing through the geometrical center of a photograph of a block of a tropical sandy topsoil; the largest distance corresponds to 51.2 mm. The fractal dimensions D_i were calculated from the slopes of the linear parts of the graph (slope $= 4 - 2D_i$).

Within a $0.4-51.2$ mm scale range (macrostructures: images of soil blocks), the spatially multifractal structured domain was characterized by periodic $0.2-0.3$ mm fluctuations around a straight line corresponding to a fractal dimension of 1.94 (Figure 14). These $0.2-0.3$ m periodic fluctuations were clearly related to the elementary surface multifractal aggregate, previously revealed within a $0.01-1.28$ mm scale range (microstructures: SEM images of soil thin sections, *i.e.*, Figure 13).

Similar fluctuations in the fractal domain have been reported by Burrough (1983a) and Armtrong (1986). Burrough (1983a) argued that partial self-similarity occurred over limited ranges of scales separated by transition zones (Mandelbrot, 1977). In this context, the last fractal domain of Figure 14 could be thought of as being characterized by lacunarities, which are the divergences from the translation invariance (fractal dimension): more or less important "holes" would exist within a fractal structure (Gouyet, 1992).

VIII Conclusion

Fractal geometry may provide an avenue toward a reliable description of soil structure, particularly in the case of heterogeneous soils.

The ultimate goal of quantitative descriptions of soil structural disorders would be the incorporation of the surface fractal dimensions and/or other characteristic dimensions of the fractal soil structure, such as the spreading and the connectivity ones into mathematical models of water transport and nutrient cycling and/or microbial dynamics in the rhizosphere, at the soil-plant interface.

In our opinion, although image analysis is carried out in two dimensions, it is a usefel technique that will gain increasing importance in the future: more research is needed to characterize the connectivity of the fractal soil structure.

In particular, special attention should be devoted to cracking patterns which are a superimposed level of heterogeneity, with a probable crack hierarchy (Dexter, 1988; Oades, 1993), and were described in terms of fractal hierarchy via numerical simulation (Perrier, 1994).

In spite of the fact that fractal geometry appears promising as a tool to quantitatively describe soil heterogeneity, it is important, in our opinion, not to forget that the field expertise of soil scientists is one of the conditions for success in soil structure analysis.

Acknowledgments

The authors would like to thank B. Burtin, M. Doirisse and Mrs. R. Philippy of the soil aggregation group (CPB-CNRS) for valuable technical assistance, and Drs. E. Perrier and P. Baveye for their constructive comments and suggestions on an early draft of the present chapter.

Financial support from the CIRAD-CA (EEC STD Tropical Agronomy 1988-1992) project, the PIREN Seine project, since 1992, and the "Méthodes, modèles et théories" CNRS Environment Committee, since 1993, are greatly appreciated.

References

Adler, P.M. 1992. *Porous media. Geometry and transports.* Butterworth-Heinemann, Boston.

Ahl, C. and J. Niemeyer. 1989. The fractal dimension of the pore-volume inside soils. *Zeitschrift für Pflanzenernährung und Bodenkunde* **152**:457-458.

Amellal, N., G. Burtin, F. Bartoli and T. Heulin. 1994. Soil aggregation in the rhizosphere of wheat: effect of inoculation with Pantoea dispersa on soil adhesion and structure. p. 7-8. In S.E. Pankhurst (ed.) *Soil Biota.* Proceedings of the Symposium on Management in substainable farming systems. Adelaide Australia (March 15-18, 1994).

Amstrong, A.C. 1986. On the fractal dimensions of some transient soil properties. *J. Soil Sci.* 37:641-652.

Avnir, D., D. Farin and P. Pfeifer. 1985. Surface geometric irregularity of particulate materials: the fractal approach. *J. Coll. Interf. Sci.* 103:112-123.

Bartoli, F., R. Philippy, M. Doirisse, S. Niquet and M. Dubuit. 1991. Structure and self-similarity in silty and sandy soils: the fractal approach. *J. Soil Sci.* 42:167-185.

Bartoli, F., R. Philippy and G. Burtin. 1992a. Poorly-ordered hydrous Fe oxides, colloidal dispersion and soil aggregation. II. Modification of silty soil aggregation by addition of Fe (III) polycations and humic macromolecule models. *J. Soil Sci.* 43:59-75.

Bartoli, F., R. Philippy and G. Burtin. 1992b. Influence of organic matter in Oxisols rich in gibbsite or in goethite. I. Structures: the fractal approach. *Geoderma* 54:231-257.

Bartoli, F., G. Burtin, R. Philippy and F. Gras. 1993. Influence of fir root zone on soil structure in a 23 m forest transect: the fractal approach. *Geoderma* 56:67-85.

Brakensiek, D.L. and W.J. Rawls. 1992. Comments on "fractal processes in soil water retention" by Scott W. Tyler and Stephen W. Wheatcraft. *Water Resour. Res.* 28:601-602.

Brewer, R. 1964. *Fabric and Mineral Analysis of Soils*. John Wiley, New York.

Brooks, R.H. and A.T. Corey. 1964. Hydraulic properties of porous media. Hydrology Paper no. 3, Colorado State University, Fort Collins.

Burrough, P.A. 1981. Fractal dimensions of landscapes and other environmental data. *Nature* 294:240-242.

Burrough, P.A. 1983a. Multiscale sources of spatial variation in soil. I. The application of fractal concepts to nested levels of soil variation. *J. Soil Sci.* 34:577-597.

Burrough, P.A. 1983b. Multiscale sources of spatial variation in soil. II. A non-Brownian fractal model and its application in soil survey. *J. Soil Sci.* 34:599-620.

Burrough, P.A. 1984. The application of fractal ideas to geophysical phenomena. *Inst. Math. Applic.* 20:36-42.

Burrough, P.A. 1989. Fractals and geochemistry. p. 383-405. In D. Avnir (ed.) *The Fractal Approach to Heterogeneous Chemistry*. John Wiley and Sons, New York.

Chepil, W.S. 1950. Methods of estimating apparent density of discrete soil grains and aggregates. *Soil Sci.* 70:351-362.

Crawford, J.W., B.D. Sleemam and I.M. Young. 1993a. On the relation between number-size distributions and the fractal dimension of aggregates. *J. Soil Sci.* 44:555-565.

Crawford, J.W., K. Ritz and I.M. Young. 1993b. Quantification of fungal morphology, gaseous transport and microbial dynamics in soil: an integrated framework utilising fractal geometry. *Geoderma* 56:157-172.

Currie, J.A. 1966. The volume and porosity of soil crumbs. *J. Soil Sci.* 17:24-35.

Daamen, C.J., Z. Xiao and J.A. Robinson. 1990. Estimation of water-retention function using scaling theory and soil physical properties. *Soil Sci. Soc. Amer. J.* 54:8-13.

Day, J.H. 1983. *Manual for Describing Soils in the Field.* LRRI 82-52. Agriculture Canada, Ottawa.

de Gennes, G. 1985. Partial filling of a fractal structure by a wetting fluid. p. 227-241. In D. Adler (Eds) *Physics of Disordered Materials*. Plenum Press, New York.

Dehoff, J. and M. Rhines. 1972. *Microscopie Quantitative*. Masson, Paris.

Dexter, A.R. 1988. Advances in characterization of soil structure. *Soil Till. Res.* 11:199-238.

Dutartre, P. 1993. *Horizons de Surface de Sols Ferrugineux Tropicaux Cultivés d'Afrique de l'Ouest (Burkina Faso et Mali). Etude des Microstructures par l'Approche Fractale.* Unpublished pH. D. dissertation, Université Henri Poincaré (Nancy I), Nancy, France.

Dutartre, P., F. Bartoli, S. Nicquet, M. Doirisse and G. Bourgeon. 1994. Etude des microstructures d'horizons de surface de sols ferrugineux tropicaux par l'approche fractale. p. 141. Proceedings of the 15th réunion des Sciences de la Terre, Nancy (April 26-28 1994).

Feder, J. 1988. *Fractals.* Plenum Press, New York.

Friesen, W.I. and R.J. Mikula. 1987. Fractal dimensions of coal particles. *J. Coll. Interf. Sci.* 120:263-271.

Fripiat, J.J. 1989. Porosity and adsorption isotherms. p. 331-340. In D. Avnir (ed.), *The Fractal Approach to Heterogeneous Chemistry.* Wiley, New York.

Gouyet, J.F. 1992. *Physique et Structures Fractales.* Masson, Paris.

Gumbs, F.A. and B.P. Warkentin. 1976. Bulk density, saturation water content and rate of wetting of soil aggregates. *Soil Sci. Soc. Amer. J.* 40:28-33.

Hatano, R., N. Kaawamura, J. Ikeda and T. Sakuma. 1992. Evaluation of the effect of morphological features of flow paths on solute transport by using fractal dimensions of methylene blue staining pattern. *Geoderma* 53:31-44.

Hermann, H. and J. Ohser. 1992. Determination of microstructural parameters of random spatial fractals by measuring chord length distributions. *J. Microscopy* 170:87-93.

Huang, C. and J.M. Bradford. 1992. Applications of a laser scanner to quantify soil microtopography. *Soil Sci. Soc. Amer. J.* 56:14-21.

Jongerius, A. 1959. The morphological soil structure classification of the Dutch soil survey institute. *Rijksfaculteit Landbouw Wetenschappen Mededelingen* 24:206-213.

Kampichler, C. and M. Hauser. 1993. Roughness of soil pore surface and its effects on available habitat of microarthropods. *Geoderma* 56:223-232.

Kaye, B.H. 1989. Image analysis techniques for characterizing fractal structures. p. 55-66. In D. Avnir (ed.) *The Fractal Approach to Heterogeneous Chemistry.* Wiley, New York.

Levitz, P. and D. Tchoubar. 1992. Disordered porous solids: from chord distributions to small angle scattering. *J. Physique I* 2:771-790.

Lovejoy, S. 1982. Area-perimeter relation for rain and cloud areas. *Science* 216:185-187.

Mandelbrot, B. 1975. *Les Objets Fractals: Forme, Hasard et Dimension.* Flammarion, Paris.

Mandelbrot, B. 1982. *The Fractal Geometry of Nature.* Freeman, San Fransisco.

Mandelbrot B.B. and J.W. Van Ness. 1968. Fractional Brownian motions, fractional noises and applications. *J. SIAM.* 10:422-437.

Mandelbrot, B.B. and J.R. Wallis. 1969. Computer experiments with fractional Gaussian Noises. Part 2. Rescaled ranges and spectra. *Water Resour. Res.* 5:228-241.

Matheron, G. 1965. *Les Variables Régionalises et leur Estimation. Une Application de la Théorie des Fonctions Aléatoires aux Sciences de la Nature.* Masson, Paris.

Matheron, G. 1967. Elments pour une thorie des milieux poreux. Masson, Paris.

McKeague, J.A., C. Wang and G.M. Coen. 1986. *Describing and Interpreting the Macrostructure of Mineral Soils: a Preliminary Report.* Research Branch Agriculture Canada, Ottawa.

Miller, E.E. and R.D. Miller. 1956. Physical theory for capillary flow phenomena. *J. Appl. Phys.* 27:324-332.

Monnier, G., P. Stengel and J.C. Fis. 1973. Une méthode de mesure de la densité apparente de petits agglomérats terreux. Application à l'analyse des systèmes de porosité du sol. *Ann. agron.* 24:533-545.

Muller, J. and J.L. McCauley. 1992. Implication of fractal geometry for fluid flow properties of sedimentary rocks. *Transp. Porous Media* 8:133-147.

Nicou, R. 1974. Contribution l'étude et l'amélioration de la porosité des sols sableux et sablo-argileux de la zone tropicale sèche. Conséquences agronomiques. *L'Agronomie Tropicale* XXIX:1100-1125.

Oades, J.M. 1993. The role of biology in the formation, stabilization and degradation of soil structure. *Geoderma* 56:377-400.

Oades, J.M. and A.G. Waters. 1991. Aggregate hierarchy in soils. *Aust. J. Soil Res.* 29:815-824.

Perrier, E. 1994. *Structure Géométrique et Fonctionnement Hydrique des Sols. Simulations Exploratoires.* Unpublished Ph. D. dissertation, Université Pierre et Marie Curie (Paris VI), Paris, France.

Pfeifer, P. 1984. Fractal dimension as working tool for surface-roughness problems. *Applied Surface Science* 18:146-164.

Pfeifer, P. and M. Obert. 1989. Fractals: basic concepts and terminology. p. 11-43. In D. Avnir (ED.) *The Fractal Approach to Heterogeneous Chemistry.* Wiley and Sons, New York.

Rawls, W.J., D.L. Brakensiek and K.E. Saxton. 1982. Estimation of soil water properties. *Trans. ASAE* 1316-1320, 1328.

Rieu, M. and G. Sposito. 1991a. Fractal fragmentation, soil porosity, and soil water properties: I. Theory. *Soil Sci. Soc. Amer. J.* 55:1231-1238.

Rieu, M. and G. Sposito. 1991b. Fractal fragmentation, soil porosity, and soil water properties: II. Applications. *Soil Sci. Soc. Amer. J.* 55:1239-1244.

Ringrose-Voase, A.J. 1990. One-dimensional image analysis of soil structure. I. Principles. *J. Soil Sci.* 41:499-512.

Serra, J. 1968. Les structures gigognes: morphologie mathmatique et interprtation mtallognique. *Mineralium Deposita* 3:135-154.

Skjeltorp, A.T. and P. Meakin. 1988. Fracture in microsphere monolayers studied by experiment and computer simulation. *Nature*, 335:424-426.

Soil Survey Staff 1981. *Revised U.S. Soil Survey Manual Draft.* U.S.D.A., Washington.

Stoyan, D., W.S. Kendall and J. Mecke. 1987. *Stochastic Geometry and its Applications.* John Wiley and Sons, New York.

Thompson, A.H., A.J. Katz and C.E. Krohn. 1987. The microgeometry and transport properties of sedimentary rock. *Adv. Phys.* 36:625-694.

Toledo, P.G., R.A. Novy, H.D. Davis and L.E. Scriven. 1990. Hydraulic conductivity of porous media at low water content. *Soil Sci. Soc. Amer. J.* 54:673-679.

Tyler, S.W. and S.W. Wheatcraft. 1990. Fractal processes in soil water retention. *Water Resour. Res.* 26:1047-1054.

Van Damme, H., P. Levitz, L. Gatineau, J.F. Alcover and J.J. Fripiat. 1988. On the determination of the surface fractal dimension of powders by granulometric analysis. *J. Coll. Interf. Sci.* 122:1-8.

Van Damme, H. and M. Ben Ohoud. 1990. From flow to fracture and fragmentation in colloidal media. II. Local order and fragmentaion geometry. p. 105-116. In J.C. Charmet, S. Roux and E. Guillin (eds.) *Disorder and Fracture.* Plenum Press, New York.

Vignes-Adler, M., A. Le Page and P.M. Adler. 1992. Fractal analysis of fracturing in two African regions, from satellite imagery to ground scale. *Tectonophysics* 196:69-86.

Warrick, A.W. 1990. Application of scaling to the characterization of spatial variability in soils. p. 39-51. In: D. Hillel and D.E. Elrick (eds.) *Scaling in Soil Physics: Principles and Applications.* SSSA Special Publ. 25, Soil Science Society of America, Madison.

Warrick, A.W., D.E. Myers and D.R. Nielsen. 1986. Geostatistical methods applied to Soil Science. p. 53-82. In A. KLute (ed.) *Methods of Soil Analysis, Part 1. Physical and Mineralogical Methods.* Agronomy Monograph 9 (2nd edition). Soil Science Society of America, Madison, Wisconsin.

Webster, R. and M.A. Oliver. 1992. Sample adequately to estimate variograms of soil properties. *J. Soil Sci.* 43:177-192.

Wittmuss, H.D. and A.P. Mazurak. 1958. Physical and chemical properties of soil aggregates in a Brunizem soil. *Soil Sci. Soc. Amer. J.* 22:1-5.

Wong, P. and J. Howard. 1986. Surface roughening and the fractal nature of rocks. *Phys. Rev. Lett.* 57:637-640.

Young, I.M and J.W. Crawford. 1991. The fractal structure of soil aggregates: its measurement and interpretation. *J. Soil Sci.* 42:187-192.

Youngs, E.G. 1990. Application of scaling to soil-water movement considering hysteresis. p. 23-37. In D. Hillel and D.E. Elrick (eds.), *Scaling in Soil Physics: Principles and Applications.* SSSA Special Publ. 25, Soil Science Society of America, Madison, Wisconsin.

The Interactions Between Soil Structure and Microbial Dynamics

J.W. Crawford and I.M. Young

Contents

I Introduction

As soil scientists, we now have a healthy respect of the pervasive nature of soil structure in mediating all soil processes. However, until recent times the complexity of structure-process interactions had led to a position of ignoring heterogeneity in favor of simpler homogenized systems, despite the recognition of the spatial-temporal variability in soil structure. The lack of any available theory which dealt adequately with the inherent structural heterogeneity, coupled with the difficulties in reproducing heterogeneous structure in the laboratory, further reinforced this position. The link between microbial processes and structure are intimately tied to the small-scale heterogeneity of the pore space, moisture conditions, and available substrate. The aquatic nature of microbes necessitate a theoretical framework that accounts for structure-moisture interactions that impact directly onto their metabolic activity, motion, and reproduction and how such actions impact on chemical cycling within the soil-plant system. The general relations between structure and microbes tend to equate the presence or absence of particular sizes of microorganisms with the presence or absence of saturated pore sizes: pores $< 20\mu$m constitute a barrier to movement for nematodes (Wallace, 1958); protozoa with a smaller body size may be inhibited where there are no water-filled pores $> 3\mu$m (Kuikman *et al.*, 1989). However, such general considerations of soil structure oversimplify the relation between pores and microbes and rarely provide a true representation of the primary mechanisms operating on the small scale: how motion and predation may be examined in terms of structural heterogeneity; how water-films

ISBN 1-56670-105-8

impede oxygen diffusion and affect motion; or how the spatial distribution of decomposable organic matter allied to ambient moisture and structure conditions impact on denitrification, root growth, etc. Additionally, while structure acts as a mediator of microbial transport and predation, microbial populations, particularly bacteria and fungi, play a major role in the genesis of that structure through exudation of polysaccharides, physical binding of soil particles, and decomposition of organic fractions which stabilize soil particles. Such a two-way interaction serves to highlight the dynamic relation between structure and microbes, a relation that perhaps is unique in soil science.

The functional quantification of structural-microbial variation in soil is the ultimate goal. Fractals offer a means by which such variability can be incorporated into theoretical frameworks that will permit structural diversity to be incorporated into models, and will also serve as a pointer to understanding the small-scale processes in soil that underpin the present and future sustainability of the soil resource. The use of fractals in microbial dynamics is in its infancy, not the least because it necessarily deals with heterogeneity of structure, moisture, gas, and substrate. The aim of this review is therefore to examine the present state of affairs and attempt to direct research into the most fruitful areas, pointing out the benefits and limitations of fractals in the context of microbial dynamics.

II Water and Gas Relations

Water, oxygen, structure and substrate are all important factors regulating the magnitude and nature of microbial activity in soil (Greenwood, 1961; Leffelaar, 1993). However, the relative importance of each is difficult to quantify due to their interrelatedness. Because of this, a number of theoretical approaches have been devised (Newman and Watson, 1977; Leffelaar, 1988; Arah and Smith, 1989; Priesack, 1991; Bazin and Prosser, 1992). Early studies of water and gas movement in soil effectively treated soil structure as a homogeneous medium (Penman, 1940; Marshall and Holmes, 1988, p. 81-92). More recently, structure is taken account of through the definition of (usually one or two) characteristic scales. This may involve a description of structure in terms of two phases: a macropore phase where fluid moves readily; and a micropore phase of lower permeability (*e.g.*, Passioura, 1971; Youngs and Leeds-Harrison, 1990, Chen and Wagenet, 1992) with exchange between the two phases described by a transfer parameter. Alternatively, it may be assumed that it is possible to define spatially averaged physical properties above some characteristic scale, and subsequently to derive equations describing the flow at such scales. These approaches have serious limitations when used in the context of understanding soil structure/microbe interactions - both in principle and in practice. Models which assume that structure exists across only a few characteristic scales (Arah and Smith, 1989; Priesack, 1991) cannot account for the broad range of pore and particle sizes evident in moisture-release curves and particle number-size distributions. This breadth in the scale of heterogeneity in soil means that transport processes will occur at rates which vary continuously from large to small scales. Thus, any assumption of homogeneity applied between the characteristic scales will employ constant rate coefficients which may over- or underestimate the rates at the scale of individual microbes or microbial communities. Furthermore, the measured values of these rate coefficients will depend on the scale at which they are measured. These same criticisms can be applied to the principle of the representative elementary volume (REV), as well as the comment that the concept is unsuitable for application to systems which interact across a range of scales, such as the soil-microbe system, where the heterogeneous distribution of microbes may be characterized by a different REV to that of hydraulic flow or gaseous diffusion (Baveye and Sposito, 1984). In order to study the significance of the simplifying assumptions regarding soil structure, as well as to develop a theory of water and gas movement at scales of relevance to microbes, which is also consistent with the broad range of scales in structural heterogeneity, we must abandon the concept of characteristic scale. Since fractals are heterogeneous structures with no characteristic scale, they potentially represent a powerful means of capturing the variability in soil structure, and providing the basis of a theoretical framework for quantifying the potential sensitivity of microbial dynamics to soil structure. In reality of course, the approximation of soil structure by a fractal will only be valid between lower and upper spatial scales, and this will be discussed further below and in Section IV. In practice therefore, the use of fractals does not circumvent the introduction of characteristic scales. The significant difference which arises from adopting the approach, is that the characteristic scale may be incorporated into the theoretical considerations as a natural consequence of the underlying hypothesized mechanisms, rather than as some assumption imposed on the framework *a priori* (see Section IV).

Structure may impose either contraints or advantages to the diffusion of oxygen and other gases through soil (Crawford *et al.*, 1993a). Such impositions have major influences over microbial processes which crucially depend on the adequate diffusion of gases to and from microbial sites (Focht, 1992). The spatial and temporal heterogeneity of pores may dramatically alter diffusive pathways and hence diffusion rates. To understand how diffusion in a fractal pore space differs from that in a homogeneous medium we begin with a consideration of the general properties of a self-similar random walk. The mean displacement of a particle assuming such a walk increases with the time interval according to

$$|x(t_2) - x(t_1)| \propto |t_2 - t_1|^H \tag{1}$$

where $x(t)$ is the position of the particle at time t, and H is a constant satisfying $0.5 < H < 1$ (Peitgen and Saupe, 1988) and $\propto$ denotes proportionality. We can interpret the significance of the value of H using the fact that the length of the particle trajectory, $L(t) \propto t$, and using Equation (1) to relate time to distance travelled, R, to give,

$$L(R) \propto R^{1/H} \propto R^{D_w} \tag{2}$$

where D_w is the fractal dimension of the particle trajectory. Thus, H is the inverse of the dimension of the particle trajectory, and as such it measures the space-filling properties of the walk. A value of $H = 0.5$ in Equation (1) yields the relation between root-mean squared displacement and time corresponding to classical Brownian motion, and the corresponding dimension of the motion is $Dw = 2$, reflecting the property that such a walk on a two-dimensional surface visits all points with probability 1. Larger values of H correspond to trajectories which are progressively less space-filling. If we assume that a self-similar fractal pore space preserves the self-similarity in the trajectories of diffusing particles, constraining the motion only via the value of H, we can calculate H in terms of structural parameters characterizing the pore space (Orbach, 1986; Haus and Kehr, 1987; Crawford *et al.*, 1993a). If S_N is the number of distinct sites visited in the random walk after N equal time steps, then since the trajectory would be expected to fill the pore volume, $S_N \propto R^{D_p}$, where D_p is the fractal dimension of the pore space. However, since $N \propto R^{D_w} \Longrightarrow R \propto N^{1/D_w}$, we can rewrite the expression of S_N in the form,

$$S_N \propto N^{D_p/D_w} \propto N^{\bar{\bar{d}}/2}, \qquad \bar{\bar{d}} = 2D_p/D_w \tag{3}$$

where $\bar{\bar{d}}$ is the spectral (or fracton) dimension of the pore space. Using Equation (1) we can now write

$$\langle R^2 \rangle \propto t^{2H} \propto t^{\bar{\bar{d}}/D_p} \tag{4}$$

corresponding to a time-dependent diffusion coefficient $D(t)$ given by $t^{(\bar{\bar{d}}/D_p)-1}$. Because $R^2 \propto t^{\bar{\bar{d}}/D_p}$, we can write the diffusion equation governing the motion in a fractal pore space in the form

$$\frac{\partial C}{\partial t} = \nabla \cdot (D(r)\nabla C) \qquad where \qquad D(r) \propto r^{-2((D_p/\bar{\bar{d}})-1))} \tag{5}$$

In this equation, $C(r,t)$ is the concentration of diffusing particles at a distance r from the source at time t. Therefore, the fractal properties of the pore space translate into a scale dependency in the diffusion coefficient in the form of a power law where the exponents depend on structural parameters of the pore space. Equation (3) can be used to determine the value of $\bar{\bar{d}}$ either analytically for certain non-random structures, or by Monte Carlo simulation for more complex random fractals (Haus and Kehr, 1987; Crawford *et al.*, 1993a). From Equation (4), $\bar{\bar{d}}$ can be interpreted as a tortuosity factor since it measures the impedance of the forward progression of a randomly diffusing particle: if $\bar{\bar{d}} < D_p$, the trajectory is forced back on itself by the obstructing structure, and the outward flow is reduced compared to diffusion in free space. In a sense, D_p and $\bar{\bar{d}}$ compete in their influence on diffusive flux, and this is reflected by their appearing as a ratio in Equation (5). A fractal pore space becomes increasingly rarefied at larger and larger length scales since porosity, $\rho(R)$, scales as $\rho(R) \propto R^{D_p-2}$. Therefore, the diffusate becomes progressively concentrated as it moves out by virtue of the self-similarity of the pore space. For $\bar{\bar{d}}$ increasingly smaller than D_p, the increased tortuosity counteracts this effect and the flow is sub-diffusive in the sense that $H < 0.5$. Structure can therefore have a

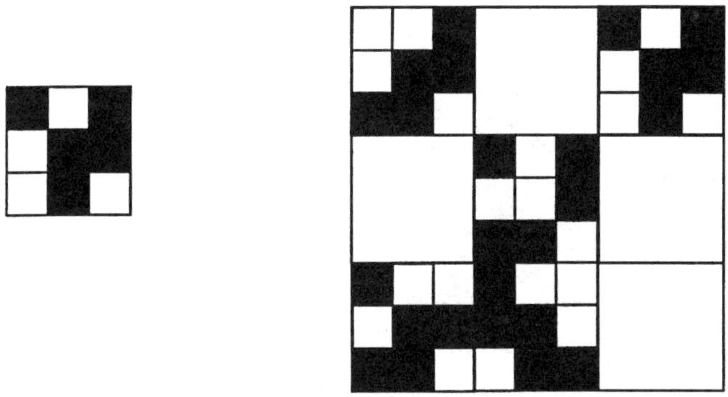

Figure 1: First two stages in the construction of a two-dimensional random recursive lattice with $m = 3$, $P = 5/9$ and $D_m = 1.46$.

very important influence on the flow rate of oxygen to microbes, and measurement of diffusion rates made at a single scale cannot lead to accurate prediction of oxygen availability. The above construction assumes that the pore volume available for free oxygen diffusion is static in time. Water films are important both for the activity and movement of microbes, as well as the connectivity of the air-filled volume in soil. One might therefore expect a variable moisture content, or more precisely, variability in the distribution of water, in soil to have an important bearing on oxygen supply; water is the major factor in determining anoxia in soil, as the diffusion rate of oxygen in water is 10^4 slower than in air (Focht, 1992). Furthermore, as discussed previously, water movement is an important vector of microbes in soil and therefore plays a vital role not only in microbial dynamics per se but also in the fate of introduced genetically manipulated organisms (Gammack *et al.*, 1992; Rattray *et al.*, 1992). It is also the means by which nitrate is transported to anoxic zones where subsequent denitrifying processes remove it from the soil-plant system. Clearly the distribution of water at microbial scales and above is crucial for the regulation of many microbial processes of ecological and agro-ecological significance. The application of fractals to water movement is discussed elsewhere in this volume where the sensitivity to structure is apparent. An independent approach (Crawford, 1994) is outlined here because of its development in the context of microbial activity and denitrification, and extension in later sections to a study of the possible influence of microbial processes on aggregate build-up

The construction is based on a generalization of the random recursive lattices developed by Clerc *et al.* (1990). As mentioned above, soil structure may only be validly approximated by a fractal between lower and upper length scales, and this approach takes this into account. We begin with a cubic lattice of unit side comprising M^3cubic cells of side $1/M$, as indicated in Figure 1, where for diagrammatic simplicity, a two-dimensional representation is shown. With probability P, the cells are filled with structure that may or may not be porous

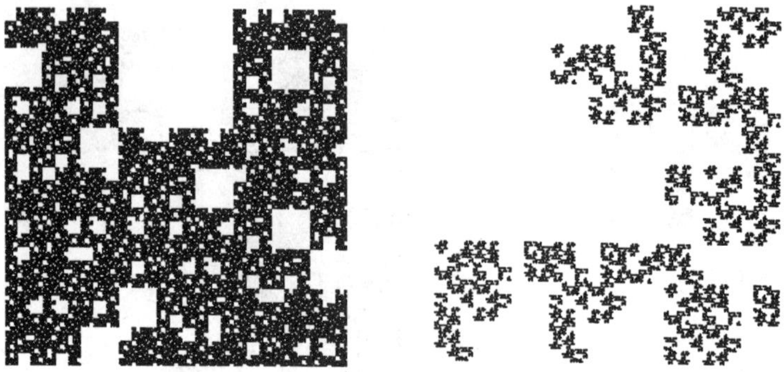

Figure 2: (a) Example of a two-dimensional lattice at the $n = 5$ iteration with $P = 8/9$, $m = 3$, $D_m = 1.89$. (b) Same as (a) except $P = 1/3$ and $D_m = 1.63$.

since at length scales below $1/M$ we assume that the structure is no longer fractal. At the next stage of the construction, a cubic lattice of side M is built from M^3 cells of unit side, and with probability P the cells are filled with random copies of the lattice generated in the previous iterate (Figure 1). At the nth stage of the iteration, the lattice has side M^n, and between this length scale and the scale of the smallest aggregated unit ($1/M$) the structure shares the properties of a fractal having dimension

$$D_m = \frac{\ln(PM^3)}{\ln M} \tag{6}$$

where D_m refers to the mass distribution of the solid matrix. Two examples with $D_m = 1.63$ and $D_m = 1.89$ are shown in Figure 2a,b (left and right, respectively). In order to proceed with an analytical treatment of the movement and distribution of water in a fractal aggregate, we approximate the pore space as a network of conductors whose conductivity is in proportion to the pore radius squared (Marshall and Holmes, 1988, p.84). At the nth iteration, the empty cells are assumed to have a resistance to flow of r_n, and the occupied cells have resistance, R_n, given by the resistance of the previous iterate. A simplification is used to approximate the complex connectivities in the network in order to derive the scaling conditions. At a particular iteration, the resistances in each cell layer perpendicular to the applied potential gradient are assumed to add in parallel, while the resistances of each layer add in series to give the total resistance of the lattice. This approximation does not capture the sensitivity of the conductivity of the lattice to the situation where unoccupied cells in a given iterate align in the direction of the flow. However, the method is sufficiently robust to discriminate between the situation where this is possible (low porosity case), and where it is not (high porosity case), and the subsequent derivation of the scaling law in terms of structure is independent of the approximation. With the above approximation, the impedance ratio $h_n = R_n/r_n$ at the nth iteration is related to the ratio at the (n-1)th iteration by

$$h_{n+1} = T_j(h_n); \qquad T_j(h_n) = \frac{r_n}{r_{n+1}} \sum_{i=0}^{M^2} a_{i,j} h_n (i + (M^2 - i) h_n)^{-1} \quad (7)$$

where $a_{i,j}$ is the number of cell layers comprising i occupied cells of resistance R_n. The T_j represent the j possible arrangements of the cells in the lattice which result in distinct lattice impedances. Iterating Equation (7) $n+1$ times from $n = 0$ we obtain

$$h_{n+1} = F_n(h_0) = T_{j_n} \circ T_{j_{n-1}} \circ \ldots \circ T_{j_0}(h_0) \quad (8)$$

where $j_r \in \{1, 2, ..., j\}$, $r = 0, 1, ..., n$, $T_j \circ T_k(h_k)$ denotes $T_j(T_k(h_k))$, and h_0 is the impedance ratio derived from the ratio of the impedance of the occupied cells to that of the pores in the $n = 0$, or primary, element. It can be shown (Crawford, 1994) that there exist special values for h_0, namely $h_0 = 0$ and $h_0 \to \infty$, where $h_{n+1} = h_0$ for all values of n. For these values of h_0, denoted h_*, the impedance ratio is independent of n, and since the length of a side of the lattice is given by M^n, this is equivalent to saying that the impedance ratio is scale invariant. The value $h_* = 0$ corresponds to the case where the resistance to flow of the pore space is effectively infinite, such as when pores are air-filled. In comparison, the case $h_* \to \infty$ corresponds to the situation where the pores are water-filled, and the impedance to flow is effectively zero. In practice, we are interested in the slightly less extreme cases where the impedance to the flow through the pore space is either much larger, or much smaller than the impedance to flow through the porous matrix in the primary elements. For such cases, the impedance ratio is no longer independent of scale and it is possible to calculate the scale dependency (Crawford, 1994).

Over the range in length scales where soil can be approximated by a fractal dimension the porosity depends on scale in a manner which is a function of the fractal dimension (Young and Crawford, 1991a). Since P determines both the porosity of the primary element, $\rho_p = 1 - P$, as well as (for fixed M) the fractal dimension of the matrix, the porosity of the primary element is a useful characterization of the porosity of the matrix at any scale. If $\rho_{air}(L)$ is the air-filled porosity of the lattice of side L, and L_a is the diameter of the smallest air-filled pore, then

$$\rho_{air}(L) = (1 - P(\frac{L}{L_a})^{D_m - 3}) \quad (9)$$

Saturated hydraulic flow corresponds to the case where $h_* \to \infty$, and the scaling properties are critically sensitive to the value of ρ_p. For the low porosity case, $\rho_p < \rho_c$, where ρ_c is the critical porosity ($\rho_c = M^{-2}$), we find that the saturated conductivity, $K_{sat}(L)$, scales according to

$$K_{sat}(L) \propto L^{2 - \Re(3 + 2(D_m/\bar{\bar{d}}) - D_m)} \quad (10)$$

where $\bar{\bar{d}}$ is the spectral dimension of the solid matrix, and $\Re$ is a constant dependent on structure. This sensitivity of the saturated conductivity to the connectivity of the matrix (expressed in terms of $\bar{\bar{d}}$) is not accounted for explicitly in other applications of fractals to hydraulic properties of soil (Tyler and Wheatcraft, 1990; Rieu and Sposito, 1991) and is clearly important for studies of the interaction of microbes and soil physical properties (Vandevivere and Baveye, 1992). For high porosity ($\rho_p > \rho_c$), the relation between conductivity and scale is complicated by the fact that there is a probability that increases with porosity, that bypass flow dominates the conductivity at any particular scale. This depends on whether large scale pores connect opposite sides of the lattice at a given scale. For $\rho_p < \rho_c$ this cannot happen.

The unsaturated conductivity, K_{unsat}, scales with L according to

$$K_{unsat}(L) \propto L^{D_m - 2(D_m/\bar{\bar{d}}) - 1} \tag{11}$$

for all values of ρ_p at a given suction (Crawford, 1994). This insensitivity of the unsaturated conductivity to porosity arises because at a given suction, the air-filled pores do not contribute to the flow, and the conductivity is governed by the connectivity and heterogeneity of the water-filled matrix, described by $\bar{\bar{d}}$ and D_m, respectively. This inherent heterogeneity in the flow path emphasizes the likely complexity in the relation between structure and passive transport of microbes in soil. Clearly, the water-filled matrix heterogeneously fills the soil volume in a manner which reflects the fractal dimension of the solid matrix. Furthermore, only the connected fraction of the water-filled matrix will contribute to the flow, giving rise to a rather heterogeneous transport of microbes down the soil profile. Therefore, measurements of transport rates of microbes through restructured or homogenized soil are likely to underestimate the rate of transport in real soil due to neglect of this factor. Additional support for this conjecture is the observation by Van Elsas *et al.* (1991) that preferential pathways generated by roots in otherwise homogenized soil increase the rate of downward movement of a bacterium. Measurements made on undisturbed cores are unlikely to circumvent the problem completely, since extrapolation to different spatial scales will be complicated because the rates of transport are unlikely to have trivial scaling properties.

The sensitive dependence of water and gas flow on structure exemplifies the importance of structure for microbial activity. It is clear that the distribution of water and gas in soil will be highly heterogeneous and that this will lead to a heterogeneous distribution of microbial activity. This is important for our understanding of soil processes in a number of ways. First, bulk measurements of respiration will scale in a manner which is a complex function of soil structure, and local rates of respiration are likely to greatly exceed rates determined from bulk measurements. There is good evidence that this is indeed the case (Rappoldt, 1992). Secondly, because of complex fluid flow patterns, it is possible that there will be a broad range of solute mixing timescales across rather small volumes of soil. As pointed out by Rappoldt (1992), together with a heterogeneous distribution of organic material, it is likely that this will result in regions of high microbial activity existing in less mobile anoxic water adjacent to regions which are rapidly replenished with percolating water of a high nitrate concentration. This conclusion is in conflict with a number of theoretical models of denitrification where a homogeneous distribution of microbes is assumed to inhabit spherical homogeneous aggregates (Leffelaar, 1988; Arah and Smith, 1989; Priesack, 1991). Such models are likely to underestimate denitrification rates. Finally, the critical role of porosity is related to another type of critical behavior, observed in a broad class of fractal lattices, which has relevance to the predictability of microbial activity. The connectivity of these lattices has a sensitive dependence on the dimension (Clerc *et al.*, 1990). Thus in two dimensions there exists a critical value for the dimension, below which the fractal is no longer connected (Falconer, 1990, p. 233). Similar, though less well developed theories correspond to the appearance of 'sheets' of structure which connect opposite sides of the lattice at a critical value of the dimension, effectively isolating the pore space in one half of the lattice from that in the other (Chayes 1991). It follows that soil having porosities or dimensions close to the critical values will have physical properties that are highly sensitive to structure and hence, microbial properties are likely to be difficult to predict. Even modest levels of clogging of the pore space by bacteria (Vandevivere and Baveye, 1992) or microbial by-products, or small changes in soil structure, could lead to significant changes in water and gas relations and hence, in the observed microbial activity.

We have assumed in the preceding arguments that the pore space and solid matrix can be approximated by fractals. While it may be true that for different soils either the pore space or the solid matrix is fractal, it cannot be the case that both can be fractal in the same soil. This can easily be seen from the fact that if both the pore space and solid are fractal in a given soil, the porosity and bulk density depend on length scale, R, according to

$$\rho_p \propto R^{D_p - 3} \qquad\qquad \rho_{mp} \propto R^{D_m - 3} \tag{12}$$

Since both fractal dimension of the pore space (D_p) and of the bulk matrix (D_m) must be less than three, Equation (12) implies that both the pore space and solid vanish at large length scales. Thus, if an understanding of gas flow and water relations is required, some additional considerations are necessary. The authors are unaware of any work that has been done on the study of the scaling properties of the complement of fractal lattices (*i.e.*, that part of the volume of space which is not occupied by the fractal —for a solid matrix which is fractal, the pore space is the complement). Using the construction in Figure 1 it is possible to calculate the asymptotic scaling properties of the complement of the fractal set describing the solid matrix. For large length scales, it can be shown (Crawford, unpublished) that the pore volume, V_p, is scaling in R according to

$$V_p(R) \propto R^{\phi}, \qquad\qquad \phi = \frac{3 - \varepsilon D_m R^{D_m - 3}}{1 - \varepsilon R^{D_m - 3}} \tag{13}$$

where the constant $\varepsilon \ll 1$. For a narrow range in R, or for $D_m \approx 3$, ϕ will be only weakly dependent on R, and so the pore space will obey an allometric relation of the form given in Equation (13).

The second ingredient in the derivation of the scaling behavior of the diffusion coefficient is the spectral dimension as defined in Equation (3). At present, an analytic counterpart corresponding to the fractal complement has to be found, but it must also be the case that the constant value be replaced by a scale-dependent value. When the scaling properties of the complement of a fractal lattice are better understood, it will be possible to derive the scaling behavior of gaseous diffusion in a soil with a fractal solid matrix.

III Physical Habitat

The composition and heterogeneity of the physical framework of soil alongside associated moisture conditions act to channel microbial movement and mediate activity. For example, adhesion of bacteria to clay particles surrounding pore walls significantly influences plasmid transfer and survival rates (Huysman and Verstraete, 1993). Under specific matric forces, rough pore channels may act to isolate pools of water in space, and provide refuge sites for bacteria from predation (Postma and Van Veen, 1990). Physical heterogeneity of the pore space suggests that in any one volume or aggregate, isolated patches of microbial populations (and activities) will be spread discontinuously, and as a consequence water will be similarly distributed according to the size and connectedness of the pore system. It is this heterogeneity, coupled with microbial processes, that may be captured by fractals. The refuge sites available to bacteria from predation by protozoa have a significant impact on the availability of a bacterial population to take part in nutrient cycling. The nature and degree of such refuge sites are related to the shape and orientation of pore walls. Following from the work by Morse *et al.* (1985) and Marquet *et al.* (1990), since the habitable surface area in a fractal landscape depends on the scale associated with an organism, if the scale is that appropriate to a microorganism (*e.g.*, a bacterium) then the surface area corresponds to the habitable pore space available to bacteria. The presence of such small-scale protected sites is vital in aiding the survival of bacteria (Heijnen and Van Veen, 1991). Crawford *et al.* (1993a) present a method whereby the potential habitable pore space for a given sized species can be measured:

$$F_{1,2} = 1 - \left(\frac{r_1}{r_2}\right)^{D_s-2} \tag{14}$$

where $F_{1,2}$ represents the space available to species 1 which provides refuge from predation by species 2, r_1 and r_2 the dimensions of species 1 and 2, and D_s the fractal dimension of the pore surface. Where species 1 represents a bacterium ($r_1 \approx 1\mu m$), species 2 a protozoan ($r_2 \approx 30\mu m$), and D_s is taken as 2.36 (Young and Crawford, 1991b), then approximately 70% of the potential habitable area available to the bacteria is inaccessible to protozoa. Where refuge sites are scale dependent, as in this case, fractals have significant advantages over traditional techniques, which may only provide very rough estimates analyzed at any one scale (Rutherford and Juma, 1992), thus providing significant underestimates of sites protected from predation. A similar investigation by

Kampichler and Hauser (1993) examined the relation between microarthropods and pore surface roughness. These authors predicted that as body size decreased by an order of magnitude, there would be a 4-fold increase in available habitat space. In a further study Kampichler (1995) takes advantage of the allometric relation between body size, B, and metabolic activity (Morse *et al.*, 1985) to connect the fractal dimension of the pore surface, D_s, with B and species abundance N:

$$N = B^{0.52-(2/3)D_s} \tag{15}$$

Using this relation, Kampichler (1995) showed that the amount of energy utilized by microarthropods was size invariant, implying that while the smaller species have access to surfaces at a small scale, the larger species, although excluded from such structure and substrate, are more efficient in utilizing substrate that is easily accessible. In effect the benefit derived by the different size classes in exploiting the soil pore space cancel out each other and thus energy consumption is similar over a large range of size classes.

Although the analysis of pore walls using fractals is a potentially powerful tool, it does not take account of the moisture films on the surface pores. In the first instance as the matric potential acting on the soil water decreases, depending on the orientation and connectivity of the pore surface, water will drain from the pores, retreating into isolated reservoirs within the small-scale structure. Therefore, at a given matric potential any one section of a pore pathway may consist of several isolated water patches, rendering any analysis of pore pathways outwith these patches inappropriate. Unfortunately the typical method of observation employed for pores, relying on the analysis of thin-sections, offers no clue to the moisture conditions at the time of sampling (see section IV).

The analysis, as presented, does not differentiate between predators with significantly different morphology or foraging strategy. In particular, Equation (14) assumes an average dimension for species, which does not change with time. In some situations, particularly protozoan predation of bacteria, this is not the case. For example, starved ciliate protozoa (*Tetrahymena*) may experience a reduction to 11.3 of their original cell volumes (Sleigh, 1989). Such a (reversible) reduction in *Tetrahymena* from an original diameter of *circa* 30 μm (Young *et al.*, 1994) would reduce refuge sites for a 1 μm bacterium from *circa* 62% to 50%. It conjures up a picture of protozoa ciliate accessing a small pore neck, consuming its fill of bacteria and then being unable to get back out: the small-scales of justice! Amoebae are also able to alter their shape when accessing small pores (Sleigh, 1989, p. 251), thereby facilitating exploitation of smaller scale structures than their spherical diameter would suggest. Bacteria can alter their size dramatically under starvation conditions, and because of this they can be found in niches which would otherwise be inaccessible (Lappin–Scott *et al.*, 1988; MacLeod *et al.*, 1988). Because of all these factors, there are limitations of the simple interpretation of Equation (14). Nevertheless, to some extent the realization that in a fractal surface, potential habitable pore space scales with microbe dimension has served as a valuable theoretical tool in reinforcing the appreciation that microbial dynamics must be tackled in the context of structural heterogeneity. Future study must clearly take account not only of complications arising from the distribution of water throughout structure, but also of the heterogeneity of active decomposable sites. An essential step, since most of the soil mineral surfaces may be thought of as biologically inert, compared to rhizosphere and 'hotspot' locations.

A promising technique to tackle the problem of reactive sites has been presented by Farin and Avnir (1987) in the context of molecules diffusing onto fractal surfaces. Farin and Avnir (1987) use the relation between reaction rate, V, particle radius, R, and the reaction dimension, D_R:

$$V \propto R^{D_R - 3} \tag{16}$$

In this case, D_R is the effective fractal dimension of the distribution of reaction sites on the surface, and the relation between D_S and D_R is governed by very similar factors that govern microbial dynamics on soil surfaces. In the context of structure-microbe processes the two most relevant governing factors are: *screening*, which describes areas of the fractal surface inaccessible to diffusing molecules and is analogous to the concept of bacterial refuge sites; and *chemical selectivity*, describing the heterogeneous distribution of reaction sites, and could be used to describe the distribution of clay and organic matter sites along the pore surface. In both of these cases $D_R < D_S$; screening acts to effectively smooth surfaces because inner parts of the surface are inaccessible to reactions, and reaction sites are a subset of all surface sites (Farin and Avnir, 1987). The degree of difference is directly related to the degree of screening and chemical selectivity.

The use of Equation (16) in microbial dynamics is untested and relies on the derivation of microbial reaction rates to surfaces, and knowledge of the distribution of chemical selectivity within the soil. When such data are available then we can begin to tease out the respective influences of a host of factors which act to control structure-microbe interactions.

IV Microbes and Soil Structure

The previous sections have highlighted fractals as a useful tool for providing a quantitative link between structure and microbial processes. Previously, the treatment of heterogeneity has required the definition of characteristic scales which compartmentalize the structure into micro-, meso- and macro-scales. Such essentially arbitrary restrictions offer no quantitative connection between structure and processes and are rendered redundant with the realization that soils may be fractal. The use of fractals to approximate soil structure has the potential to significantly increase our understanding of the interactions which take place across scales, and which are ultimately manifest at the scale of the plant and above. However, in practice, fractals do not circumvent the requirement for a definition of a characteristic scale, since the correlations implied by fractal scaling cannot persist over all spatial scales. If they did, then the porosity of a soil with a fractal matrix would tend to unity as the scale of observation increased, and the density to infinity at the smallest scales. In order to be able to scale above the fractal regime, it is crucial to know this cut-off scale, or "correlation length". Furthermore, the picture as presented is rather static, since it does not take into account the fact that structure is changing in time. As Jastrow and Miller (1991) point out, the soil-microbial system is both dynamic and interactive. A full description of soil structure must therefore be based on a dynamical theory, and such a theory must lead to an understanding of the origin of fractal scaling. As a consequence, it may also be possible to identify the factors which influence the loss of fractal scaling above the correlation length, and the possible relation through these factors, between the correlation length and the structure of soil. Such understanding would complete the fractal picture.

Microbes and roots play an important role in the aggregation process (Elliot and Coleman, 1988), which is correlated with mycorrhiza content and fine roots (Miller and Jastrow, 1990), and microbial polysaccharide products (Martens and Frankenberger, 1992; Dinel *et al.*, 1992). Although microbial activity is usually associated with increased stability, the converse may sometimes be the case (Lynch, 1981). In field soils, there is likely to be a dynamic balance between stabilizing and destabilizing processes (Lynch, 1981; Elliot and Coleman, 1988; Guidi *et al.*, 1988; Jastrow and Miller, 1991).

Tisdall and Oades (1982) have proposed a hierarchical model for aggregation, where the aggregates at each level of the hierarchy are characterized by distinct binding agents. The model is of rather general applicability, although the details will vary from soil to soil (Elliot and Coleman, 1988). At the smallest aggregate scale ($<2\ \mu m$) particles are bound by inorganic agents such as H-bonding or coulomb attraction between individual clay plates to form highly stable crystals, with some second-order modification of bond strength by organic materials. At the next level, the smallest aggregates are bound together by very stable, persistent organic bonds to form structures of diameter between $2\ \mu m$ and $20\ \mu m$. These particles are stable to ultrasonic vibration and seem to be associated with bacterial cells. Several bonding agents, including persistent organics, crystalline oxides and aluminosilicates, bind these particles together to produce stable aggregates having diameters in the range $20\ \mu m$ to $250\ \mu m$. Although persistent, these aggregates are less stable than the smaller classes, and can be disrupted by sonication. Finally, organic material such as hyphae and fine roots bind these aggregates to form large agglomerations having diameters $>2000\ \mu m$. These aggregates are the least stable and are to a large extent influenced by agricultural practices.

Key features of Tisdall and Oades' (1982) conceptual model are a scale-dependent aggregate stability, and hierarchical clustering. In light of the hierarchical construction in Section II, dynamical cluster-cluster aggregation models may be of some applicability in conjunction with fractal scaling. It is known that various aggregation processes lead to fractal structure due to the feedback between structure and the conformation of aggregating particles (Meakin, 1988). Diffusion-limited aggregation (DLA) is the simplest example (Witten and Sander, 1981) in which a randomly diffusing field of non-interacting particles stick with probability 1 when they contact a cluster growing from a central seed. Theoretical studies of colloidal dynamics examine the effects of various diffusion laws on colloidal growth and aggregate structure and sizes (Meakin, 1988), where it is assumed that the bonds are irreversible. In the context of soil aggregation, the effect of scale-dependent cohesion, aggregate disruption and particle-size distribution on aggregate structure, growth rate and length scale of the largest coherent structures, may be examined. The results (Crawford *et al.*, 1994) are based on a Monte Carlo simulation of a system of particles moving randomly on a 100 × 100 homogeneous grid. The simulation is performed in two dimensions due to limitations of computer memory and run time, but this will not affect the general conclusions. Cohesion is possible when two particles occupy adjacent sites and to begin with, the primary structural units (*i.e.*, the smallest particles from which all other aggregates are made) are assumed to be identical, and mass is conserved. The random movement of aggregates is assumed to approximate the fact that soil aggregates will bind in random orientations. The influence of scale-dependent cohesion is examined by parameterizing cohesion on the basis of a probability, P_c, that a collision leads to cohesion when

$$P_c = 1/m^\theta \tag{17}$$

where m is the number of primary particles comprising the smaller of the two aggregates (*i.e.*, the aggregate's mass), and θ is a constant. Such a probability law can be interpreted by examining the influence of particle size on the rate of cohesion if aggregates contact at a rate which is independent of mass, since then, $1/P_c$ is proportional to the timescale for cohesion of particles of mass m. A large value of θ implies that small particles bind more rapidly than large ones. To include the effect of a size-dependent stability, as featured in Tisdall and Oades' (1982) conceptual model, we introduce a probability law to describe aggregate fragmentation of the form

$$P_f = e^{-n_b/m} \tag{18}$$

where P_f is the fragmentation probability, m is again the number of primary particles comprising the smaller aggregate of the two which form the parent (*i.e.*, the mass of that aggregate) and n_b is the number of bonds between the two aggregates. Thus, larger aggregates (*i.e.*, smaller n_b/m) are less stable than smaller aggregates. The precise choice of the functional forms for Equations (17) and (18) is not important, but the choice presented here guarantees that the aggregates will eventually stop growing when the rate of disruption equals the rate of aggregation. The simulation was run for values of ranging from 0 to 5, and for P_f as given in Equation (18), or set equal to 0 for all m. The consequences for the rate of increase in mean aggregate size, fractal dimension and correlation length are examined.

Figure 3a and b (left and right, respectively) shows the structures that result when $P_f = 0$, $\theta = 0$ and $\theta = 5$. These structures are fractal, and the fractal dimension corresponding to $\theta = 0$, $\theta = 2$ (not shown in Figure 3) and $\theta = 5$ were calculated to be 1.61±0.015, 1.64±0.008 and 1.64±0.007, respectively.

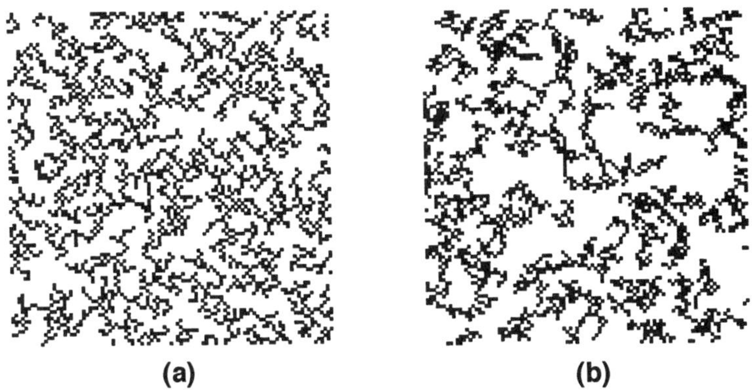

Figure 3: (a) Aggregate ensemble with $P_f = 0$ and $\theta = 0$. (b) Same as (a) except $\theta = 5$.

Furthermore, simulations carried out with P_f given by Equation (18) result in structures which have fractal dimensions not significantly different from the $P_f = 0$ case. From this we can conclude that the fractal dimension is more sensitive to whether or not there is scale dependency in the cohesion law, rather than to the details of that law. Furthermore, the dimension is insensitive to fragmentation. Although the study is conducted in two dimensions, we can make a comparison with observed values for the fractal dimension of soil if we assume that an upper bound to the dimension of aggregates in a three-dimensional simulation is given by adding one to the values in the present study (Sander, 1986; Meakin, 1988). Observations of soil structure (see above) suggest that $2.7 < D_m < 3$ is a reasonable bound for the fractal dimension of soils. Therefore, the simulation with $\theta = 0$ yields structures which are too open to be of relevance to soil aggregation. A scale-dependent cohesion law, consistent with Tisdall and Oades' model, implying that smaller aggregates bind more readily than larger ones seems to be required to produce structures with higher fractal dimensions. However, it is clear that even this is probably not enough on its own to lead to aggregates having dimensions close to those observed. Additional factors such as an initial particle-size distribution and mechanical re-packing, which maximizes the total bonding energy, are necessary to produce aggregates of sufficiently high fractal dimension (Crawford *et al.*, 1994).

Figure 4 shows the average aggregate mass as a function of time for the cases $\theta = 2$ and $\theta = 5$, both with $P_f = 0$ and with P_f varying with aggregate size according to Equation (18). Clearly, fragmentation slows down the growth rate of the aggregates, but the value of θ has a far greater influence. Fragmentation is the process that ultimately limits the aggregate size, but the final size depends on the balance between fragmentation and aggregation. Since this final aggregate size defines the correlation length above which we would not expect soil to obey fractal scaling, it is of crucial importance to understand the sensitivity to the parameters of the dynamical model. To do this we study the aggregate ensemble

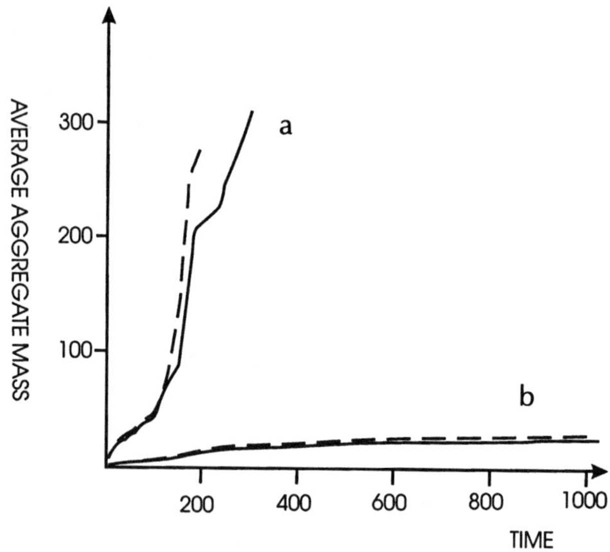

Figure 4: Plot of mean aggregate "mass" (*i.e.*, number of primary particles comprising the aggregate) as a function of time in arbitrary units. Curve (a) is for the $\theta = 2$ case, and curve (b) is for the $\theta = 5$ case. Broken lines correspond to the dynamics with $P_f = 0$, and solid lines correspond to P_f, as given in Equation (17).

at steady state, where the rate of aggregation equals the rate of fragmentation. If $N(m)dm$ is the mass distribution function, τ_f is the fragmentation timescale and τ_c is the cohesion timescale, all defined at the steady state, then the rate of cohesion equals the rate of fragmentation when

$$\frac{1}{2\tau_c}\int_m N(m^*)P_c(m^*)dm^* = \frac{1}{\tau_f}\int_m N(m^*)P_f(m^*)dm^* \tag{19}$$

Using Equations (17) and (18), and assuming that $n_b/m \ll 1$ at steady state (*i.e.*, the number of contact points between aggregates is much less than the total number of particles in the smaller of the two), this can be rewritten:

$$\frac{1}{2\tau_c}\int_m N(m^*)/(m^*)^\theta dm^* = \frac{1}{\tau_f}\int_m N(m^*)\left[1 - \frac{n_b}{m^*}\right] dm^* \tag{20}$$

The solution gives

$$\frac{1}{\tau_c}\left\langle\frac{1}{m^\theta}\right\rangle = \frac{1}{\tau_f}\left[1 - \left\langle\frac{n_b}{m}\right\rangle\right]$$

from which a characteristic mass for the steady-state particle ensemble can be defined (to zeroth order in the ratio n_b/m):

$$m \propto \left(\frac{\tau_f}{\tau_c}\right)^{1/\theta} \tag{21}$$

This illustrates the sensitivity of the characteristic mass to the parameters controlling the dynamics. The value of θ is clearly critical and the characteristic mass decreases as θ is increased. For large values of θ, the resulting aggregate sizes will be less sensitive to the details of fragmentation, since the fragmentation timescale appears raised to a small power. These results explain the trends illustrated in Figure 4. Thus, while the details of the cohesion law do not strongly influence the resulting structure of the aggregates, they do play a crucial role in determining the range of aggregate sizes created. This conclusion is independent of the choice of function in Equation (18), but arises from the assumption that P_f is a function of n_b/m, and that $n_b/m \ll 1$ at steady state. In such a circumstance, the relative change in fragmentation probability with mass is insensitive to the details of the process, and so the correlation length is controlled by the cohesion probability, which changes according to θ as aggregate size increases. In order to test these conclusions for real soil, it will be necessary to determine the scale dependency of cohesion and fragmentation quantitatively in field samples under various management treatments.

Another soil process that provides indirect clues to the nature of the binding forces in soil is cracking. Crack patterns, which develop in thin layers of slurried soil as it dries, have been used as an indicator of an effect of adding soil conditioners, or crop treatment, on soil structure (O'Callaghan and Loveday, 1973; Painuli and Pagliai, 1990; Steinberger and West, 1991; Dasog and Shashidhara, 1993). However, the derived parameters such as surface area to total length, or crack surface area to total surface area do not relate directly to the crack-forming mechanisms. Meakin (1987) presents a theoretical model for crack-pattern formation in terms of the detailed nature of the non-linear forces binding particles confined to a thin layer. The form of the resulting pattern depends on the parameters of the forces, and it has been proposed that the crack pattern can be approximated by a fractal (Meakin, 1987; Skjeltorp and Meakin, 1988). This suggests that fractal geometry in combination with an appropriate theoretical model may provide insight into the forces responsible for binding soil particles.

Figure 5 shows an example of a crack pattern formed after a 4-mm deep soil slurry was left to dry for 24 hrs (Preston, Young and Griffiths, 1995, unpublished). Three treatments were imposed: soil and water with no nutrient added; soil and water with glucose added; soil, water and glucose with nutrient solution added. The glucose was added to promote microbial polysaccharide production, and the addition of nutrients was intended to optimize the carbon-to-nitrogen ratio. Images of the crack patterns were processed and the fractal dimensions estimated, and these are shown in Figure 6. There was a significant decrease in the fractal dimension as glucose and nutrient were added. This can be interpreted as a reduction in the degree of branching in the crack pattern. This reduction is most likely due to an increase in stability due to microbial polysaccharide production as carbon and nitrogen were added to the slurry. Clearly, although the effect is significant, it is not large and a greater exploration of the influence of different functional groups of microbes is required. Furthermore, in order to understand the change of dimension in terms of the nature of the

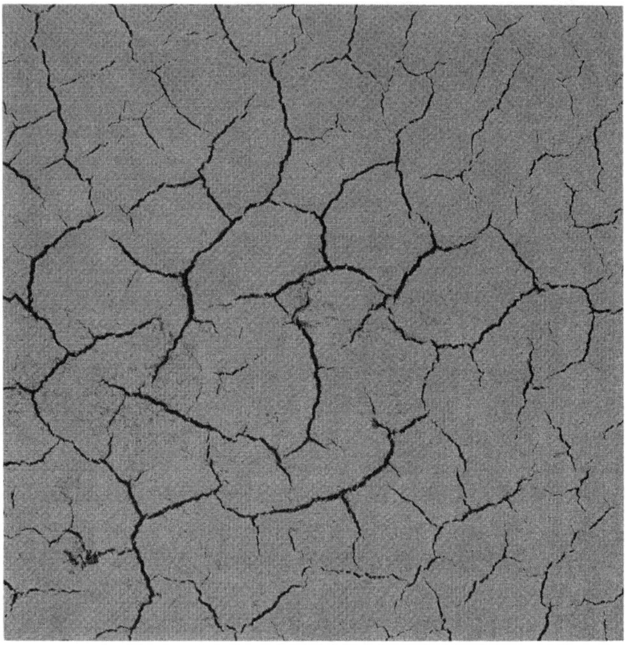

Figure 5: Example of a crack pattern formed in a Craibstone soil slurry poured to 4-mm depth and dried over 48 hrs.

binding forces, the results need to be interpreted using a theoretical model of cracking. Such considerations are currently under development.

Despite the obvious importance of this topic, there has been comparatively little quantitative work done on the influence of microbial activity on soil structure. Perhaps some of the reason is due to the difficulty in dealing quantitatively with soil structure in a manner which can be readily related to microbial activity. The use of fractals to describe the structure may therefore be a promising step forward. It is clear from the previous three sections that soil structure can play a very important role in determining microbial activity and population structure. In this section it was argued that microbes may in turn play an important role in determining both soil structure and the upper limit, where the mass distribution is correlated and fractal. This confirms that the soil-microbe system is both dynamic and interactive (Jastrow and Miller, 1991) and suggests the potential for self-organization. Multi-component systems with feedback have a propensity for complex dynamics which may lead to stable or highly unstable behavior (Procaccia, 1988; Sleeman, 1989; Schroeder, 1990), depending on the nature and parameters of the dynamics. Thus, there may be no such thing as a stable soil-microbe system, and indeed the variability may be an important manifestation of the processes which maintain soil in a persistently viable state.

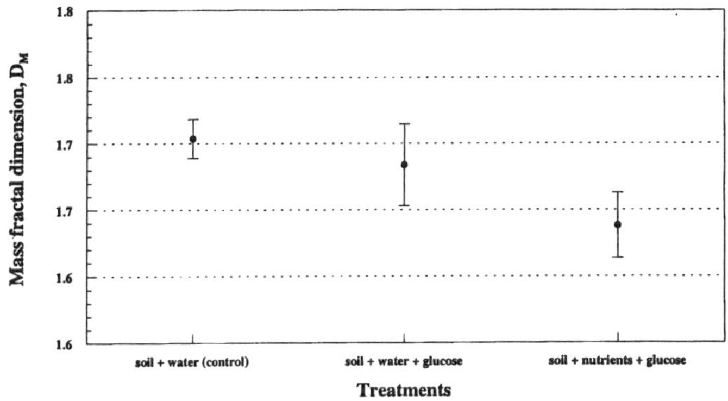

Figure 6: Plot of fractal dimension of the crack pattern as a function of treatment (see text for treatment details).

V Observational Techniques

The small-scale dynamics of structure-microbe relations make it virtually impossible to directly observe the important interactions; it is recognized that only a tiny fraction of the total number of organisms in soil can be cultured in the laboratory (Smith and Stribley, 1994). Further, direct non-destructive observation of structure has been, until relatively recently, impossible.

One of the most common methods of indirectly inferring structural parameters in relation to microbial dynamics has been the drying limb of the moisture release curve (*e.g.*, Rutherford and Juma, 1992; Heijnen and Van Veen, 1991). Given that the fractal dimension, D_m, may be approximated from the relation between air porosity, V_p, and matric potential ψ_m (Ahl and Niemeyer, 1989),

$$V_p \propto \psi_m^{D_m - 3} \tag{22}$$

there was some hope that the moisture-release-curve would provide a method whereby fractals and microbial dynamics might be linked. However, as discussed by Young and Crawford (1991a), Equation (22) tends to underestimate the fractal dimension as it assumes that large pores do not drain through small pores, which is clearly unrealistic. Further, a power-law exponent, as expressed in Equation (22), may be a consequence of a fractal pore volume, a fractal solid volume, a fractal pore wall, or a non-fractal self-similar pore wall (Crawford *et al.*, 1995). One cannot infer from the measurement which is the case. Therefore, whereas a moisture release curve is relatively simple and inexpensive to obtain, it is unlikely to be of much practical use in connection with fractals and microbial dynamics. Similar criticisms can be levelled at the approximation of a fractal dimension from particle- and aggregate-size distributions (Crawford, Sleeman and Young, 1993); such number-size distributions are widely used in relation to microbial dynamics (Juma, 1993).

At present the most reliable method of directly observing soil structure is through resin impregnated soil blocks/thin sections (see Fitzpatrick, 1993 for review). Whereas slight changes in structure through sampling, acetone replacement, and resin injection may occur, thin sections have the potential to reveal structural mineralogical and biological diversity through the judicious use of staining techniques from scales of cms to nm. Images captured from thin-sections have been successfully identified as exhibiting fractal structures (Bartoli *et al.*, 1991; Crawford *et al.*, 1993a; Bartoli *et al.*, this volume). Typically, the fractal dimension is estimated by the box-counting technique. However, as highlighted in Section II, there are problems in differentiating between a fractal distribution of solid matrix and a fractal distribution of pore space from thin sections (Crawford and Matsui, 1996). Any fractal distribution of particles in space consists of the component comprising the fractal, and the remaining component of space. As discussed in Section II (Equation (13)) the pore volume that complements a fractal mass distribution can have scaling properties similar to that of a fractal. This clearly has important repercussions when analyzing images of soil. An improper characterisation of the heterogeneity of the pore space leads to errors in scaling the processes that occur within, *e.g.*, moisture content, gas movement. The scaling exponent ϕ in Equation (13) is itself scale-dependent, whereas the fractal dimension is not. In theory, analysis of such scale-dependency should be able to distinguish between fractal and non-fractal scaling in soil.

However, there may be situations where the measured fractal dimension is different over different ranges in spatial scale. In reality this may be so particularly where the measurements are indicative of the mechanisms controlling soil structural genesis. Where scale-dependent aggregate stability exists due to different aggregate-forming mechanisms operating across scales (Tisdall and Oades, 1982) then, if the fractal dimension does indeed offer some insight into the origin of genesis, it may be expected to vary similarly across scale. In the future we may well find soils with a fractal pore space (varying across scale) at a limited range of scales and a fractal solid (varying across scale) at a limited range outwith that defined by the fractal pore space. The essential points are that the pore and solid space cannot be fractals within the same scale range and the measured power-law exponent must always be between 2 and 3 for a 3-D object.

Clearly, a major disadvantage of thin sections is that structure may only be resolved in 2 dimensions under the assumption of structural isotropy (*i.e.*, for a thin-section structure where say, $D = 1.36$, the addition of 1 provides an estimate of $D = 2.36$ for the 3-D object). Although techniques exist that allow a 3-D picture of structure from a series of thin sections taken from one sample to be compiled, they are expensive and extremely time-consuming (Scott, Webster and Nortcliff, 1988). A more practical alternative is Computer-Aided Tomography (CAT). CAT permits 3-D, non-destructive imaging of the internal structure of soil (see Aylmore, 1993, for review). It works on the principle of the attenuation of energy beams (typically γ or X-ray in soil studies) through absorbtion by soil/water bodies, and permits optical slices of soil to be resolved at *circa* 1 mm^3 resolution. Numerical techniques allow 3-D images to be reconstructed. For soil studies γ–ray sources are preferred over X-ray. One of the main reasons for this is that there are readily available γ–ray sources providing large differentials in energy levels, which are necessary if variations in water content and bulk density are to be separately analyzed. Whereas several workers have extensively examined soil structure using CAT (Anderson and Gantzer, 1989; Aylmore and Hamza, 1990), little effort has been employed in investigating the fractal nature of 3-D images. At present the relative coarse resolution provided by the most accessible CAT scanners coupled with the expense in using them limits the possible applications. With the future developments of high resolution (*i.e.*, 25 μm and 1 μm; Aylmore, 1993), CAT scanners may have potential use in structure-microbial investigations.

VI Summary and Concluding Remarks

Fractals are nothing more than a partial descriptive tool unless they are linked to some understanding of the underlying processes. From this and other chapters of this volume, it is clear that variability of soil structure must be accounted for and that fractals offer the possibility of a unified framework for studying a very broad range of soil processes and their interrelations. Since it could well be the case that these interrelations are more important than the details of the individual processes acting in isolation, the possibility of a unified approach offered by fractals is potentially very important. Even if it proves to be the case that fractals provide only a first-order model for any particular process, this pervasiveness should be an overriding consideration until the interactions are better understood.

The activity of microbes is governed by structure at correspondingly small scales, and is related to processes occurring across scales from microns to metres and above. Water relations and oxygen supply are of primary importance, and these are sensitively dependent on soil structure. The fractal dimension alone is not sufficient to characterize structure from the point of view of transport processes and the spectral dimension must be introduced as an additional parameter that is related to connectivity. The diffusion of oxygen depends on the fractal dimension of the pore space, which characterizes pore heterogeneity, and the spectral dimension. The effect of structure is to reduce the effective diffusion coefficient progressively at larger and larger scales. Thus, diffusion measurements made on a soil at a single scale and with no account of structure, cannot be related to the supply of oxygen to microbes without introducing a potentially substantial error. The same is true in connection with water relations, where hydraulic conductivity and the distribution of water in soil are governed by the latter's structure. In this instance the fractal and spectral dimensions of the bulk solid are the key structural parameters, since by virtue of its surface tension, water adheres to the solid matrix rather than moving freely through the pore space. Porosity in a fractal soil is scale dependent, whether it is the pore space or solid matrix that is fractal, and the fractal approximation can only be valid between lower and upper bounds in length scale. The porosity at the smallest spatial scale where the approximation is valid is a critical parameter in the scaling relations for hydraulic properties. Below the critical porosity, both the saturated and unsaturated hydraulic conductivities are scale dependent (although with different scaling exponents), while at values of porosity above the critical value only the unsaturated conductivity obeys a simple scaling relation. For soils with such porosities, the scaling of the saturated conductivity is influenced by bypass flow, which intermittently dominates as scale increases.

Under given moisture, oxygen and nutrient regimes, microbial dynamics is strongly influenced by interactions between species. For organisms which feed on others, physical protection and other constraints on movement are important. The fractal dimension of the pore wall describes how the surface area is partitioned between structures of different length scale, and as such provides information on the pore surface area that is protected from predation by larger organisms. For bacteria preyed upon by protozoa, the fraction of surface area that is potentially habitable by bacteria and is inaccessible to protozoa can exceed 50. Constraints imposed by structure on the movement of randomly moving microbes can be treated in a similar manner to diffusion. Since not all of the pore structure is accessible to larger microbes, the diffusion rate depends not only on the structure of the pore space, but also on the dimensions of the organism.

Since fractal structure has important consequences for understanding and scaling microbial activity, it is important to appreciate the factors which influence the upper cut-off to fractal scaling, *i.e.,* the correlation length above which fractal scaling no longer applies. Since the soil-microbe system is dynamic, it is important also to understand how structure changes in time, and the extent to which feedback between activity and structure promotes the potential for self-organization. An active and complex microbial population can directly or indirectly influence the structure through the production of bound aliphatic compounds, or by the binding action of hyphae. The aggregation process can lead to fractal structure, although the structure may be rather insensitive to the details of the binding properties resulting from microbial activity. However, the upper length scale which bounds the fractal regime can be very sensitive to microbial activity. The presence of strong feedback, and the fact that microbial processes may strongly influence structure, suggests that the scale at which the soil system may be managed is rather small (*i.e.,* below the correlation length).

The fact that fractals are relevant to the soil-microbe system only over a limited range in length scale presents a problem for scaling above the correlation length. In practice it will be necessary to average over fractal patches where each patch is characterized by the relevant fractal dimensions, correlation length, and critical porosity. If the correlation between patches decays as a power law, then it may be that some new fractal regime emerges, and there is some evidence that soil properties can be described by fractals at the field scale (Neuman, 1990). If this is not the case, then some other geostatistical tool may be relevant (McBratney and Webster, 1986; Burrough, 1993). The distinction is that if the patches are strongly correlated, then an understanding of the correlation may lead to a management strategy that impinges on the system at a larger scale than would be implied by the studies below the correlation length.

While structure may be the most important control over microbial processes (Van Veen and Kuikman, 1990) it is not the only constraint on microbial activity. The distribution of sorption sites will govern the mobility of bacteria and their susceptibility to predation (Gammack *et al.*, 1992; Kuikman *et al.*, 1991; Tan *et al.*, 1991). The distribution of organic matter in soil is of crucial importance for microbial activity but very little quantitative work exists on this. This may be especially important in the context of denitrification, as has been pointed out by Rappoldt (1992), since denitrification rates in a soil with heterogeneous distribution of activity could be orders of magnitude greater than would be anticipated under the assumption that activity is homogeneously distributed. Clearly this is an important area for future work.

Another important area which has been neglected to a significant extent is the dynamics of multi-component microbial systems in a heterogeneous environment. The behavior of such systems is very sensitive to the existence of protected habitat space (Kuno, 1987) and to the relative activity of the different functional groups (May, 1986). There is very little quantitative information concerning the dynamics of the individual components of the microbial community in soil, and even less about the interaction with structure. Thus a truly dynamic theory of the soil-microbe system is still some way off.

Soil microorganisms are a vital component of the ecosystem, linking the top and bottom of the food chain to close the cycle. They are therefore central to the sustainability of the global ecology. We have shown that microbial activity is sensitively dependent on the structure of soil, and that feedbacks between structural generation and microbially mediated processes may lead to self-regulation. While the latter may account for apparent overall stability of the system, it is also the key to catastrophic instability, since we cannot tamper with any part of the system without affecting the whole. Ultimately, this sensitivity limits our ability to predict the outcome of some impact on the system, either through modification of existing components, or addition of new (*e.g.*, genetically engineered microorganisms). The impact is unlikely to be linearly dependent on the magnitude of the imposed change if that change is large. Our knowledge, as it exists, is not yet sufficiently complete to understand the meaning of "large", or to know whether we have reached the limit of predictability in any particular system, or not.

References

Ahl, C. and J. Niemeyer. 1989. The fractal dimension of the pore-volume inside soils. *Z. Pflanzenernhr. Bodenk.* 152: 457-458.

Anderson, S.H. and C.J. Gantzer. 1989. Determination of soil water content by X-ray computed tomography and magnetic resonance imaging. *Irrig. Sci.* 10:63-71.

Aylmore, L.A.G. 1993. Use of computer-assisted tomography in studying water movement around plant roots. *Adv. Agron.* 49:1-54.

Aylmore, L.A.G. and M. Hamza. 1990. Water and solute movement to plant roots. *Trans. 14th Int. Congr. Soil Sci.* II, 124-129.

Arah, J.R.M. and K.A. Smith. 1989. Modelling denitrification in aggregated soils: relative importance of moisture tension, soil structure and oxidizable organic matter. p. 271-286. In J.Aa. Hansen and K. Henriksen (eds.), *Nitrogen in Organic Wastes Applied to Soils.* Academic Press, London.

Bartoli, F, R. Philippy, M. Dorisse, S. Niquet and M. Dubuit. 1991. Structure and self- similarity in silty and sandy soils: the fractal approach. *J. Soil Sci.* 42: 167-186.

Baveye, P. and G. Sposito. 1984. The operational significance of the continuum hypothesis in the theory of water movement through soils and aquifers. *Water Resour. Res.* 22:521-530.

Bazin, M.J. and J.I. Prosser. 1992. Modelling microbial ecosystems. *J. App. Bact. Sym. Supp.* 73:89S-95S.

Burrough, P.A. 1993. Soil variability: a late 20th century view. *Soils Fert.* 56:529-562.

Chayes, J.T., L. Chayes, E. Grannan and G. Swindle. 1991. Phase transitions in Mandelbrots percolation process in 3 dimensions. *Prob. Th. Rel. Fields* 90:291-300.

Chen, C. and R.J. Wagenet. 1992. Simulation of water and chemicals in macropore soils. Part I. Representation of the equivalent macropore influence and its effect on soil water flow. *J. Hydrol.* 130:105-126.

Clerc, J.P., G. Giraud, J.M. Laugier and J.M. Luck. 1990. The electrical conductivity of binary disordered systems, percolation clusters, fractals and related models. *Adv. Phys.* 39:191-309.

Crawford, J.W. 1994. The relationship between structure and the hydraulic properties of soil. *Eur. J. Soil Sci.* 45:493-502.

Crawford, J.W. and N. Matsui. 1996. Heterogeneity of the pore and solid volume of soil: distinguishing a fractal space from its non-fractal complement. *Geoderma* 73:183-195.

Crawford, J.W., N. Matsui and I.M.Young. 1995. The relation between the moisture release curve and the structure of soil. *Eur. J. Soil Sci.* 46:369-375.

Crawford, J.W., K. Ritz and I.M. Young. 1993a. Quantification of fungal morphology, gaseous transport and microbial dynamics in soil: an integrated framework utilising fractal geometry. *Geoderma* 56: 157-172.

Crawford, J.W., B.D Sleeman and I.M. Young. 1993b. On the relation between number-sized distributions and the fractal dimension of aggregates. *J. Soil Sci.* 44: 555-565.

Crawford, J.W., S. Verrall and I.M. Young. 1994. The origin and loss of fractal scaling in soil. Unpublished manuscript.

Dasog, G.S. and G.B. Shashidhara. 1993. Dimension and volume of cracks in a vertisol under different crop covers. *Soil Sci.* 156:424-428.

Dinel, H., P.E.M. Lvesque, P. Jambu and D. Righi. 1992. Microbial activity and long-chain aliphatics in the formation of stable soil aggregates. *Soil Sci. Soc. Am. J..* 56:1455-1463.

Elliot, E.T. and D.C. Coleman. 1988. Let the soil work for us. *Ecol. Bull.* 39:23-32.

van Elsas, J.D., J.T. Trevors and L.S. Van Overbeek. 1991. Influence of soil properties on the vertical movement of genetically marked Pseudomonas fluorescens through large soil microcosms. *Biol. Fertil. Soils* 10:249-255.

Farin, D. and D, Avnir. 1987. Reactive fractal surfaces. *J. Phys. Chem.* 91:5517-5521.

Falconer, K. 1990. *Fractal Geometry*. John-Wiley and Sons, London, England.

FitzPatrick, E.A. 1993. *Soil Microscopy and Micromorphology*. John Wiley and Sons, London, England.

Focht, D.D. 1992. Diffusional constraints on microbial processes in soil. *Soil Sci.* 154:300-307.

Gammack, S.M., E. Paterson, J.S. Kemp, M.S. Cresser and K. Killham. 1992. Factors affecting the movement of microbial inocula in soils. p. 263-305. In J-M. Bollag and G. Stotzky (eds.) *Soil Biochemistry*. Marcel Dekker, New York.

Greenwood, D.J. 1961. The effect of oxygen concentration on the decomposition of organic materials in soil. *Plant Soil* 14:360-376.

Guidi, G., A. Pera, M. Giovannetti, G. Poggio and M. Bertoldi. 1988. Variations of soil structure and microbial population in a comost amended soil. *Plant Soil* 106:113-119.

Haus, J.W. and K.W. Kehr. 1987. Diffusion in regular and disordered lattices. *Phys. Rep.* 150(5-6):263-416.

Heijnen, C.E. and J.A. Van Veen. 1991. A determination of protective microhabitats for bacteria introduced into soil. *FEMS Microbiol. Ecol.* 85:73-80.

Huysman, F. and W. Verstraete. 1993. Effect of cell surface characteristics on the adhesion of bacteria to soil particles. *Biol. Fertil. Soils* 16:21-26.

Jastrow, J.D. and R.M. Miller. 1991. Methods for assessing the effects of biota on soil structure. *Agric. Ecos. Env.* 34:279-303.

Juma, N.G. 1993. Interrelationships between soil structure/texture, soil biota/soil organic matter and crop production. *Geoderma* 57, 3-30.

Kampichler, C. and M. Hauser. 1993. Roughness of soil pore surface and its effect on available habitat space of microarthropods. *Geoderma* 56:223-232.

Kampichler, C. 1995. Biomass distribution of a microarthropod community in spruce forest soil. *Biol. Fertil. Soils* 19:263-265.

Kuikman, P.J., J.D. Van Elsas, A.G. Jansen, S.L.G.E. Burgers and J.A. Van Veen. 1990. Population dynamics and activity of bacteria and protozoa in relation to their spatial distribution in soil. *Soil Biol. Biochem.* 22:1063-1073.

Kuikman, P.J., M.M.I. Van Vuuren and J.A. Van Veen. 1989. Effect of soil moisture regime on predation by protozoa of bacterial biomass and the release of bacterial nitrogen. *Agric. Ecosys. Env.* 27:271-279.

Kuikman, P.J., A.G. Jansen and J.A. Van Veen. 1991. N-nitrogen mineralisation from bacteria by protozoan grazing at different soil moisture regimes. *Soil Biol. Biochem.* 23:193-200.

Kuno, E. 1987. Principles of predator-prey interaction in theoretical, experimental and natural population systems. *Adv. Ecol. Res.* 16:249-337.

Lappin–Scott, H.M., F. Cusack and J.W. Costerton. 1988. Nutrient resuscitation and growth of starved cells in sandstone cores: a novel approach to enhanced oil recovery. *App. Environ. Microbiol.* 54:1373-1382.

Leffelaar, P.A. 1988. Dynamics of partial anaerobiosis, denitrification and water in a soil aggregate: simulation. *Soil Sci.* 146:427-444.

Leffelaar, P.A. 1993. Water movement, oxygen supply and biological processes on the aggregate scale. *Geoderma* 57:143-165.

Lynch, J.M. 1981. Promotion and inhibition of soil aggregate stabilization by micro-organisms. *J. Gen. Microbiol.* 126:371-375.

MacLeod, A., H.M. Lappin–Scott and J.W. Costerton. 1988. Plugging of a model rock system by using starved bacteria. *App. Environ. Microbiol.* 54:1365-1372.

Marshall, T.J. and J.W. Holmes. 1988. *Soil Physics*. Cambridge University Press, Cambridge, England.

Marquet, P.A., S.A. Navarrete and J.C. Castilla. 1990. Scaling population density to body size in rocky intertidal communities. *Science* 250:1125-1127.

Martens, D.A. and W.T. Frankenberger. 1992. Decomposition of bacterial polymers in soil and their influence on soil structure. *Biol. Fertil. Soils* 13:65-73.

May, R.M. 1986. When two and two do not make four: nonlinear phenomena in ecology. *Proc. R. Soc. Lond.* B228:241-266.

McBratney, A.B. and R. Webster. 1986. Choosing functions for semi-variograms of soil properties and fitting them to sampling estimates. *J. Soil Sci.* 37:617-639.

Meakin, P. 1987. A simple model for elastic fracture in thin films. *Thin Solid Films* 151:165-190.

Meakin, P. 1988. Models for colloidal aggregation. *Ann. Rev. Phys. Chem.* 39:237-267.

Miller, R.M. and J.D. Jastrow. 1990. Hierarchy of root and mycorrhizal fungal interactions with soil aggregation. *Soil Biol. Biochem.* 22:579-584.

Morse, D.R., J.H., Lawton, M.M. Dodson and M.H. Williamson. 1985. Fractal dimension of vegetation and the distribution of arthopod body lengths. *Nature* 314: 731-733.

Newman, E.I. and A. Watson. 1977. Microbial abundance in the rhizosphere: a computer model. *Plant Soil* 48:17-56.

Neuman, S.P. 1990. Universal scaling of hydraulic conductivities and dispersivities in geologic media. *Water Resour. Res.* 26:1749-1758.

O'Callaghan, J.F. and J. Loveday. 1973. Quantitative measurement of soil cracking patterns. *Pattern Recognition* 5:83-98.

Orbach, R. 1986. Dynamics of fractal networks. *Science* 231:814-819.

Painuli, D.K. and M. Pagliai. 1990. Effect of polyvinyl alcohol, dextran and humic acid on some physical properties of a clay and loam soil. I. Cracking and aggregate stability. *Agrochimica* 34:117-130.

Passioura, J.B. 1971. Hydrodynamic dispersion in aggregated media. I. Theory. *Soil Sci.* 111:339-344.

Peitgen, H. and D. Saupe. (eds.) 1988. *The Science of Fractal Images.* Springer–Verlag, Berlin.

Penman, H.L. 1940. Gas and vapour movement in the soil. I. The diffusion of vapours through porous solids. *J. Agric. Sci.* 30:210-234.

Postma J. and J.A. Van Veen. 1990. Habitable pore space and survival of rhizobium Leguminosorum bivor trifolii introduced into soil. *Microb. Ecol.* 19:149-161.

Priesack, E. 1991. Analytical solution of solute diffusion and biodegradation in spherical aggregates. *Soil Sci. Soc. Am. J.* 55:1227-1230.

Procaccia, I. 1988. Universal properties of dynamically complex systems: the organisation of chaos. *Nature* 333:618-623.

Rappoldt, C. 1992. *Diffusion in Aggregated Soil.* Unpublished Ph.D. Dissertation. Wageningen Agricultural University, Wageningen, The Netherlands.

Rattray, E.A., J.I. Prosser, L.A. Glover and K. Killham. 1992. Matric potential in relation to survival and activity of a genetically modified microbial inoculum in soil. *Soil Biol. Biochem.* 24:421-425.

Rieu, M. and G. Sposito. 1991. Fractal fragmentation, soil porosity and soil water properties: I. Theory. *Soil Sci. Soc. Am. J.* 55:1231-1238.

Rutherford, P.M. and N.G. Juma. 1992. Influence of texture on habitable pore space and bacterial-protozoan populations in soil. *Biol. Fertil. Soils* 12, 221–27.

Sander, L.M. 1986. Fractal growth processes. *Nature* 322:789-793.

Schroeder, M. 1990. *Fractals, Chaos and Power Laws. Minutes from an Infinite Paradise.* W.H. Freeman Company, New York.

Scott, G.J.T., R. Webster and S. Nortcliff. 1988. The topology of pore structure in cracking clay soil. II. Connectivity density and its estimation. *J. Soil Sci.* 39: 315-326.

Skjeltorp, A.T. and P. Meakin. 1988. Fracture in microsphere monolayers studied by experiment and computer simulation. *Nature* 335:424-426.

Sleeman, B.D.S. 1989. Complexity in biological systems and Hamiltonian dynamics. *Proc. R. Soc. Lond.* 425:17–47.

Sleigh, M. 1989. *Protozoa and other Protists.* Edward Arnold, London.

Smith, N.C. and D.P. Stribley. 1994. A new approach to direct extraction of microorganisms from soil. Chapter 5. In Ritz, K., Dighton, J. and Giller, K.E. (eds.) *Beyond the Biomass, Compositional and Functional Analysis of Soil Microbial Communities.* Wiley and Sons, London.

Steinberger, Y. and N.E. West. 1991. Effects of polyacrylamide on some biological and physical features of soil: preliminary results from a greenhouse study. *Arid Soil Res. Rehab.* 5:77-81.

Tan, Y., W.J. Bond, A.D. Rovira, P.G. Brisbane and D.M. Griffin. 1991. Movement through soil of a biological control agent, *Pseudomonas fluorescens. Soil Biol. Biochem.* 23:821-825.

Tisdall, J.M. and J.M. Oades. 1982. Organic matter and water-stable aggregates in soils. *J. Soil Sci.* 33:141-163.

Tyler, S.W. and S.W. Wheatcraft. 1990. Fractal processes in soil water retention. *Water Resour. Res.* 26:1047-1054.

Vandevivere, P. and P. Baveye. 1992. Saturated hydraulic conductivity reduction caused by aerobic bacteria in sand columns. *Soil Sci. Soc. Am. J.* 56:1-13.

Van Veen, J.A. and P.J. Kuikman. 1990. Soil structural aspects of decomposition of organic matter by micro-organisms. *Biogeochemistry* 11:213-233.

Wallace, H.R. 1958. Movement of eelworms. I. The influence of pore size and moisture content of the soil on the migration of larvae of the beet eelworm *Heterodera schachtii* Schmidt. *Ann. Appl. Biol.* 46:74-85.

Witten, T.A. and L.M. Sander. 1981. Diffusion limited aggregation, a kinetic critical phenomenon. *Phys. Rev. Lett.* 47:1400-1403.

Young, I.M. and J.W. Crawford. 1991a. The fractal structure of soil aggregates: its measurement and interpretation. *J. Soil Sci.* 42:187-192.

Young, I.M. and J.W. Crawford. 1991b. The analysis of fracture profiles using fractal geometry. *Aus. J. Soil Res.* 30:291-295.

Young, I.M, A. Roberts, B.S., Griffiths and S. Caul. 1994. Growth of ciliate protozoan in model ballotini systems of different particle sizes. *Soil Biol. Biochem.* 26:1173-1178.

Youngs, E.G. and P.B. Leeds-Harrison. 1990. Aspects of transport processes in structured soils. *J. Soil Sci.* 41:665-675.

Using Fractal Dimension Of Stained Flow Patterns In Clay Soils To Predict Bypass Flow

R. Hatano and H.W.G. Booltink

Contents

ISBN 1-56670-105-8

I Introduction

The flow of water and solutes in structured soil is strongly affected by macropores. Water often moves through vertically continuous macropores, such as cracks, worm holes and root channels, bypassing the soil matrix within peds. Bouma (1984) defined bypass flow as the movement of free water along macropores through an unsaturated soil matrix.

Although this phenomenon was already known in the 19th century, only recently, with the increased awareness for water use efficiency and leaching of chemicals, have studies been initiated to better understand bypass flow. Detailed reviews on the effects of macropores on water- (Beven and Germann, 1982) and solute flow (White, 1985) have already been published. These reviews emphasize the fact that standard equations predicting water flow and solute transport in homogeneous soils are inadequate to describe water and solute movement in structured field soils. To discuss the direction of future research, the American Society of Agronomy held a symposium in 1988 on "Transport of water and solutes in macropores" (*Geoderma* 46, 1990, special issue). As we shall again briefly mention in section V, most bypass flow simulation models proposed up to this time did not consider the morphological features of soil structure. In the symposium, Bouma (1990), however, emphasized that the use of morphometric data should be considered to find realistic boundary conditions in simulation models of water regimes in structured soils. For this purpose, a technique to quantify soil morphology in combination with the measurement of water and solute transport in representative soil samples should be developed. Pedotransfer functions relating easy obtainable soil characteristics to more complex ones play an important role in this quantification (Bouma and Van Lanen, 1987).

To obtain a realistic pedotransfer function predicting bypass flow, we have studied the use of fractal dimensions of water-conducting macropores (Hatano *et al.*, 1992; Hatano and Booltink, 1992; Booltink *et al.*, 1993). In the present chapter we shall introduce a procedure to estimate fractal dimensions of stained flow patterns. The fractal dimensions are used in regression equations explaining various processes related to bypass flow. Results of these regression analyses are incorporated in a deterministic water flow simulation model.

II Bypass Flow, Principles and Applications

Bouma (1990) summarized factors governing bypass flow as: 1) rain intensity, 2) rain quantity, 3) surface microrelief, 4) surface texture, 5) surface water repellency and 6) surface water content. More bypass flow occurs at: 1) high rain intensities, 2) high rain quantities, 3) little surface relief, 4) high clay contents, 5) high water repellency and 6) high surface water contents. Furthermore, he emphasized the contribution of soil morphology to bypass flow. However, a clear relation between soil morphology and bypass flow was not available

Evidence of bypass flow has been observed in undisturbed soil cores in the laboratory and in the field (Beven and Germann, 1981). Three types of experimental procedures, using adequate sample size, are available: 1) breakthrough experiments, 2) tensiometer monitoring and 3) dye-staining and salt-marking. Concerning the adequacy of the sample size used in these experiments, Bouma (1985) suggested the use of the Representative Elementary Volume (REV), reducing the coefficient of variance associated with the measurements, a concept first defined by Beven and German (1981). Lauren *et al.* (1987) argued that a sample area for flux measurement should contain at least 30 peds in cross section in order to be representative.

II.A Breakthrough Experiment

Breakthrough curves with a rapid elevation of tracer concentration within a short period after initial breakthrough are indications of preferential flow through macropores (Anderson and Bouma, 1977a, Bouma *et al.*, 1983; White *et al.*, 1984; Hatano *et al.*, 1985; Andreini and Steenhuis, 1990; Ishiguro, 1991). Rose and Passioura (1971) presented a procedure to obtain a dispersion coefficient by curve fitting. However, this technique sometimes does not allow the evaluation of the dispersion coefficient in well-structured soils. On the basis of the assumption that dispersion in aggregated porous media depends on two elements, dispersion in macropores and interdiffusion between the macropores and soil matrix, Passioura (1971) developed an appropriate boundary condition for the evaluation of the dispersion coefficient. Interdiffusion was assumed to depend on pore geometry, macropore-water velocity and representative aggregate size in the sample. He substituted the ratio of microporosity within aggregates to total porosity for the value of the pore geometry factor and obtained the following equation:

$$\frac{D_A}{U \cdot a^2} > 0.3 \cdot \frac{\theta_T}{\theta_A} \tag{1}$$

where D_A is the diffusion coefficient within aggregates, approximated to be equal to 5×10^{-10} m^2 s^{-1} by Hatano *et al.* (1993), L is the length of the sample, U is the macropore water velocity (U = darcian flow rate/macroporosity), a is the radius of aggregates, and θ_T and θ_A are the porosity and microporosity within aggregates, respectively. By applying Equation (1), the sample length required for obtaining a reasonable dispersion coefficient can be estimated. For a sample with loosely packed, fine aggregates ($a = 1$ mm, $\theta_T/\theta_A = 2$, $T = 0.5$, $U = 10^{-4}$ m s^{-1} corresponding to a darcian flow rate of 2.5×10^{-4} m s^{-1}), the required sample length is 0.12 m. As often reported, water-conducting macropore is less than 1 vol % and aggregate size in structured soil is often more than 20 mm (*e.g.*, Bouma and Wosten, 1977). On the basis of these latter values, relative to structure element size and macroporosity, the required sample length is 153 m, which is, however, practically impossible. This difficulty may be partly resolved by having recourse to the hypothetical mobile-immobile analysis. Bouma and Wosten (1979) considered that 60% of the water inside peds act as immobile water during an infiltration in a clay soil with a large number of vertically continuous channels.

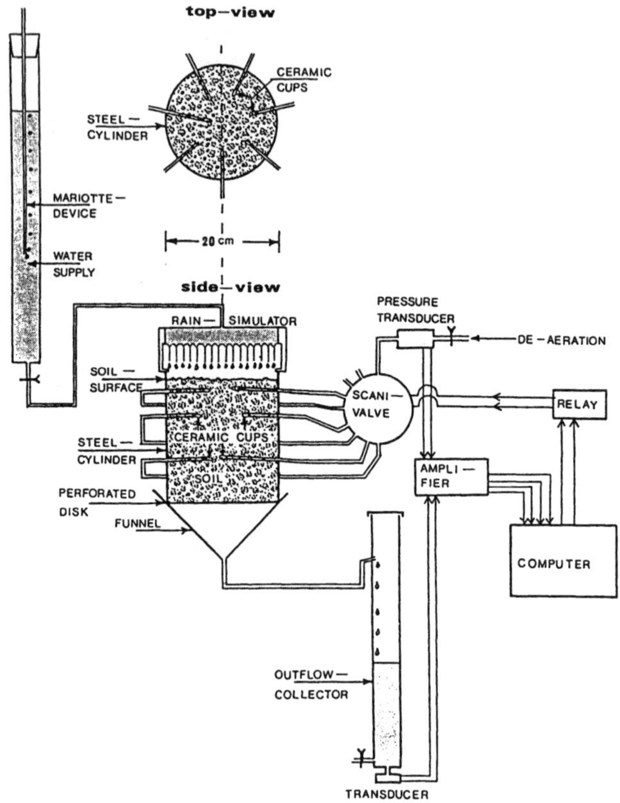

Figure 1: Laboratory set-up for measuring bypass flow.

II.B Tensiometer Monitoring

Booltink and Bouma (1991) presented an experimental set-up for tensiometer monitoring in undisturbed soil cores of a 6-L volume during a simulated rain event (Figure 1). The outflow at the bottom of the core and pressure heads were continuously monitored by a computer. Methylene blue powder was spread on the surface of the core before the application of simulated rainfall. The outflow always had a blue color. This provided evidence for the occurrence of bypass flow. Tensiometer monitoring showed a clear rise of the matric potential in the second half of the soil core, whereas at the same time a slight rise could be observed in the upper part of the core. This also evinces the occurrence of bypass flow and suggests the accumulation of water at the end of discontinuous macropores. Similar observations were reported by Anderson and Bouma (1977b). Van Stiphout *et al.* (1987) defined subsurface infiltration of bypassing water at the end of discontinuous macropores as "Internal Catchment". Moreover, Booltink and Bouma (1992) categorized the reactions of tensiometers in

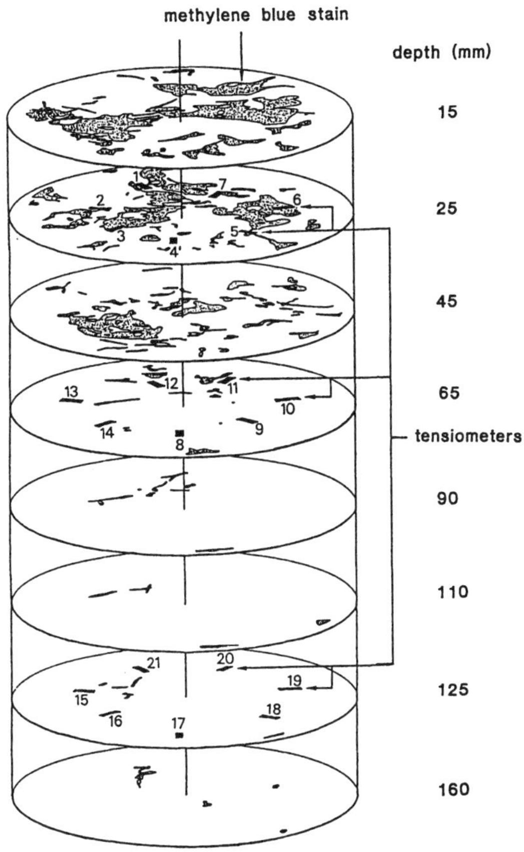

Figure 2: Reconstruction of methylene blue staining patterns of a 6-L core sample. Patterns were drawn at a depths of 15, 25, 45, 65, 90, 110, 125 and 160 mm in the sample. The tensiometer locations at the three depths are also shown.

combination with observations of methylene blue staining patterns (Figure 2). Taking the distance between the positions of tensiometers and stains into consideration, they estimated that 33 vol % of sample was in direct contact with continuous water-conducting macropores and 42 vol % was influenced by internal catchment. The experimental set-up in Figure 1 can be easily extended to measure breakthrough curves of chemicals by measuring the concentration in the outflow or by direct measurement of concentration at the outlet, using specific ion electrodes (Booltink *et al.* 1993).

II.C Dye Staining

Dye staining techniques provide direct visual information on pore continuity, as shown in Figure 2. This technique has been used to obtain evidence of the occurrence of preferential water flow along macropores, sometimes combined with salt-marking techniques (Ritchie *et al.*, 1972; Kissel *et al.*, 1973). Bouma and Dekker (1978) used a 0.03% (by weight) methylene blue solution as a tracer to demonstrate infiltration patterns of water into cracked clay soils under different rainfall intensities from 4 to 80 $mm.hr^{-1}$ and quantities ranging from 2.5 to 40 mm. They measured the area of stains on the wall of vertically continuous macropores to a depth of 1 m in 0.5 m^2 plots. The stained area was defined as contact area where water infiltrates laterally into peds. Contact area could be presented as a function of rain intensity and duration. Hoogmoed and Bouma (1981) used this relationship as a boundary condition in a model simulating bypass flow.

Utilization of image analysis techniques has allowed a more precise analysis of the hydraulic function of stains. Bouma *et al.* (1977) used the Quantimet image analyzer to measure number, length, area and perimeter of macropores stained and not stained by methylene blue in large thin sections of 70 cm^2. Based on the value of the ratio Area/Perimeter^2 (denoted hereafter A/Pe^2), they proposed to classify macropores into three categories: 1) planes defined by $A/Pe^2 < 0.015$, 2) vughs defined by $0.015 < A/Pe^2 < 0.04$, and 3) channels defined by $0.04 < A/Pe^2$. Average percentages of stains in each type of macropore, and the surface-roughness of stains, explained the differences in measured saturated hydraulic conductivity of clay soils. Murphy *et al.* (1977a,b) and Ringrose-Voase and Bullock (1984) proposed similar morphometric techniques.

Bouma *et al.* (1979) found that saturated hydraulic conductivities of clay soils were often restricted by pore necks smaller than 30 μm. They improved a model to calculate saturated hydraulic conductivity based on Poiseuille's law using the neck size. But Poiseuille's law does not consider the irregularity or roughness of shapes, such as the smoothness of pore wall, which seems to be related to pore continuity, as indicated by Bouma *et al.* (1977). Therefore, Bouma *et al.* (1979) suggested that studies to obtain empirical functions describing hydraulic conductivity using visible morphological features are also required.

III Application of Fractal Theory to Stained Flow Patterns

Fractal theory describes the self-similarity of morphological features, such as surface roughness and shape (Mandelbrot, 1982). A macropore system consists of a connection of different types of macropores of various sizes (Bouma *et al.* 1977). Hatano *et al.* (1992) studied the application of fractal theory to

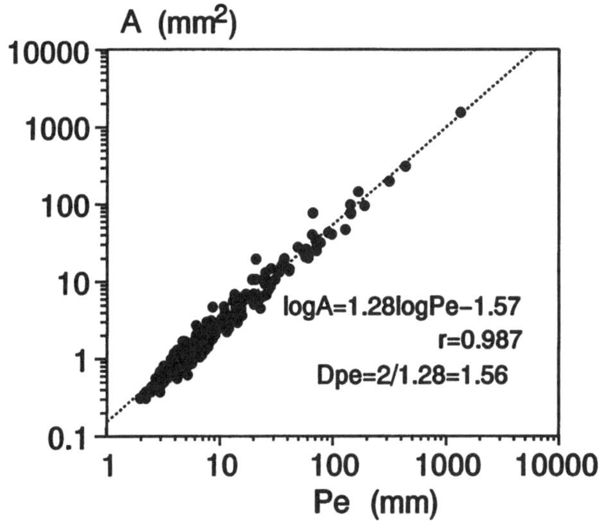

Figure 3: Double logarithmic plots of the area and the perimeter of methylene blue stains in a cross section of soil core sample.

characterize stained flow patterns. They proposed procedures to estimate the fractal dimensions of surface roughness and of staining patterns.

III.A Fractal Dimension of Surface Roughness

Fractal dimensions of surface roughness are estimated from the relationship between area and perimeter of stained parts (Hatano *et al.*,1992). This technique was used to determine the fractal dimension of the perimeter of rain area by Lovejoy (1982). As a general concept, the length of a smooth line, which has a dimension of 1, is proportional to the area raised to the power of 1/2 and to the volume raised to the power of 1/3. If the fractal dimension of the perimeter is D_{pe}, the perimeter (Pe) raised to the power of $1/D_{pe}$ is proportional to the length, to the area raised to the power of 1/2, and to the volume raised to the power of 1/3, respectively (Mandelbrot, 1982). In Figure 3, the double logarithmic plots of area (A) versus perimeter (Pe) of stained parts on a cross section are regressed to a straight line. D_{pe} is estimated from the slope of the regression line by:

$$\log A = \left(\frac{2}{D_{pe}}\right) \log Pe + C \tag{2}$$

where C is a constant. The D_{pe} value ranges from 1, for completely smooth boundary, to 2, for completely rough boundary. Hatano *et al.* (1992) reported that the D_{pe} values within a soil horizon remained almost constant and ranged from 1.30 for spherical ped faces to 1.65 for the surface of porous pumice

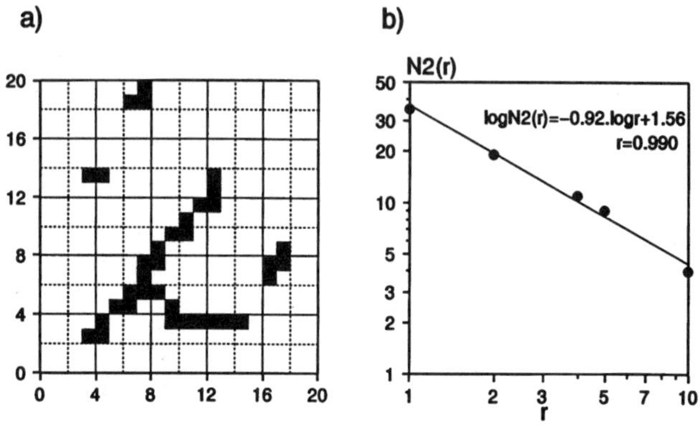

Figure 4: Box-counting method to measure the 2-dimensional fractal dimensions of stain patterns. (a) Fractionalization of stained patterns into squares and (b) double logarithmic plots of the number of squares containing the stains ($N_2(r)$) versus the square width (r).

stones. The average value of D_{pe} for angular blocky ped faces was 1.58. For the surface of large blocky peds or prisms D_{pe} was equal to 1.45. A smaller value of D_{pe} indicates that the soil has a smoother ped face.

III.B Fractal Dimension of Shape

This fractal dimension is generally called the Hausdorff dimension (Mandelbrot, 1982) or box-counting dimension (Edgar, 1992). It is related to the basic shape of an object in a plane, for example a dot or line. Figure 4 illustrates the procedure to estimate the fractal dimension on a cross section. The fractal dimension of a shape is derived by fractionalizing a stained cross section into squares (pixels) of side r and subsequently counting the number of stained squares $N_2(r)$. If the image data of the staining pattern have been prepared as a binary dataset on a small square cell mesh, the stained squares are easily counted. By repeating this process with increasing r values, a range of $N_2(r)$ values is obtained. As shown in Figure 4b, the fractal dimension of a staining pattern is estimated from the slope of the regression line of the double logarithmic plots of $N_2(r)$ against r:

$$\log N_2(r) = -D_{s2} \log r + C \tag{3}$$

where C is a constant. The value of D_{s2} ranges from 0 to 2. When D_{s2} is 0, the staining pattern is a dot pattern. Figure 5 shows that the $\log N_2(r)$ versus $\log r$ plots for dots, arranged according to a grid pattern with section width h, fall into two straight lines that intersect at r equal to h. In the range of r smaller than h, D_{s2} is calculated to be 0, which is the dimension of a dot. In the range of r larger than h, D_{s2} becomes 2, which is the dimension of a completely filled area. This allows us to use D_{s2} values in the range of

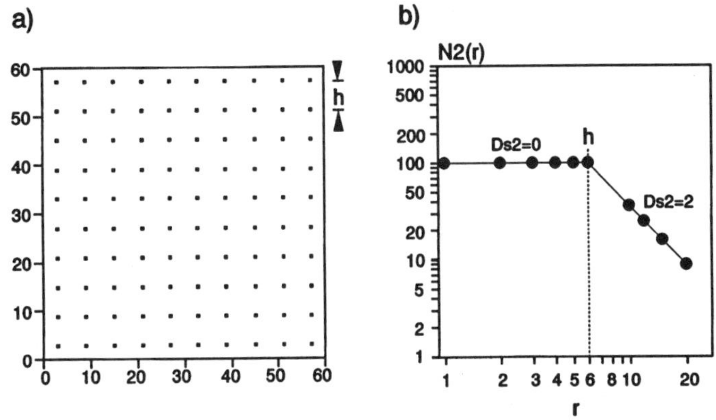

Figure 5: (a) Dots disposed at the nodes of a grid, with spacing h, (b) double logarithmic relationship between square width (r) and the number of squares containing the dots ($N_2(r)$).

$r < h$ as unique properties of the shape of objects distributed on a plane. An example of double logarithmic plots between $N_2(r)$ and r for the image of the methylene blue staining patterns is presented in Figure 6. The soil was taken from the B horizon of a Brown Forest soil with well-developed subangular blocky peds (Hatano *et al.* 1992). The image data are composed of 460 × 460 pixels, which are produced by a Nexus 6400 image analyzer. The width of a pixel is 0.221 mm. The double logarithmic plots also fall into two straight lines, which intersect at r equal to 13 pixels. In the range of r larger than 13, the slope of the line is -2. On the other hand, the slope of the line, in the range of r smaller than 13, is -1.47. Therefore, D_{s2}, as a unique characteristic of staining patterns, is equal to 1.47. Figure 7 shows an example of the vertical distribution of D_{s2} through five large soil cores (200 mm in diameter and 200 mm long) taken at a depth of 15-35 cm in a structured clay soil (Hatano and Booltink, 1992). D_{s2} values for all cores ranged from 0.85 to 1.67. The average values of D_{s2} for the upper half (25-90 mm) and lower half (110-160 mm) of the core as well as for the whole core are shown in Table 1. Staining patterns in the upper half are relatively similar. Those in the lower half show considerable variation.

To relate D_{s2} values to staining patterns, images of methylene blue staining patterns in horizontal cross sections at depths of 45 and 110 mm in three cores are presented in Figure 8. Although three images at a depth of 45 mm in the cores (a, b and c in Figure 8) have almost the same stained area, the staining pattern in (a) is longer and more slender, the staining pattern in (b) is more extensive and the staining pattern in (c) is more crowded than the others. D_{s2} values of 1.18 for staining pattern (a), 1.35 for pattern (b) and 1.30 for pattern (c) reflect these differences very well. The three images at a depth of 110 mm (d, e and f in Figure 8) have smaller stained areas than at a depth of 45 mm,

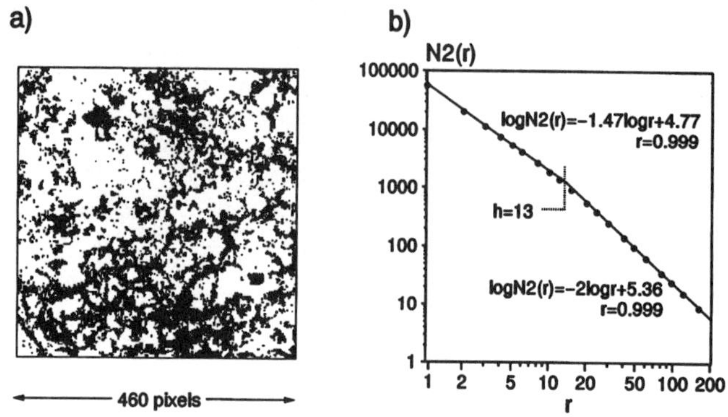

Figure 6: (a) Staining patterns on a horizontal cross section at a depth of 3.5 cm in a core taken from the B horizon of a Brown Forest soil with a well developed subangular blocky peds, (b) double logarithmic relation between square width (r) and the number of stained squares ($N_2(r)$).

Sample	Whole core	Upper half of core	Lower half of core
A	1.19	1.24	1.13
B	1.20	1.31	1.06
C	1.31	1.33	1.29
D	1.43	1.33	1.63
E	1.17	1.27	1.03

Table 1: Average values of 2-dimensional fractal dimensions of methylene blue staining patterns (D_{s2}) in a structured clay soil. (Data from Hatano and Booltink, 1992.)

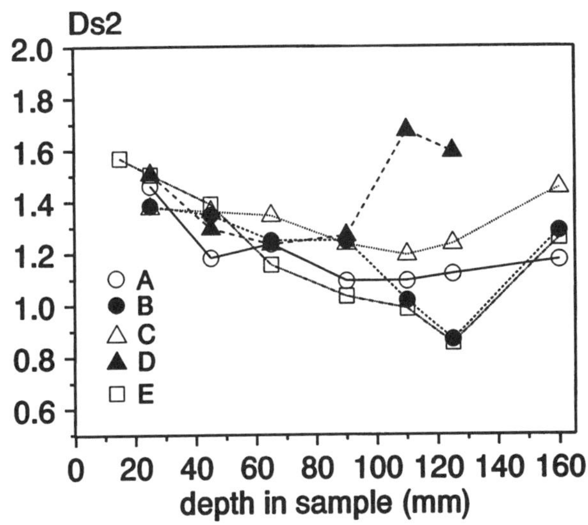

Figure 7: Vertical distribution of 2-dimensional fractal dimensions of staining patterns in horizontal cross sections of 6-L cores.

but the staining patterns are characterized very well by the D_{s2} values. The D_{s2} value is 1.09 for the very slender pattern in (d), 1.02 for the fragmentary pattern in (e) and 1.67 for the extensive pattern in (f).

III.C Three-Dimensional Fractal Dimension from a Reconstructed Sample

Theoretically, the 3-dimensional fractal dimension (D_{s3}) of an object with an isotropic structure is equal to $D_{s2} + 1$. But structures of stained flow patterns are anisotropic, as demonstrated in the above example. In general, the fractal dimension D_{s3} can be obtained in a similar way as D_{s2}. In this case, cubes (voxels) with a width r are used instead of squares (pixels), as used to derive D_{s2}. The model for D_{s3} is:

$$\log N_3(r) = -D_{s3} \log r + C \tag{4}$$

where C is a constant.

For the evaluation of D_{s3}, 3-dimensional image data are needed, which requires vertical sampling on horizontal cross sections at an interval with adequate resolution, generally 30 μm to 1 mm. But, it is impossible to sample a horizontal cross section at such a thin interval. Therefore, Hatano and Booltink (1992) developed a procedure to estimate D_{s3} values of staining patterns in cores using the values of D_{s2} and $N_2(r)$ measured in horizontal cross sections at various depths in the core.

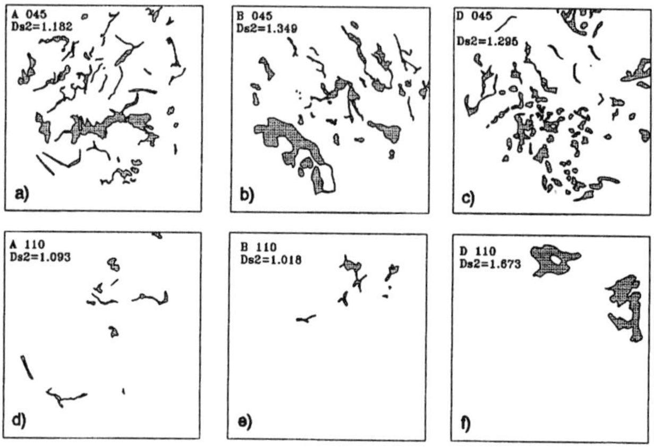

Figure 8: Typical examples of staining patterns in horizontal cross sections of 6-L cores at depths of 45 mm (a-c) and 110 mm (d-f)

The total stained volume (V) in a core sample is approximated by:

$$V = N_3(r) \cdot r^3 \tag{5}$$

At the smallest value of r, Equation (5) provides the best approximation of the actual stained volume. On the other hand, the value of V is also obtained by vertical integration of the products of thickness and stained area (A) in each horizontal cross section. The stained area in the jth cross section is approximated by:

$$A(j) = N_2(r, j) \cdot r^2 \tag{6}$$

Taking the interval thickness as r, the total stained volume is approximated by:

$$V = r \cdot \sum_{j=1}^{L/r} A(j) \tag{7}$$

where L is the total thickness of soil considered. Inserting Equations (5) and (6) into Equation (7), one obtains:

$$N_3(r) = \sum_{j=1}^{L/r} N_2(r, j) \tag{8}$$

which allows calculation of $N_3(r)$ for every r value. D_{s3} is estimated as the absolute value of the slope of the regression line of the double logarithmic plot of $N_3(r)$ against r.

This procedure is applied to various depths in a core, for example the upper half and the lower half of the core, as well as to the whole core. Figure 9 shows an example of double logarithmic plots of $N_3(r)$ versus r for the whole

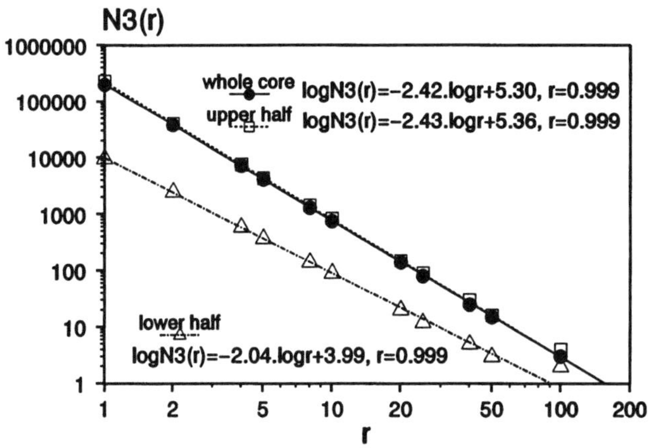

Figure 9: Double logarithmic relationship between cube width (r) and the number of stained cubes $N_3(r)$, which were obtained by the reconstruction of horizontal cuts.

Sample	Whole core	upper half of core	Lower half of core
A	2.27	2.29	2.08
B	2.32	2.36	2.20
C	2.33	2.39	2.28
D	2.39	2.38	2.58
E	2.42	2.43	2.04

Table 2: Three-dimensional fractal dimensions of stain patterns (D_{s3}). (Data from Hatano and Booltink 1992.)

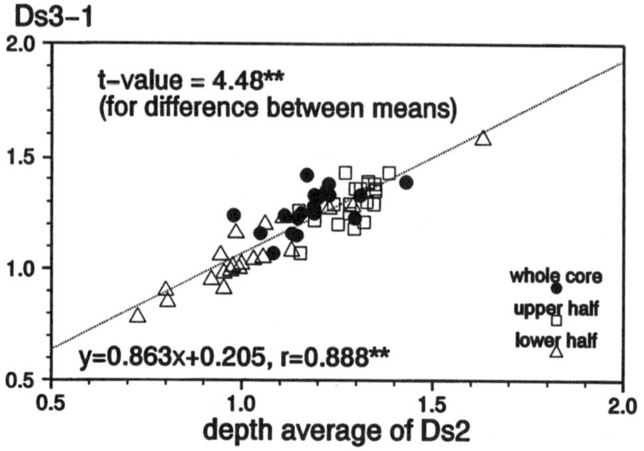

Figure 10: Comparison between 2-dimensional fractal dimensions (D_{s2}) and 3-dimensional fractal dimensions (D_{s3}). The relation between D_{s2} and $D_{s3} - 1$ and the result of t-test for the difference between means of D_{s2} and $D_{s3} - 1$ are presented.

core and for the upper and lower halves of the core whose staining patterns are illustrated in Figure 2. All the plots follow straight lines. D_{s3} values corresponding to the depth-averaged values of D_{s2} in Table 1 are presented in Table 2. D_{s3} values do not equal the sum of the depth average of D_{s2} plus 1. Figure 10 shows a relationship between the average values of D_{s2} and $D_{s3} - 1$ values. This figure includes other data which we have obtained. D_{s3} values were significantly correlated to D_{s2} values. On the other hand, t-test suggested that at the 1% level there was a significant difference between the means of D_{s2} and $D_{s3} - 1$ values. Therefore, we would recommend using $D_{s3} - 1$ values instead of the depth-weighted average of the 2-dimensional fractal dimension of staining patterns.

IV Pedotransfer Function to Predict Bypass Flow using Fractal Dimensions

IV.A Bypass Flow

Figure 11 shows a typical example of outflow curves in undisturbed unsaturated soil cores (200 mm in diameter and 200 mm long) taken from a structured clay soil in Eastern Flevoland in the Netherlands (Booltink *et al.*, 1993). One 15-mm rainfall containing a 0.01 M methylene blue was applied to each of the 15 soil cores with an intensity of 10, 20, 30 or 40 mm.hr^{-1}. Outflow started after a significant time lag from the start of the rainfall. This time lag was strongly

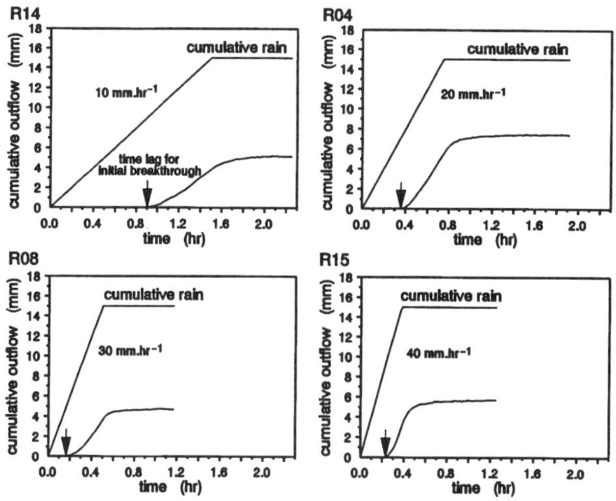

Figure 11: Outflow curves in 6-L soil cores following application of a 15 mm rain shower with a rain intensity of 10 mm.hr^{-1} (R14), 20 mm.hr^{-1} (R04), 30 mm.hr^{-1} (R08) and 40 mm.hr^{-1} (R15).

related to the rain intensity applied. However, as shown in Table 3, there were large variations among the values of time lags within one rain intensity. The outflow rate during a rainfall was a little lower than the applied rain intensity (Table 3). As soon as the rain ceased, the outflow reduced remarkably and stopped after a few minutes.

Total outflow was not related to the applied rain intensity (Table 3). However, as shown in Figure 12, there was a significant relationship between the time lag (T_l) and the mean water-retention time during infiltration. Mean water-retention time was estimated in terms of the ratio of the difference between the total amount of rain (R) and outflow (O) to the rain intensity (R_i) (Hatano and Booltink, 1994). The regression equation is:

$$\frac{(R-O)}{R_i} = 1.060 \cdot T_l + 0.0729 \tag{9}$$

with a correlation coefficient of 0.907, which is significant at the 1% level. This indicates that the time lag for outflow is induced by internal catchment (storage of water in vertically discontinuous macropores) and surface storage.

Water absorption into peds during infiltration depends on the soil moisture status in the core (Booltink and Bouma, 1993). Taking the mean water pressure head (P) into consideration in the regression analysis, an improved regression equation is obtained:

$$\frac{(R-O)}{R_i} = 1.044 \cdot T_l + 0.00501 \cdot P - 0.199 \tag{10}$$

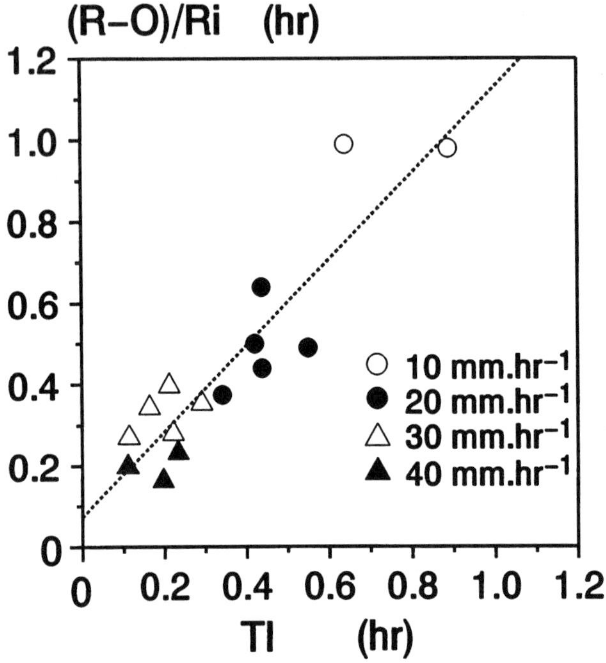

Figure 12: Relationship between the time lag for initial breakthrough (T_l) and the mean water-retention time estimated in terms of the ratio of the difference between total amount of rain (R) and total outflow (O) to rain intensity (R_i).

with a correlation coefficient of 0.933, which is significant at the 1% level. The significant regression coefficient of the pressure head term (P) suggests that a low pressure head caused by high initial water content shortens the time required for water retention. The outflow rate (O_r) appears more strongly related to a combination of the rain intensity (R_i) and the water retention $(R - O)$, which is the reciprocal of the mean water-retention time, than the rain intensity only, as indicated by the following equations:

$$O_r = 0.691 \cdot R_i - 0.768 \tag{11a}$$

$$O_r = 5.082 \cdot \frac{(R - O)}{R_i} + 2.416 \tag{11b}$$

The correlation coefficients for these equations are 0.883 and 0.956, respectively, which are significant at the 1 % level. It seems that the amount of water retention $(R - O)$ acts as a resistance to water flow.

Sample	R_i	T_l	O	O_r	P	Vs_u	Vs_l	D_{s3u}	D_{s3l}
	mm. hr^{-1}	hr	mm	mm. hr^{-1}	kPa				
R13	10	0.638	5.1	6.7	60.8	0.0652	0.0019	2.26	1.78
R14	10	0.888	5.2	7.4	61.6	0.0357	0.0016	2.07	1.98
R03	20	0.420	5.0	13.8	54.7	0.0748	0.0026	2.25	1.95
R04	20	0.342	7.5	15.6	27.9	0.0887	0.0033	2.35	2.05
R05	20	0.438	6.2	15.4	48.6	0.1199	0.0099	2.43	2.02
R06	20	0.437	1.9	7.8	61.0	0.0304	0.0058	2.20	2.00
R07	20	0.550	5.2	14.0	56.6	0.1124	0.0013	2.36	2.23
R08	30	0.163	4.7	16.8	61.9	0.0975	0.0212	2.36	1.99
R09	30	0.222	6.6	21.3	55.7	0.0806	0.0096	2.29	1.90
R10	30	0.113	6.9	25.2	58.2	0.0605	0.0152	2.18	2.06
R11	30	0.292	4.4	16.8	61.2	0.1064	0.0076	2.29	1.91
R12	30	0.212	3.1	11.4	63.9	0.0711	0.0093	2.22	1.85
R15	40	0.233	5.7	28.4	58.9	0.0836	0.0207	2.38	2.16
R16	40	0.197	8.5	30.8	37.3	0.0608	0.0132	2.30	2.01
R17	40	0.110	7.1	26.4	63.5	0.0730	0.0791	2.21	2.27

† R_i represents the rain intensity, T_l represents the time lag for initial breakthrough, O represents the amount of outflow, O_r represents the outflow rate, P represents the initial soil water pressure head, Vs represents the volumetric fraction of stained parts, D_{s3} represents the 3-dimensional fractal dimension of staining pattern, and the subscripts u and l represent the upper half and the lower half of the core, respectively.

Table 3: Physical and morphological results characterizing bypass flow measured in 6-L core samples at an application of 15 mm rain (Data from Booltink *et al.* 1993.)

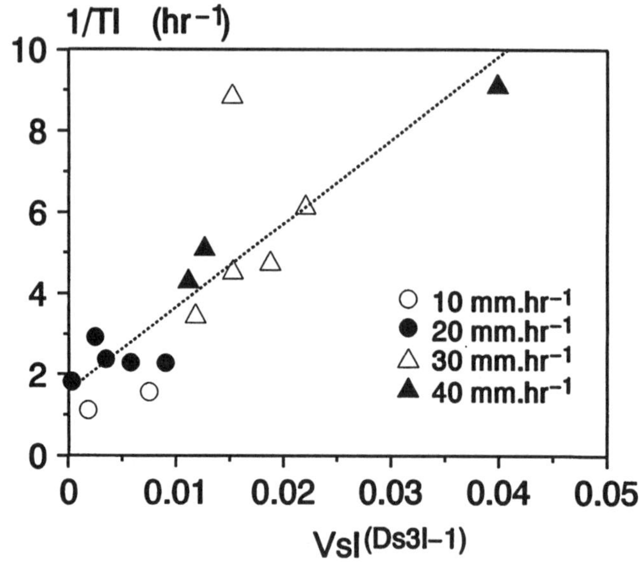

Figure 13: Reciprocal of the time lag for initial breakthrough as a function of the volumetric fraction of stained parts (V_{sl}) raised to the power of fractal dimensions ($D_{s3l} - 1$) in the lower half of the soil cores.

IV.B Macropore Geometry and Bypass Flow

After the breakthrough experiments, all cores were cut into 20-mm horizontal sections. Methylene blue staining patterns on each horizontal cross section were analyzed with a Nexus 6400 image analyzer. Area and perimeter of all individual stains were measured. Binary image data were used for the calculation of the fractal dimensions and of the volumes of stained parts. Three dimensional fractal dimensions (D_{s3}) and volumetric fractions of stained parts (V_s) in the upper and lower halves of the core are presented in Table 3. The time lag until initial breakthrough (T_l) could be significantly regressed on the volumetric fraction of stained parts (V_{sl}) and the depth-weighted fractal dimension ($D_{s3l} - 1$) of the lower half of the core, as shown in Figure 13. This regression equation is:

$$\frac{1}{T_l} = 196.0 \cdot V_{sl}^{(D_{s3l}-1)} + 1.761 \qquad (12)$$

with a correlation coefficient of 0.804, which is significant at the 1% level. The geometry of the upper half of the core could not explain the time lag.

Although this pedotransfer function is strictly empirical, it has a clear physical meaning. An increase of volumetric fraction of water-conducting macropores (V_{sl}) and a decrease of the fractal dimension ($D_{s3l} - 1$) shorten the time lag. A decrease of the fractal dimension indicates that the shape of the macropores tends to be more straight than tortuous.

As shown in Table 3, the time lag was related to the rain intensity (R_i). Taking this into consideration, the following improved regression equation could be derived:

$$\frac{1}{T_l} = 0.0984 \cdot R_i + 130.5 \cdot V_{sl}^{(D_{s3l}-1)} - 2.416 \tag{13}$$

with a correlation coefficient of 0.850, which is significant at the 1% level. This indicates that a high rain intensity shortens the time lag.

These equations suggest that more bypass flow occurs at high rain intensities, high soil water contents, and large volumes of vertically continuous straight macropores. It is clear that the 3-dimensional fractal dimension of stained flow patterns is a useful tool to quantitatively describe measured physical phenomena related to bypass flow.

IV.C Adequate Sample Size for the Measurement of Bypass Flow

Following Bouma's (1990) suggestion that the sample volume for flux measurement should contain at least 30 peds to be representative, for this experiment we used soil cores with a volume of 6 L. There were nevertheless large variations among the values of bypass flow characteristics within one rain intensity (Table 3). However, as shown in Equation (13), the variation of time lag was explained by the geometry of water-conducting macropores. The Representative Elementary Volume for bypass flow measurement may therefore be the volume of a sample with representative fractal dimension and the volume of water-conducting macropores.

By combining different samples into one larger sample, we investigated the effect of an increase in sample volume on the values of D_{s3} and V_s (Figure 14). The value of D_{s3} was estimated from the slope of a line obtained by plotting the $N_3(r)$ versus r on a double logarithmic paper. At the same time, the value of V_s was estimated as the ratio of the sum of $N_3(1)$ values and of the sample volume. The mean values and 95% confidence limits of the values of D_{s3} and V_s for 15 samples were also calculated.

As shown in Figure 14a, the value of D_{s3} was within the 95% confidence interval when 4 or more samples were combined. After the combination of 5 samples or more, the average value of D_{s3} was almost constant. On the other hand, the value of V_s reached the 95% confidence interval when 5 samples were combined (Figure 14b). Therefore, the Representative Elementary Volume for the measurement of bypass flow may be estimated as 30 L with a cross-sectional area of 0.15 m^2 (volume of one sample is 6 L and the area of the cross section is 0.03 m^2). However, this sample volume is too large to use in laboratory experiments because the weight of a 30 L sample may be over 45 Kg. It is not necessary to use a sample as large as 30 L if bypass flow is related to soil morphology, as in Equation (13). We should rather choose a sample size and a sample number adequate to relate bypass flow to soil morphology. In view of the results mentioned above, the appropriate sample size may be about 6 L and the optimal number of samples may be above 5.

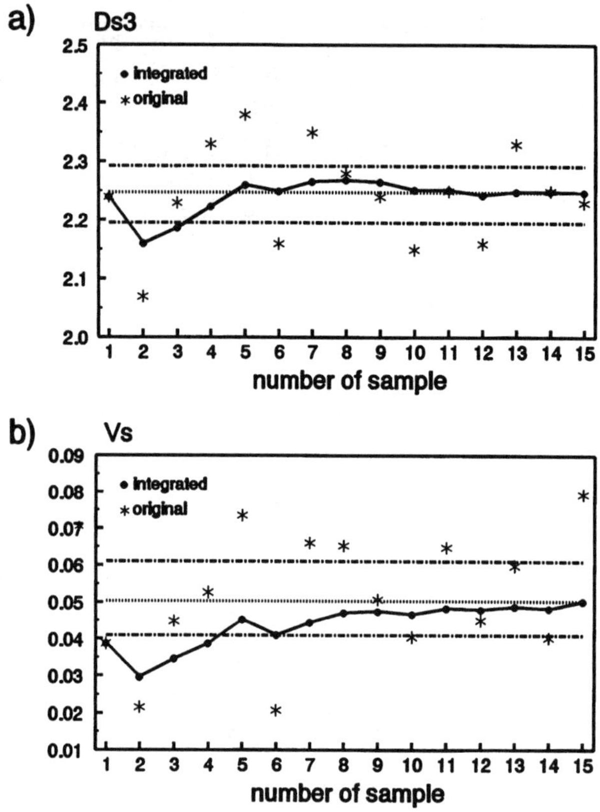

Figure 14: Variations of the values of 3-dimensional fractal dimension (D_{s3}) and volumetric fraction of water-conducting macropores (V_s) with integration of samples.

V Using Soil Morphology Characterized with Fractals in a Model to Simulate Bypass Flow

V.A Modeling Bypass Flow

Two main approaches for simulating bypass flow can be distinguished: the one- and two-domain models. One-domain models treat the soil matrix, including macropores, as one continuous flow system, whereas two-domain models describe water flow in micro- and macropores separately.

Kinematic wave theory, applied by Germann and Beven (1985), is an example of a one-domain model approach. Water is added to the soil surface as a rectangular pulse that is transported through the soil as a kinematic wave with a slowly decreasing speed due to loss of energy. Germann (1990) further modified kinematic wave theory by applying a sink term to describe absorption of water in the surrounding soil from the macropores. The recently developed technique of lattice gasses (Frisch, 1986; DiPietro, 1993) allows direct simulation of the Navier-Stokes equation in a two-dimensional, non-isotropic, porous medium. Due to the high data- and computational demands of the lattice-gasses approach, it can only be applied on small samples and is not suitable for simulation of bypass flow at the pedon or field scale.

Most bypass flow simulation models follow the two-domain approach. Edwards *et al.* (1979) used a two-dimensional model, simulating infiltration and redistribution of water into a soil surface containing cylindrical macropores. Beven and Clark (1986) used a system of uniformly distributed channels with different widths and depths to simulate infiltration into a homogeneous soil matrix containing macropores. Chen and Wagenet (1992) simulated water and chemical transport by combining Darcian flow for transport in the micropore or soil matrix domain with Chezy-Manning equations for macropore transport. Hatano and Sakuma (1991) used a capacity model to simulate the transport of water and solutes in an aggregated soil. They distinguished a mobile, an immobile and a mixing phase for water and chemical interactions. Van Genuchten and Wierenga (1976), De Smedt *et al.* (1986) and Vanclooster *et al.* (1992) developed two-region models to describe convective dispersive transport in a mobile and immobile phase. The use of this type of models in structured soils is, however, limited since non-Darcian flow modes, such as film-flow along macropore walls, cannot be described accurately by means of convective dispersive flow. In all these modeling approaches, "real" soil structure defined, as the physical constitution of a soil matrix as expressed by the size, shape and arrangements of the elementary particles and voids, is not considered.

Bronswijk (1988) calculated macropore sizes by means of swelling and shrinkage characteristics measured on structure elements in the laboratory. In his approach, the excess water at the soil surface during a rain event was immediately transported to the ground water without interaction with the surrounding soil. Jarvis and Leeds (1987, 1990) developed a model that calculates water balances in the micro- and macropore domains. Soil structure is represented by means of cube-shaped soil aggregates and transport of water through the macropores is calculated using an empirical tortuosity factor. Jarvis (1991) included transpiration and water flow through the soil matrix in the model.

Hoogmoed and Bouma (1980) used a Darcian flow model for simulation of water transport in the micropore domain combined with measured soil morphological characteristics to simulate the absorption of water along macropore walls during bypass flow events. Although this model is based on soil morphological observations it does not include the effects of tortuous water transport trough macropores.

Fractal theory was used by Hatano and Booltink (1992) and by Booltink *et al.* (1993). They quantified soil structure in terms of fractal dimensions of methylene blue stained macropores and linked these dimensions to soil physical characteristics such as macropore flow velocity, initial breakthrough of bypass flow water and cumulative amounts of bypass flow.

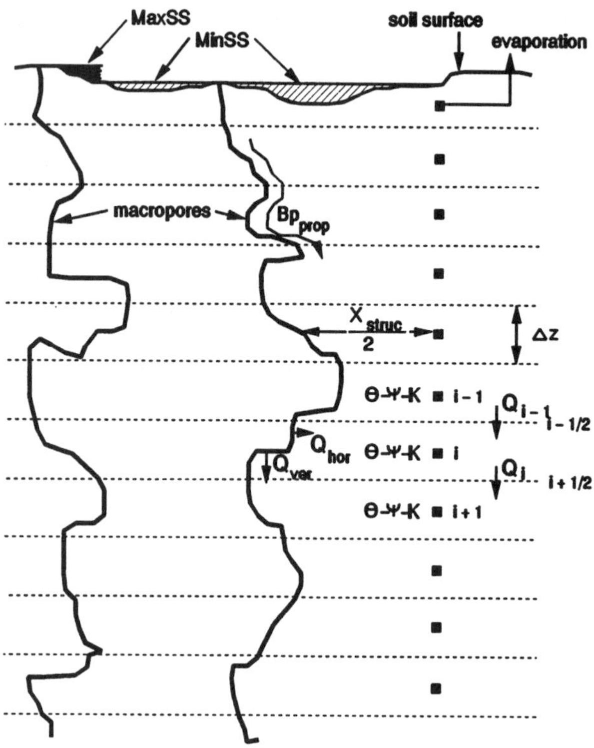

Figure 15: Schematic overview of the two-domain simulation model.

V.B Simulation of Water Flow in the Micropore Domain

Figure 15 shows the general modeling approach followed by Booltink *et al.* (1993). The soil is partitioned into several compartments of equal thickness (Δz). Water flow (Q) from compartment $i-1$ to i is simulated using a finite-difference solution of Richards' equation:

$$\frac{\partial \theta}{\partial t} = \frac{\partial}{\partial z}[K(\theta)\frac{\partial H}{\partial z}] - U(z,t) \tag{14}$$

where θ denotes the volumetric water content (m^3 m^{-3}) at the nodes in the center of each compartment, t is time (day), K is the hydraulic conductivity (m day^{-1}) at the boundary of the nodes (indicated by $i + 1/2$ and $i - 1/2$), the hydraulic head H (m) is equal to the sum of the matrix potential Ψ and of the depth z and is defined at the centers of the nodes, and U is a sink term representing water uptake by plants (day^{-1}). Evaporation is calculated from the top node only.

V.C Simulation of Water Flow in the Macropore Domain

For every rain or irrigation event, the amount of water that could not infiltrate through the soil surface (during the period available) is calculated. The surplus of water is stored on the soil surface and when a threshold value for surface storage (MinSS in Figure 15) is exceeded, water flows into the macropores. Not all existing macropores present are equally accessible, due to small differences in microrelief (Booltink *et al.*, 1993). A maximum surface storage (MaxSS in Figure 15) is therefore defined. Beyond this level, water flows directly into the macropores. Between MinSS and MaxSS, excess water is divided proportionally between surface storage and bypass flow. Water remaining on the surface continues to infiltrate after the rain has stopped. In the macropore domain, water transport is simulated using a tipping-bucket approach. Propagation of the water front in the macropores (Bp_{prop} [mm day^{-1}]) is calculated using Equation (15), based on rain intensities (R_i) and measured soil morphological characteristics (Booltink *et al.*, 1993):

$$Bp_{prop} = a_{trv} \cdot R_i + d_{trv} \cdot T_l + b_{trv} \tag{15}$$

Parameters a_{trv}, b_{trv} and d_{trv} are empirical constants. The dimensionless parameter T_l represents the time lag for initial breakthrough at a specified depth and depends on the volumetric fraction V_s of stained macropores and the fractal dimension of these stains ($D_{s3} - 1$) (Equation (12)). High values of V_s, which are an indication of either a large number of small pores, or few big water-conducting macropores, will shorten the time lag for initial breakthrough. Although fractal dimensions do not give any information on the connectivity of a macropore system, they can provide information on the geometry of water-conducting macropores. In a horizontal plane, strict vertical macropores appear as line-shape entities; the fractal dimension ($D_{s3} - 1$) will, therefore, be close to 1, the dimension of a line. On the other hand, horizontally-oriented macropores will, in that horizontal plane, appear as stain-like spots with a typical $D_{s3} - 1$ value close to 2. Low fractal dimensions will result in rapid macropore flow and short time lags for breakthrough of bypass flow water. Parameters for Equation (15) were derived from laboratory experiments with large (200 mm in diameter and 200 mm long) soil cores, as described in section IV. The soil cores were taken at the boundary between the plough layer and the subsoil since this appeared to be the most restricting layer for bypass flow. Parameters may, therefore, not be correct at smaller scales but they are representative for the system considered.

V.D Interaction Between Micro- and Macropore Domain

Water flowing into macropores is absorbed into macropore walls. Horizontally and vertically oriented macropores behave differently in that respect.

Lateral transport of water from vertically continuous macropores is based on absorption, with diffusion as the most important driving force, as described by a diffusivity equation:

$$Q_{hor} = \frac{Av_{act}}{Av_{pot}} D(\theta) \frac{\Delta\theta}{\Delta[\frac{X_{struc}}{2}]} \tag{16a}$$

where Q_{hor} represents the horizontal absorption flux of water (m day^{-1}), D is the soil water diffusivity in m^{2}day^{-1}, $\Delta\theta$ was calculated as a geometrical mean of saturation (macropore wall) and the actual value of θ in the adjacent soil matrix, as simulated by Richards' equation at nodal points. X_{struc} is the structure element diameter (m). Nodal points for simulating water flow in the micropore domain are located in the center of the structure elements, as depicted in Figure 15. Since macropores were not completely filled with bypass flow water, the potential available vertical contact area (Av_{pot}) between vertically oriented macropores and bypass flow had to be reduced in proportion to their actual, measured, values (Av_{act}).

Vertical absorption of water on horizontal pedfaces of structural elements is dominated by gravity forces. As described by the Darcy equation:

$$Q_{ver} = \frac{Ah_{act}}{Ah_{pot}} K(\Psi) \left[\frac{\Delta\Psi}{\Delta[\frac{X_{struc}}{2}]} + 1 \right] \tag{16b}$$

where Q_{ver} represents the vertical absorption flux (m day^{-1}), and $\Delta\Psi$ was also calculated as a geometrical mean of saturation at the macropore wall and the actual value of θ in the adjacent soil matrix, as simulated by Richards' equation at nodal points. The vertical absorption flux Q_{ver} was reduced for the area of soil in contact with bypass water similarly as in the case of the horizontal absorption flux Q_{hor}. The absorption fluxes were added as an additional sink term to Richards' equation. Contact areas for horizontal and vertical absorption were determined using dye tracers in laboratory experiments. This procedure, previously described by Booltink *et al.* (1993), is based on stratification of methylene blue stained macropores into sets of horizontally- and vertically-oriented macropores using the ratio between area and perimeter of individual stains: $A/Pe^2 < 0.015$ (Bouma *et al.*, 1977). The total contact area Av_{act} for the vertically-oriented cracks was calculated using the following equation:

$$Av_{act} = \sum_{s=1}^{n} \Delta z \cdot Pe \tag{17}$$

in which s is the crack number, n is the total numbers of cracks within a compartment, Δz is the compartment thickness and Pe is the perimeter of the crack. Actual contact areas for horizontally-oriented cracks Ah_{act} were calculated by summing up the measured areas of the horizontally-oriented macropores after stratification.

V.E Model Application to Laboratory Experiments

At the Kandelaar experimental farm in Eastern Flevoland in the Netherlands fifteen soil samples (0.20 m diameter by 0.20 m length) were taken and brought

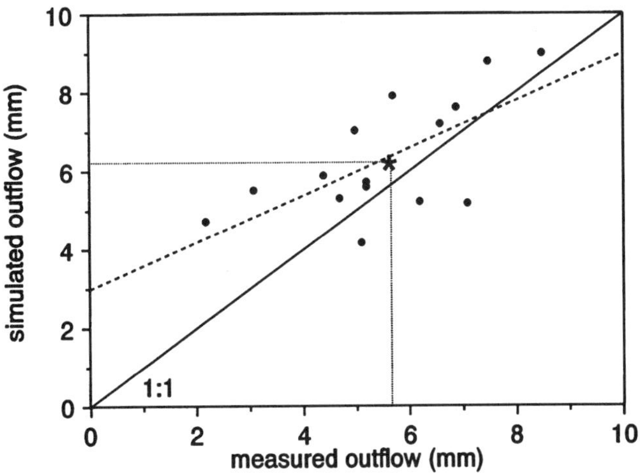

Figure 16: Scatter diagram of measured versus simulated outflow for the simulation of the laboratory experiments. The asterisk indicates the average measured versus the average simulated point. The dashed line represents the regression line through the data points.

to the laboratory. The soil samples were subsequently placed on a funnel, which was connected with an outflow collector equipped with a transducer.

Rain showers of 10, 20, 30 and 40 mm.hr^{-1} were applied on the soil samples using a small rain simulator consisting of an adjustable tube pump and a needle irrigator. All measurements were controlled by a computer equipped with a 14-bit 16-channel A/D-convertor that measured with an interval of 10 s. Additional soil physical characteristics such as the retentivity and the conductivity curves were determined using the one-step outflow method (Doering, 1965, Kool and Parker, 1987). No additional parameter adjustment or calibration was carried out. Booltink *et al.* (1993) describe the experiment and measurement set-up in detail. Model performance was tested by comparing simulation results with measured amounts of outflow for the various rainfall intensities.

V.F Simulation Results: Laboratory Experiments

Figure 16 presents simulated and measured outflow as a scatter diagram. The model overestimates the outflow slightly since data points are not quite evenly distributed around the 1:1 line. The asterisk (*) in Figure 16 represents the average measured and simulated outflow for the 15 samples.

The regression line (R = 0.664) of measured versus simulated outflow (dashed line in Figure 16) deviates from the 1:1 line. This indicates that simulation results show a less dynamic behavior than measured data. Especially at low outflow rates the model tends to overestimate outflow. Table 4 presents average values for the simulation results. Outflow is the most important term

	Average (mm)	Standard Deviation (mm)
Rain	15.0	-
Surface Infiltration	5.4	1.15
Absorption		
Vertical cracks	2.6	1.20
Horizontal cracks	0.7	0.64
Bypass Flow	6.3	1.49

Table 4: Simulated average mass balance for the model simulations on the laboratory experiments for all 15 samples. (Data from Booltink *et al.*, 1993.)

in the mass balance. Due to the relatively dry initial condition and the water remaining on top of the sample, surface infiltration is high compared to previous experiments (Booltink and Bouma, 1991). Absorption into horizontal pedfaces seems hardly relevant.

V.G Model Application to Field Experiments

At the experimental farm "De Kandelaar" in Eastern Flevoland in The Netherlands a tile-drained research site of approximately 100 x 300 m was selected. Tile drains in this site are 150 m long and 48 m apart. A CR10-datalogger (Campbell) continuously monitored rain intensity, ground water level and drain discharge from one tile drain in the center of the study site for 4 years. The simulation model was calibrated using these data. For this purpose, a Monte Carlo simulation procedure was used in which uncertainty of model inputs is characterized in terms of distributions with or without correlations among the various parameters (Jansen *et al.*, 1993a; 1993b). In this procedure the first step is the application of a sensitivity analysis to select the most relevant parameters for further calibration. An iterative Monte Carlo search procedure, which divides simulation outputs in acceptable and unacceptable results, was used to reduce the parameter spaces. With this so-called Rotated Random Scan procedure, one gradually and efficiently zooms in on the acceptable parameter set. This set was used for the final simulations. The sensitivity analysis and the Rotated Random Scan procedure are described in detail by Booltink (1994).

V.H Simulation Results: Field Experiments

In Figure 17, measured and simulated ground water levels are presented in a scatter diagram. The regression coefficient (R = 0.963) of measured versus simulated ground water levels (dashed line in Figure 17) did not significantly differ from the 1:1 line, which represents a perfect simulation. In Table 5, simulated terms of the mass balance are presented.

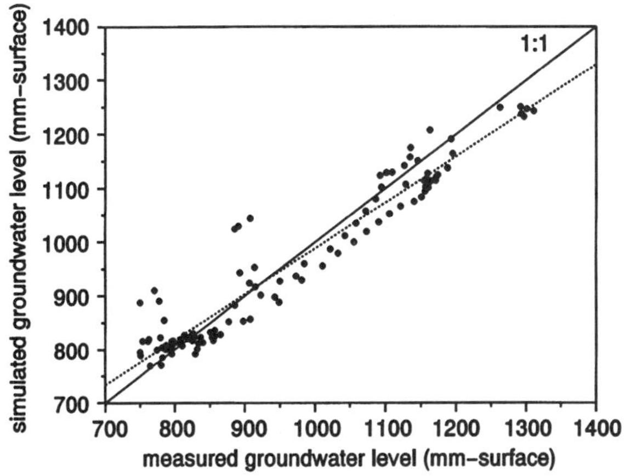

Figure 17: Scatter diagram for measured and simulated ground water levels for the field experiment for the period 1989-1990. The dotted line indicates the regression line through the data points.

Cumulative surface infiltration is the most important term in the water balance. Bypass flow is also substantial and was calculated to be 48.1 mm. Absorption along macropores in this period was small (8.7 mm) due to the permanent wet soil conditions during winter time. Mass balance errors during simulations were small, 0.6 mm average with only a few outliers.

VI Conclusions

This chapter shows that morphology of pathways is an important factor in various processes related to bypass flow. Methylene blue is useful as a tracer to define flowpaths. The fractal analysis shown in this paper can simplify the complex interpretation of staining patterns in terms of fractal dimensions of stains. A regression analysis of the relationship between bypass flow and fractal dimension showed that the depth-weighted fractal dimension of a two-dimensional structure ($D_{s3} - 1$), as well as the volumetric fraction of stains, are useful tools to describe outflow. Although the various regression equations that were obtained are empirical, they can explain reasonably well the various phenomena governing bypass flow. In particular, the macropores where vertical continuation is dominant induce significant bypass flow.

An empirical function predicting the time of initial breakthrough of bypass flow was incorporated in a simulation model to describe bypass flow. Since contact time of water and soil is important in calculating absorption processes along macropore walls this function can be used to describe the propagation of tortuous water transport through a macropore system. Stratification of macropores in terms of vertical continuous cracks and horizontal pedfaces of structure

	Average (mm)	Standard Deviation (mm)
Surface infiltration	131.5	37.5
Evaporation	38.0	3.9
Drainage	79.9	11.5
Absorption	8.7	11.7
Bypass flow	48.1	27.9
Leachate	66.8	10.9
Profile storage	2.9	1.2

Table 5: Simulated cumulative average mass balance for simulations on the field data for the period 1991-1992. Absorption includes horizontal and vertical absorption. Discharge of water to deeper soil layers is considered as leachate. (Data from Booltink, 1994.)

elements, together with the use of fractal dimensions to describe soil geometry derived from measured methylene blue staining patterns, is a promising operational method for the morphological characterization of soil structure. The results of both simulation studies, in the laboratory and for the field situation, were satisfactory. However, due to the different time scales and the different boundary conditions in the field compared to the laboratory, a comparison between the two simulation studies cannot be made.

Further collection of data and validation of the empirical function developed in different soil types is recommended for the future. Coupling this data to soil profile and structure descriptions in existing soil databases may then enhance the usefulness of existing soil data.

References

Andreini, M.S. and T.S. Steenhuis. 1990. Preferential paths of flow under conventional and conservation tillage. *Geoderma* 46: 85-102.

Anderson, J.L. and J. Bouma. 1977a. Water movement through pedal soils. I. Saturated flow. *Soil Sci. Soc. Am. J.* 41: 413-418.

Anderson, J.L. and J. Bouma. 1977b. Water movement through pedal soils. II. Unsaturated flow. *Soil Sci. Soc. Am. J.* 41: 419-423.

Beven, K. and R.T. Clarke. 1986. On the variation of infiltration into a homogeneous soil matrix containing a population of macropores. *Water Resour. Res.* 22: 383-388.

Beven, K. and P. Germann. 1981. Water flow in soil macropores. II. A combined flow model, *J Soil Sci.* 32: 15-29

Beven, K. and P. Germann. 1982. Macropores and water flow in soils. *Water Resour. Res.* 18: 1311-1325.

Booltink, H.W.G. 1994. Field scale distributed modelling of bypass flow in a heavily textured clay soil. *J. Hydrol.* 163: 65-84.

Booltink, H.W.G. and J. Bouma. 1991. Physical and morphological characterization of bypass flow in a well-structured clay soil. *Soil Sci. Soc. Am. J.* 55: 1249-1254.

Booltink, H.W.G. and J. Bouma. 1993. Sensitivity analyses on processes affecting bypass flow. *Hydrol. Process.* 7: 33-43.

Booltink, H.W.G., R. Hatano and J. Bouma. 1993. Measurement and simulation of bypass flow in a structured clay soil: a physico-morphological approach. *J. Hydrol.* 148:149-168.

Bouma, J. 1984. Using soil morphology to develop measurement methods and simulation techniques for water movement in heavy clay soils. In J. Bouma and P.A.C. Raats (Eds.), *Water and Solute Movement in Heavy Clay Soils.* Int. Soc. Soil Science (ISSS) Symp. Inst. Land Reclamation and Irrigation (ILRI), Wageningen, Netherlands, Publ. No.37, pp.298-316.

Bouma, J. 1985. Soil variability and soil survey. In J. Bouma and D.R. Nielsen (Eds.), *Soil Spatial Variability.* Int. Soc. Soil Science (ISSS) and Soil Sci. Soc. Am. (SSSA) Symp. PUDOC, Wageningen, Netherlands, pp.130-149.

Bouma, J. 1990. Using morphometric expressions for macropores to improve soil physical analyses of field soils. *Geoderma* 46: 3-11.

Bouma, J., C. Belmans, L.W. Dekker and W.J.M. Jeurissen. 1983. Assessing the suitability of soils with macropores for subsurface liquid waste disposal. *J. Environ. Qual.* 12: 305-310.

Bouma, J. and L.W. Dekker. 1978. A case study on infiltration into dry clay soil. I. Morphological observations. *Geoderma,* 20: 27-40.

Bouma, J., A. Jongerius, O. Boersma, A. de Jager and D. Schoonderbeek. 1977. The function of different types of macropores during saturated flow through four swelling soil horizons. *Soil Sci. Soc. Am. J.* 41: 945-950.

Bouma, J., A. Jongerius and D. Schoonderbeek. 1979. Calculation of saturated hydraulic conductivity of some pedal clay soils using micromorphometric data. *Soil Sci. Soc. Am. J.* 43: 261-264.

Bouma, J. and H.A.J. Van Lanen. 1987. Transfer functions and threshold values: from soil characteristics to land qualities. pp. 106-111. In *Quantified Land Evaluation.* Proceedings of a workshop. ISSS/SSSA, Washington, D.C. ITC Publ. Enshede, the Netherlands.

Bouma, J. and J.H.M. Wosten. 1979. Flow patterns during extended saturated flow in two, undisturbed swelling clay soils with different macrostructures. *Soil Sci. Soc. Am. J.* 43: 16-22.

Bronswijk, J.J.B. 1988. Modeling of water balance, cracking and subsidence of clay soils. *J. Hydrol.* 97: 199-212.

Chen, C. and R.J. Wagenet. 1992. Simulation of water and chemicals in macropore soils. Part I. Representation of the equivalent macropore influence and its effect on soil water flow. *J. Hydrol.* 130: 105-126.

De Smedt, F., F. Wauters and J. Sevilla. 1986. Study of tracer movement through unsaturated sand. *Geoderma* 38: 223-236.

DiPietro, L. 1993. *Transferts d'Eau dans des Milieux à Porosité Bimodale:*

Modélisation par la Méthode de Gaz sur Réseaux. Unpublished Ph.D. Dissertation. University of Montpellier. Montpellier, France.

Doering, E.J. 1965. Soil-water diffusivity by the one-step method. *Soil Sci.* 99: 332-326

Edgar, G.A. 1992. *Measure, Topology and Fractal Geometry*. Springer-Verlag, New York.

Edwards, W.M., R.R. van der Ploeg and W. Ehlers. 1979. A numerical study on the effects of non-capillary sized pores upon infiltration. *Soil Sci. Soc. Am. J.* 43:851-856

Frisch, U. 1986. A new strategy for hydrodynamics: lattice gasses. In: Proceedings II European turbulence conference. Springer Verlag, Berlin.

Germann, P.F. 1990. Preferential flow and the generation of runoff. I. Boundary layer flow theory. *Water Resour. Res.* 26: 3055-3063.

Germann, P.F. and K. Beven. 1985. Kinematic wave approximation to infiltration into soils with sorbing macropores. *Water Resour. Res.* 21: 990-996.

Hatano, R. and H.W.G. Booltink. 1992. Using fractal dimensions of stained flow patterns in a clay soil to predict bypass flow. *J. Hydrol.* 135: 121-131.

Hatano, R. and H.W.G. Booltink. 1994. A morphological approach to describe solute movement influenced by bypass flow in a structured clay soil. *Soil Sci. Plant Nutr.* 40: 573-580.

Hatano, R., N. Kawamura, J. Ikeda and T. Sakuma. 1992. Evaluation of the effect of morphological features of flow paths on solute transport by using fractal dimensions of methylene blue staining pattern. *Geoderma* 53: 31-44.

Hatano, R. and T. Sakuma. 1991. A plate model for solute transport through aggregated soil columns. I. Theoretical description. *Geoderma* 50: 13-23.

Hatano, R., T. Sakuma and H. Okajima. 1985. The source-sink effect of clayey soil peds on solute transport. *Soil Sci. Plant Nutr.* 31: 199-213.

Hatano, R., A. Tomita and T. Sakuma. 1993. Diffusion processes in water saturated spherical soil aggregates. *Soil Sci. Plant Nutr.* 39: 245-255.

Hoogmoed, W.B. and J. Bouma. 1980. A simulation model for predicting infiltration into cracked clay soil. *Soil Sci. Soc. Am. J.* 44: 458-461.

Ishiguro, M. 1991. Solute transport through hard pans of paddy fields. I. Effect of vertical tubular pores made by rice roots on solute transport. *Soil Sci.* 152: 432-439.

Jansen, P.H.M., P.S.C. Heuberger and R. Sanders. 1993a. *UNCSAM 1.1: a Software Package for Sensitivity and Uncertainty Analysis*; manual. RIVM report nr. 959101004, RIVM, Bilthoven, The Netherlands.

Jansen, P.H.M., P.S.C. Heuberger and R. Sanders. 1993b. *Rotated-Random Scanning: a Simple Method for Set Valued Model Calibration*. RIVM internal report, in preparation, RIVM, Bilthoven, The Netherlands.

Jarvis, N.J. 1991. *Macro: a Model of Water Movement and Solute Transport in Macroporous Soils*. Dept. of Soil Sciences, Swedish Univ. of Agric. Sci., Uppsala, Sweden. pp 1-58.

Jarvis, N.J. and P.B. Leeds-Harrison. 1987. Modelling water movement in drained clay soil. I. Description of the model, sample output and sensitivity analysis. *J. Soil Sci.* 38: 487-498.

Jarvis, N.J. and P.B. Leeds-Harrison. 1990. Field test of a water balance model of cracking clay soils. *J. Hydrol.* 112: 203-218.

Kissel, D.E., J.T. Ritchie and E. Burnett. 1973. Chloride movement in undisturbed swelling clay soil. *Soil Sci. Soc. Am. Proc.* 37: 21-24.

Kool, J.B. and J.C. Parker. 1987. *Estimating Soil Hydraulic Properties from Transient Flow Experiments: SFIT user's guide.* Electric Power Research Institute, Palo Alto, California, USA.

Lauren, J.G., J. Wagenet, J. Bouma and J.H.M. Wosten. 1988. Variability of saturated hydraulic conductivity in a Glossaquic Hapludalf with macropores. *Soil Sci.* 145: 20-28.

Lovejoy, S. 1982. Area-perimeter relation for rain and cloud areas. *Science* 216: 185-187.

Mandelbrot, B.B. 1982. *The Fractal Geometry of Nature.* W.H. Freeman and Company, New York.

Passioura, J.B. 1977. Hydrodynamic dispersion in aggregated media. I. Theory. *Soil Sci.* 111: 339-344.

Ritchie, J.T., D.E. Kissel and E. Burnett. 1972. Water movement in undisturbed swelling clay soil. *Soil Sci. Soc. Amer. Proc.* 36: 874-879.

Ringrose-Voase, A.J. and P. Bullock. 1984. The automatic recognition and measurement of soil pore types by image analysis and computer programs. *J. Soil Sci.* 35: 673-684.

Rose, D.A. and J.B. Passioura. 1971. The analysis of experiments on hydrodynamic dispersion. *Soil Sci.* 111: 252-257.

Vanclooster, M., H. Vereecken, J. Diels, F. Huysman, W. Verstraete and J. Feyen. 1992. Effect of mobile and immobile water in predicting nitrogen leaching from cropped soil. *Modelling Geo-Biosphere Processes* 1: 23-40.

Van Genuchten, M.Th. and P.J. Wierenga. 1976. Mass transfer studies in sorbing porous media. I. Analytical solutions. *Soil Sci. Soc. Am. J.* 40: 473-480.

Van Stiphout, T.P.J., H.A.J. Van Lanen, O.H. Boersma and J. Bouma. 1987. The effect of bypass flow and internal catchment of rain on the water regime in a clay loam grassland soil. *J. Hydrol.* 95: 1-11.

White, R.E., G.W. Thomas and M.S. Smith. 1984. Modelling water flow through undisturbed soil cores using a transfer function model derived from ^{3}HOH and Cl transport. *J. Soil Sci.* 35: 159-168.

White, R.E. 1985. The influence of macropores on the transport of dissolved and suspended matter through soil. *Adv. Soil Sci.* 3: 95-120.

2-D and 3-D Fingering in Unsaturated Soils Investigated by Fractal Analysis, Invasion Percolation Modeling and Non-Destructive Image Processing

S. Crestana and A. N. D. Posadas

Contents

ISBN 1-56670-105-8

I Introduction

"Soil is one of the most complex systems known. Highly nonlinear and highly variable, the soil system contains an infinite number and variety of chemical and biological phenomena. They ensure that the system is never static, never at equilibrium. Soils, if they are to be understood, must be studied at the atomic, human, and global scales. They are as worthy of study in and of themselves as are the heavens and the depths of the oceans. Past research has shown how to manage soil for a specific use, but today there is societal pressure to do this and at the same time maximize the long-term productivity of the soil and minimize environmental pollution. The tools of modern science now allow us to deal with surfaces at the atomic level. Concepts such as fractals and chaos help us to deal with the mind-boggling variability of soils" (Sposito and Reginato, 1992).

One of the physical phenomena considered to be of great relevance and practical importance in soil physics, as well as in petroleum engineering, is the fingering phenomenon or hydrodynamic instability, which occurs under certain physical conditions, when a fluid flows through an unsaturated medium.

The hydrodynamic instability of the flow through the unsaturated zone of the soil may take place due to soil stratification or to imposed initial conditions (Glass *et al.*, 1991). The occurrence of flow instability may be easily observed in a soil when the hydraulic conductivity increases with depth, as in the case of stratified soils, for instance, when the fluid meets, during its movement, an interface of great variation of hydraulic conductivity, from a smaller value (*e.g.*, fine texture) to a greater value (coarse texture). The equations that usually describe the transport of water and solutes in saturated media, assume valid the hypothesis of a stable wetting front. Therefore, because they have not been conceived to deal with unstable flow conditions, they usually do not apply satisfactorily to such conditions.

The fluid flow through preferential paths or fingers is extremely important in agriculture in the hydrological processes of infiltration and in the transport of agrochemicals through the soil profile (Sposito and Reginato, 1992). For example, the role that the preferential flow plays in the transport of pesticides, heavy metals, radioactive waste and other contaminants is particularly interesting. These chemicals may, as a consequence of preferential flow, reach the ground water and compromise the quality of water resources that will later be used for irrigation or human use.

The fingering phenomenon in soils has been extensely studied in the laboratory (Hill and Parlange, 1972; White *et al.*, 1977; Tamai *et al.*, 1987; Glass *et al.*, 1989a,b; Baker and Hillel, 1990; Glass *et al.*, 1990; Selker *et al.*, 1992) as well as in the field (Starr *et al.*, 1978; Glass *et al.*, 1988; Ritsema and Dekker, 1993). These studies have been carried out basically in two dimensions through visual observations (Chang, 1990), light transmission (Glass *et al.*, 1989) and digitizing and image processing (Posadas and Crestana, 1993).

Experiments were recently carried out to learn more about the physical mechanisms of the occurrence of fingering in soils. These experiments showed that fingering is dominated by gravity, that it occurs occasionally during infiltration and fluid redistribution, and that initial soil-water content conditions are key factors determining its appearance (Glass *et al.*, 1988). Consequently, it may take place even in homogeneous soils (Selker et al, 1992; Posadas and Crestana, 1995), where the determining factor is the rate of fluid application at the soil surface.

In relation to theoretical models for the explanation of the fingering phenomenon, some efforts have been made toward quantification, the most relevant ones being the analytical and semi-analytical models of Raats (1973), Philip (1975), Hillel and Baker (1988), Glass *et al.* (1989a,b), Chang *et al.* (1994), and those of Diment and Watson (1983) and Hosang (1993), which involve numerical simulation. Due to the complexity of the three-dimensional description of the phenomenon, all of these models were developed in two dimensions, despite the fact that, on the whole, the phenomenon presents three-dimensional characteristics.

Recently, Glass *et al.* (1991) and Selker *et al.* (1992) have extended their original model from two to three dimensions, validating it experimentally in the laboratory, which is a very interesting result. The sample of sand used was frozen after development of the fingers and was sliced in order to describe the fingers quantitatively. In spite of the precautions that were taken, the experimental procedures employed may still have introduced disturbances that may have affected the validation process of the model. Besides, the quantification was done for just one instant of the same sample.

Thus, the phenomenon of fingering in soils, similar in many ways to that occurring in petroleum reservoirs, presents great challenges to scientific research, as much from the perspective of conception and validation of theoretical models as from that of representative experiments.

In recent years, great attention has been given to the study of fluid transport in porous media. The main reasons for this are the economic interest in improving the secondary and tertiary recovery of petroleum and the possible water and soil contamination due to intensive use of agrochemicals. In order to understand this extremely complex problem, some researchers have devoted themselves to the study of fluid transport in connection to the geometry of porous media. Many experiments performed in this way have shown that the fluid transport-porous medium coupling has self-similarity or fractal characteristics in a range of defined scales (Katz and Thompson, 1985; Hansen and Skjelter, 1988). Those characteristics have led to the creation of some simulation models such as Diffusion-Limited Aggregation (*DLA*) (Witten and Sanders, 1983), which simulates the viscous-fingering phenomenon (Maløy *et al.*, 1985a,b; Chen and Wilkinson,1985), and invasion percolation (Wilkinson and Willemsen, 1983), which simulates the capillary-fingering phenomenon. Fingering in soils, which is basically of capillary character, has fractal characteristics (Chang, 1990; Chang *et al.*, 1994; Posadas and Crestana, 1993; Posadas and Crestana, 1995), suggesting that the use of a modified theory of invasion percolation is an appropriate model of simulation (Posadas *et al.*, 1993; Posadas and Crestana, 1995; Onody *et al.*, 1995).

The purpose of this chapter is to describe the state of the art related to the

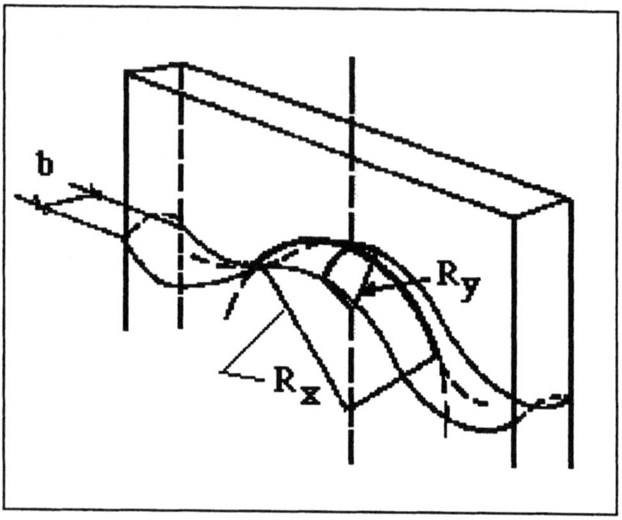

Figure 1: Hele-Shaw cell showing the geometry of the fluid-fluid interface and the two curvature radii R_x and R_y.

research on the "fingering" phenomenon, and to describe applications of fractal theory (Mandelbrot, 1982) and of digitizing techniques, image processing, and X-ray and nuclear magnetic resonance tomography (Crestana, 1992; Posadas and Crestana, 1995; Tannús *et al.*, 1995; Posadas *et al.*, 1996) to the characterization of fingering phenomena in soils, as well as introducing a modified theory of invasion percolation to the simulation of fingers (Posadas and Crestana, 1995; Onody *et al.*, 1995).

II Viscous Fingering

The problem of viscous fingering in porous media is of vital importance in petroleum recovery (Berry and Christie, 1991; Nittman *et al.*, 1985; Lenormand *et al.*, 1988). It is also an interesting problem in view of hydrodynamics and physics of porous media, because fingering is one of the growth phenomena that happens far from thermodynamic equilibrium and it possesses three-dimensional characteristics. In this respect, viscous fingering presents invariance of scale (auto-similarity), making it amenable to the application of theories such as that of renormalization group (*RG*), "scaling" and disordered systems in general (Bensimon, 1986; Dupont *et al.*, 1993).

II.A Fingering in the Hele-Shaw Cell

The Hele-Shaw cell, as seen in Figure 1, consists of two parallel transparent plates of size L separated by a distance b. Hele-Shaw (1898) studied the flow of water through several objects used as cells. He observed the lines of currents of flow by injecting coloring substances that traced in colors these lines of current. His experiments showed that for a Hele-Shaw cell with a small value of b ($b << 1.0$), the flow had characteristics of a "flow potential" for a low Reynolds number (Bear, 1972). When b was increased, the fluid became turbulent, with confusing current lines. Thus, considering the Hele-Shaw cell in Figure 1, the equation of velocity of flow V, obtained from the Navier-Stokes equation (Bear, 1972) governing the flow, is given as (Feder, 1988):

$$\mathbf{V} = -\frac{K}{\mu}\boldsymbol{\nabla}(p + \rho g z) \tag{1}$$

where $\mathbf{V}$ is the average velocity on the cell thickness, p is pressure, ρ the fluid density, g the gravity acceleration, z the coordinate on z-axis, μ the fluid viscosity and $K = b^2/12$ the cell permeability.

In the case of incompressible fluids, using the continuity equation and applying $\boldsymbol{\nabla} \cdot \boldsymbol{\nabla}$ in Equation (1), the Laplace equation is obtained:

$$\nabla^2 \phi = 0 \tag{2}$$

where $\phi = -\frac{K}{\mu}(p + gz)$ is a potential.

Equation (2) is characteristic of problems of potential found in electrostatics and diffusion, among others, studied in physics. Consequently, the flow controlled by the Laplace equation (2) is called potential flow (Feder, 1988).

In order to find a solution for the flow velocity, the boundary conditions of the system should be specified. For instance, some known pressure is applied to both ends of the cell and it is assumed that the fluid velocity is equal to zero, when in contact with the walls of the container.

Also, the interface between two fluids (*i.e.*, water and air) is controlled by capillary forces when at rest, and the difference of pressure between both of them is expressed as:

$$\Delta p = (p_1 - p_2) = \gamma \left(\frac{1}{R_x} + \frac{1}{R_y} \right) \tag{3}$$

where γ is the interfacial tension of the interface between the two fluids, R_x and R_y are the main curvature radii. Typically $R_x >> R_y$, when $b << 1.0$ (Feder, 1988).

Now let us inject the fluid (1) at a constant rate V at $z = -\infty$ and withdraw fluid (2) at the same rate at $z = \infty$. The interface between the two fluids will then move with a velocity $\mathbf{V} = (0, 0, V)$ along the z-axis. However, it turns out that the interface is unstable if the viscosity of the driving fluid (1) is smaller than the viscosity of the fluid being driven (2). Engeberts and Klinkeberg (1951) coined the term **viscous fingering** in relation to their observation of such instabilities when water drives oil out of a porous medium (Feder, 1988).

Flow in porous media also follows Equations (1) and (2), and therefore the flow in Hele-Shaw cells is often used to model the flow in porous media. However, as we shall see, there are important differences and the validity of using the Hele-Shaw cell as a model of flow in porous media is questionable. (Feder, 1988; Vicsek, 1992).

The theory of viscous fingering was developed and compared to experiments independently by Saffman and Taylor (1958) and by Chuoke *et al.* (1959). Recently there has been a growing interest in the field and many new theoretical and experimental results have been published, *e.g.*, those of Bensimon *et al.* (1986a), Jensen *et al.* (1987) and DeGregoria and Schwartz (1987). A recent review is given by Homsey (1987).

In order to test the stability of the advancing interface we follow the standard practice (Saffman and Taylor, 1958; Chuoke *et al.*, 1959) and assume that the straight interface is perturbed by a sinusoidal displacement so that in the moving frame of reference the position of the interface is given by the real part of

$$\xi = \varepsilon \exp\left(2\pi n t + \frac{2\pi x}{\lambda} i\right) \tag{4}$$

where ε is the wave amplitude, n is the instability coefficient, t is time, λ is the perturbation wavelength, x is the horizontal axis and i is $\sqrt{-1}$. If $n < 0$, ξ decreases with time and the displacement front remains stable; if $n > 0$, ξ then grows exponentially, making the displacement front unstable.

Solving Equations (1) to (3), including only linear terms, one finds that ξ in the front is unstable with respect to perturbations that have a wavelength λ longer than a critical wavelength λ_c given by (Feder, 1988):

$$\lambda_c = 2\pi \sqrt{\frac{\gamma}{\left(\frac{\mu_2}{K_2} - \frac{\mu_1}{K_1}\right)(V - V_c)}} \tag{5}$$

where μ_1, μ_2, K_1, K_2 and V_c are the viscosity, the hydraulic conductivity of fluid (1) and fluid (2) and the critical velocity, respectively.

Perturbations with shorter wavelength λ_c are stabilized by interfacial tension. Thus, for all wavelengths $\lambda > \lambda_c$, the displacement front will be unstable, and perturbations with a wavelength $\lambda_{\max}$ expressed as:

$$\lambda_{\max} = \sqrt{3}\lambda_c \tag{6}$$

will present the greatest growth rate and dominate the dynamics of the displacement front (Feder, 1988).

On the other hand, introducing the viscosity μ of the fluid and making $g \approx 0$ (neglecting gravitational effects), it is found that (Feder, 1988):

$$\lambda_{\max} \propto \sqrt{\frac{\gamma}{\overline{V}\mu}} = \frac{1}{\sqrt{N_{ca}}} \tag{7}$$

where N_{ca} is called the capillary number and is defined as:

$$N_{ca} \equiv \frac{\overline{V}\mu}{\gamma} \tag{8}$$

The capillary number measures the relation between the viscous and capillary forces. If we fix μ and γ, then the capillary number N_{ca} governs the growth

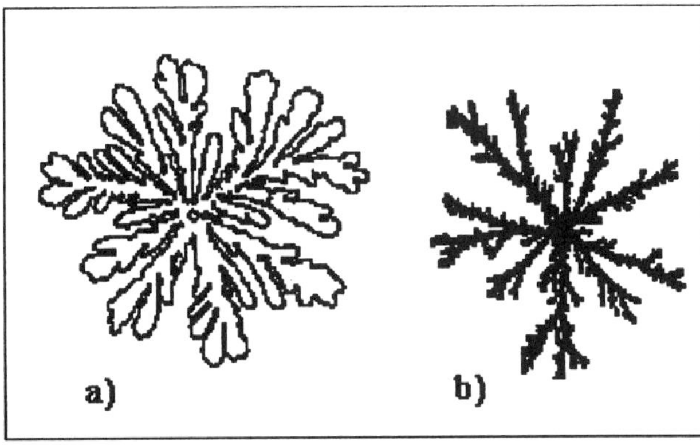

Figure 2: Viscous fingering generated in a radial Hele-Shaw cell. In (a) air was used as a fluid to dislocate glycerol. The capillary number obtained was $N_{ca} = 0.1$. The generated structure is not fractal (Maløy *et al.*, 1985b). (b) represents a fractal structure. In this case water was used to dislocate a "non-Newtonian" and highly viscous fluid ($N_{ca} >> 1.0$). The fractal dimension found for this structure was $D \approx 1.7$ (Daccord *et al.*, 1986).

dynamics of fingers, since it will only depend on the velocity module $\overline{V}$. If we vary this $\overline{V}$, we can control the capillary number N_{ca} and also the domination of the viscous or capillary forces for the front displacement. For instance, if the velocity $\overline{V}$ is great, then the viscous forces will govern the system. On the contrary, if $\overline{V}$ is small or slow, the capillary forces will govern the process.

Theories were also demonstrated by Paterson (1981) and Maløy *et al.* (1985), for different values of the capillary number N_{ca}, using a circular Hele-Shaw cell with the liquid ejected in the center of the cell. Nittmann *et al.* (1985), using a circular Hele-Shaw cell and a capillary number $N_{ca} >> 1.0$, found that the structure of generated fingers had fractal characteristics. Its fractal dimension was measured and found to be equal to 1.7, similar to the structures obtained through the theoretical model of DLA (Witten and Sanders, 1981).

Figure 2 presents two experiments of the viscous fingering phenomenon in a radial Hele-Shaw cell. In Figure 2a, a non-fractal structure is shown, obtained by using the capillary number $N_{ca} = 0.1$ (Maløy *et al.*, 1985a), whereas in Figure 2b, a fractal structure is shown, obtained through a capillary number $N_{ca} >> 1.0$ (Daccord *et al.*, 1986). Thus, the viscous fingering, generated through a Hele-Shaw cell, may only present fractal characteristics under certain physical conditions (Vicsek, 1992).

II.B Viscous Fingering in a Porous Medium

The flow in a porous medium is also controlled by the same equations, (2) and (3), deduced for the Hele-Shaw cell. However, K now represents the porous

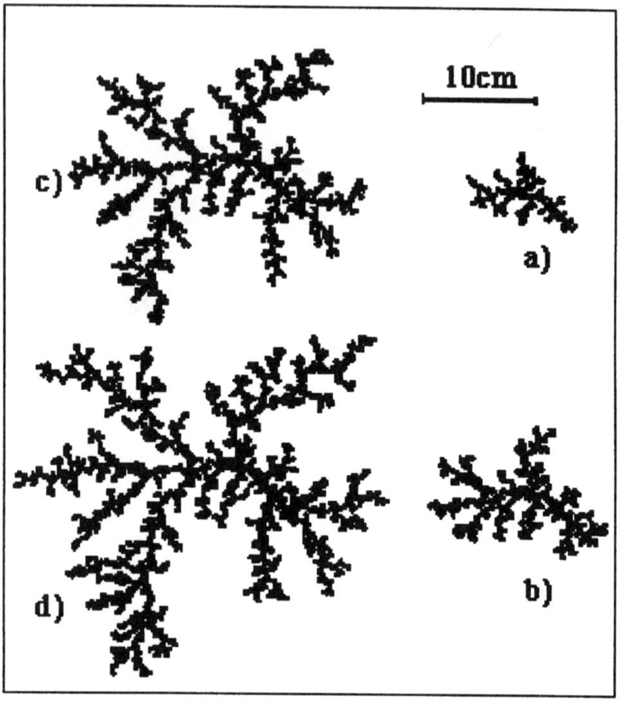

Figure 3: Structure of fingers generated in a radial cell with a 40-cm diameter, containing glass spheres representing a porous medium. The experimental conditions used were the same ones employed in the experiment of Figure 2a. This structure has fractal characteristics, as was demonstrated by Maløy *et al.* (1985a).

medium permeability (not $K = b^2/12$ anymore). Also, the experiments showed that the fingering dynamics in a porous medium is quite different. For instance, Figure 3, which is a fractal object, represents the fingering obtained in a porous medium in conditions similar to those obtained in Figure 2a (Maløy *et al.*, 1985a,b). Both figures are completely different. The most important difference between the Hele-Shaw cell and the cell containing a porous sample lies in the contour conditions. For a porous medium, the flow is controlled through microscopic wavelengths defined by the size of the pores in all directions of space. In the Hele-Shaw cell the flow is controlled only through the geometric parameter of scale b that represents the separation between the plates. The velocity of flow $\mathbf{V}$ in a porous medium given by Equation (1) remains the same, but it is now called Darcy's equation. Also, the Laplace equation is still valid for a porous medium, but with a different physical meaning. In the Hele-Shaw cell the length scale is indeed given by the critical wavelength λ_c, whereas in the porous medium the length scale is associated to the size of the pores. The dynamics of fluid displacement front is also different. For the Hele-Shaw cell it only matters that the distributions of pressure satisfy the contour conditions between the two plates (absolute difference of pressure between both fluids). For a porous medium, the front displacement is associated to the capillary pressure of the pore throat, which transmits the fluid from one pore to another. This introduces a randomness in the problem, because the pore diameters are distributed randomly in the porous system.

The dynamics of the viscous-fingering front in porous media therefore has two main components: the global pressure distribution controlled by Darcy's law and therefore the Laplace equation, and the local fluctuations in pore geometry. The result of these two factors is a growing fractal structure (Feder, 1988).

Thus, for a capillary number $N_{ca} >> 1.0$ the fractal structure of the fingers in a porous medium (case of viscous fingering) follows the DLA growth model, with fractal dimension $D \approx 1.7$. For a capillary number $N_{ca} << 1.0$ (capillary fingering), the observed fingering structure has fractal characteristics pertaining to the model of invasion percolation (Willkinson and Willensen, 1983; Daccord, 1987), with fractal dimension $D \approx 1.89$ (Dias and Wilkinson, 1986).

In the last years, the DLA model (Tamai *et al.*, 1987; Lenormand, 1986; Lenormand *et al.*, 1988) has been extensively used, from both experimental and theoretical contexts, in the study of the viscous fingering phenomenon. The same applies to studies of capillary fingering, using the model of invasion percolation (Wilkenson and Willemsen, 1983; Meakin *et al.*, 1992; Dias and Wilkinson, 1986). These studies were basically done in two dimensions and neglected gravitational effects. In three dimensions, theoretical work has been done with some validation experiments of Frette *et al.* (1990). However, these experiments were done under ideal conditions and, apparently, far from real conditions (Frette *et al.*, 1991). The reason for this basically lies in the difficulty that the phenomenon presents in 3D, when the gravitational effects can no longer be disregarded. In addition, there is a lack of methodologies capable of measuring finger dynamics within the porous material without perturbing it significantly. Nonetheless, some efforts and encouraging results have been obtained by using tomographic techniques, such as magnetic resonance imaging (MRI) (Chen *et al.*, 1988), but also in experiments done in static conditions and using quite idealized materials, such as glass spheres. Recently, Crestana (1992), Posadas *et al.* (1993), and Posadas and Crestana (1995) have introduced the magnetic resonance and X-ray tomography techniques to the study of the fingering phenomenon in soils. The results obtained showed that both techniques are potentially promising to the study of the fingering phenomenon in porous media, in both dynamic and static conditions. These results open new ways and point to new frontiers concerning scientific research on the fingering phenomenon, which will be presented at the end of this chapter.

III Fingering in Soils

The occurrence of the fingering phenomenon in soils, according to the definition of the capillary number N_{ca} given by Equation (8), is basically driven by capillary forces (Posadas and Crestana, 1993). That is to say that capillary forces control the dynamics of the phenomenon. However, one should remember that Equation (8) was obtained for $g \approx 0$, *i.e.*, gravitational effects were neglected. The experimental results of the fingering phenomenon in soils show that the phenomenon dynamics is controlled by gravity (Hill and Parlange, 1972; Glass *et al.*, 1988, 1989; Baker and Hillel, 1990). Thus, the effect of gravity cannot be disregarded when we study the fingering phenomenon in soils, even in two dimensions.

When gravitational forces are important, instead of using the capillary number N_{ca} it is better to introduce another parameter that will take this effect into consideration. This parameter is called the "bond number" (B), and represents the relation between gravitational and capillary force, and is expressed as (Blumberg *et al.*, 1980; Roux and Wilkinson, 1988; Wilkinson, 1986; Jacquin and Adler, 1985):

$$B = \frac{\Delta \rho g \overline{r}^2}{\gamma} \tag{9}$$

where $\Delta\rho$ is the difference in density between the two fluids, g is acceleration due to gravity, $\overline{r}$ is the average diameter of the pore and γ is the surface tension. For example, in the typical case of the two fluids water and air, and considering a pore with average diameter $\overline{r} = 20\mu m$, $\Delta\rho = 1.0$ g/cm^3, $g = 980$ cm/s^2 and $\gamma = 72.25$ dyne/cm in Equation (9), we have $B \sim 10^{-4}$. This value is very small, indicating that, on a microscopic level, gravitational effects may be disregarded. However, in bigger samples gravitational pressure may be equal to or greater than the capillary pressure. For instance, for a system N pores in height, these pores will contribute cumulatively, each with its gravitational pressure, giving a total contribution of pressure equal to NB. Thus, for a system that is 10 cm long, $N = 5000$ pores will fit in it, which multiplied by $B = 10^{-4}$ gives $NB \sim 0.5$. This value is of the same order of magnitude as the capillary forces. Then, on a macroscopic level, the gravitational effects are of great importance to the development of the capillary fingering phenomenon, which at a given instant may dominate the phenomenon dynamics. This is what usually happens in the case of soils. It must be mentioned also that in soils, besides gravitational pressure, hydraulic pressure acts, which is a macroscopic parameter. In this way, gravitational pressure and hydraulic pressure, which are two macroscopic parameters, will play a very important role in the study of the fingering phenomenon dynamics in soils, especially after the transient and in conditions of constant injection of fluid (as in the case of irrigation, for instance). In conditions of drainage or fluid redistribution in the soil matrix after interrupting its application, the situation may be different.

As noted above, fingering phenomena in soils have been intensively studied by researchers of soil physics. For example, Philip (1975), on the basis of results obtained by Saffman and Taylor (1958) and Chuoke *et al.* (1959), determined the critical wavelength λ_c for the occurrence of fingers in soils in two dimensions, taking into consideration the effects of gravity and the water

Sandy soil sample number	Particle diameter: d (mm)	Sieve diameter
1	0.106 < d < 0.149	140 - 100
2	0.212 < d < 0.500	70 - 35
3	0.297 < d < 0.500	50 - 35
4	0.500 < d < 1.000	35 - 18
5	1.000 < d < 2.000	18 - 10
6	0.149 < d < 1.000	100 - 18

Table 1: Samples of sandy soil selected for the fingering phenomenon experiments, in 2D.

content of the soil sample. This relation is expressed as:

$$\lambda_c = 2\pi\sqrt{\frac{\gamma}{\rho g(1 - \frac{\theta}{K}\overline{V})}} \tag{10}$$

where ρ is the fluid density, $\overline{V}$ is the average velocity of the fluid, K is the soil hydraulic conductivity, g is the gravity acceleration and γ is surface tension of the fluid-air interface.

Most theoretical work on fingering in soils was developed from Philip's (1975) work with variants for specific situations for a given type of soil, with special attention given to the work done by Glass *et al.* (1989a,b), Baker and Hillel (1990) and Glass *et al.* (1990).

The only accounts in the literature about the use of fractal theory in the study of the fingering phenomenon in soils were introduced by Chang (1990), Chang *et al.* (1994), Posadas and Crestana (1993) and Posadas and Crestana (1995). Chang *et al.* (1994) characterize the dynamics of the fingers' front through the fractal dimension D and try to describe the phenomenon by introducing the concept of "effective" surface tension. They define a surface tension that depends on front dynamics. This surface tension is related to the fractal dimension D. It is expressed as:

$$\gamma^* = C\gamma\varepsilon^{1-D} \tag{11}$$

where γ^* is the effective surface tension that depends on the front dynamic, γ is the surface tension under static conditions, C is a dimensionless proportionality constant, ε is the scale length and D is the fractal dimension.

If we analyze Equation (11), it appears that the "effective" surface tension, besides depending on the fractal dimension D, also depends on the scale ε, a variable used to determine D.

From a mathematical point of view, ε may be infinitely small, but the physical meaning is only considered in a specific range of values. Thus, the smallest value of ε is assumed to be equal to the average pore diameter of the porous system. Through this concept, the "effective" surface tension is estimated as

Sandy soil sample number	Saturated hydraulic conductivity K (cm/s) x 10^{-3}	Total porosity Φ (%)	Bulk density ρ_b (g/cm^3)
1	6.3 ± 0.5	44.8 ± 3.0	1.50 ± 0.08
2	26.7 ± 1.5	31.2 ± 3.0	1.60 ± 0.08
3	30.5 ±2.0	32.4 ± 1.6	1.60 ± 0.08
4	47.8 ± 3.5	18.2 ± 4.0	2.18 ± 0.20
5	52.7 ± 3.7	16.0 ± 3.0	2.24 ± 0.20
6	14.2 ± 1.5	19.0 ± 3.0	2.16 ± 0.20

Table 2: Values of saturated hydraulic conductivity (K), of total porosity (ε) and of bulk density (ρ_b).

well as the maximum wavelength capable of generating fingers. However, it is still necessary to do more research to clarify this concept, especially the experimental validation of the γ^* value found.

The results obtained by Posadas *et al.* (1993) and Posadas and Crestana (1995) will be presented in the next section of this chapter.

IV Fractal and Imaging Analysis to Characterize Fingering Phenomenon in 2D

In order to characterize the fingering phenomenon, fractal theory and imaging techniques, both in dynamic (Chang *et al.*, 1994; Posadas and Crestana, 1993; Posadas *et al.*, 1993) and in static conditions (Posadas, 1994), were employed. Experiments were performed in the laboratory, using planar columns (2D), similar to the experiments done by Glass *et al.* (1989). In order to perform those experiments, six samples of quartz sand used in construction, passing through US standard sieves number10, 18, 35, 50, 70, 100 and 140 respectively, were selected, as shown in Table 1.

The samples in Table 1 were previously characterized by having their hydraulic conductivities in saturation (K) measured, as well as their global (ρ_b) densities and their total porosities (Φ), as shown in Table 2.

IV.A Double-Layer Sand Column

The experiment regarding water infiltration in double-layer soils was set up at the EMBRAPA-CNPDIA laboratory, São Carlos, SP, Brazil. An acrylic column (height: 100 cm, width: 30 cm, thickness: 1.0 cm) was built, in order to have the results of this work compared to those obtained by Glass *et al.* (1989a,b), and Chang *et al.*(1994). At the bottom of the column, ten holes of 0.5 cm diameter were made to let out the air. The top section of the column remained open to the environment. The material used consisted of six samples of quartz sand, listed in Table 1. Five different experiments were conducted: each one was repeated five times, employing five different packed columns. The filling of the acrylic column with each sample of employed soil was accomplished by placing it in two layers inside the column, the top layer being of fine sand, sample number 1 in Table 1, and being approximately 10 cm high (Posadas and Crestana, 1993). The bottom layer filled up to 80 cm of the column (this layer is considered coarse when compared to the top one), and consisted of samples number 2, 3, 4, 5 or 6 from Table 1, making five different columns. Both layers were filled into the container so as to have the most homogeneous distribution as possible. Through a tube placed on the top of each column of soil, 45 ml of water were applied on the surface during three seconds, by means of an acrylic plate with 1.0-mm diameter holes, in order to have the water uniformly spread on the surface. Once the infiltration was begun, the water pressure was kept constant through a 1.5-cm high layer of water during the experiment. In this situation, the appearance and the later development of the fingers were observed through the column.

IV.B Experiment Filming and Image Digitization

As the infiltration of water started through the double-layer column, the images began to be collected by utilizing a commercial video camera. The filming was done until the end of the experiment, at the instant when the fastest finger reached the bottom of the box, as shown in Figure 4. The obtained images were digitized for different instants during the infiltration process by means of a micro-computer, supplied with a frame-grabber with 256 gray levels and resolution of 512×512 pixels. In order to have the preferential infiltration process quantified, the images, after being digitized, were processed using the Khoros image processing package (Donohoe, 1992).

In three out of the five samples employed (samples 2, 4 and 5), the appearance of fingers was observed. In the remaining two (numbers 3 and 6), the flow displacement remained stable, without presenting fingers or preferential paths in all repetitions done.

Figures 4 through 11 show the results of the experiments where the fingering phenomenon was observed. The digitized images represent the development of fingers at the end of the experiment, for each sample utilized, when the first finger reached the bottom of the acrylic column. For instance, Figure 4 represents the development of rather compact and wide fingers. A transition region occurs at the interface between the two layers, approximately 7 cm high, signaling the limit up to where the fluid was dislocating uniformly. After overcoming this transition region the fingers began to develop. This transition zone, also observed by Hill and Parlange (1972) and Glass *et al.* (1989), is called the induction zone. Figure 5a represents the result of another test for the same sample as in Figure 4. This test was repeated using similar conditions. In Figure 5a we can observe the development of two fingers up to the end of the experiment, while the third one joined the finger on the left almost midway. These fingers are similar to those in Figure 4, but a bit more compact and wider.

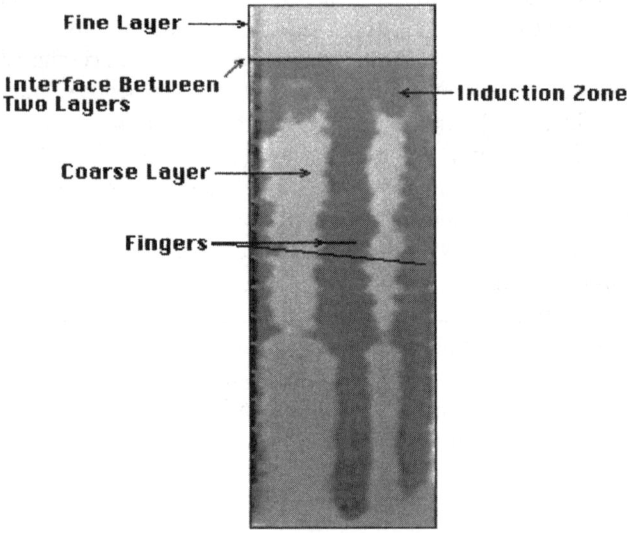

Figure 4: Image digitized without pre-processing. This picture represents the formation of fingers for the sample of sandy soil number 2. Three fingers may be observed, in dark gray, as being quite compact and wide. We can observe also the induction zone. The fine layer refers to the fine texture sandy soil and the coarse layer to the coarse texture sandy soil.

Nonetheless, the induction zone is less noticeable. In this experiment, a piece of filter paper was placed on the interface between the two layers, in order to prevent the small particles of the fine layer from penetrating the second layer. This procedure was not adopted in the first experiment. Thus, according to the experimental condition, the small particles, existing just under the interface of the fine layer, may or may not block the large pores of the coarse layer, being one of the determining factors of the size of the induction zone.

In the other three tests, repeated with this same soil, the results were similar to those presented in Figures 4 and 5a. On average, there were three fingers with geometrically similar characteristics, as shown in Table 3. The filter paper was always placed on the interface between the two layers. The induction zone was reduced when compared to the case without filtering paper.

Although the developed fingers had similar geometric characteristics, their spatial localization was aleatory, as shown in Figures 4 and 5a. In Figure 4, two fingers dislocated along the borders of the column and one along the middle, whereas in Figure 5a the two fingers that reached the bottom of the column dislocated along the middle. The transit time for the fastest finger to reach the bottom of the box was approximately 15 minutes on the average. This time is of the same order of magnitude as the time that the fluid would take to travel the same distance in saturation conditions, as will be discussed later (Table 3).

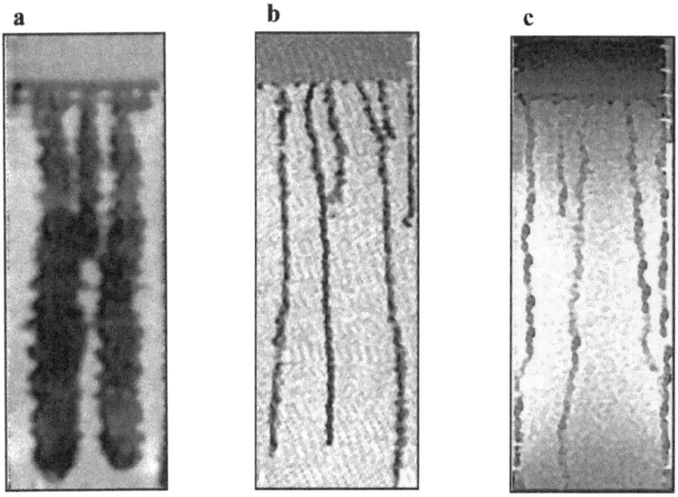

Figure 5: (a) Digitized image representing another test done with soil sample number 2. We may observe two quite wide and compact fingers, similar to those in Figure 4. However, this time the third finger, the middle one, joined the finger on the left halfway; (b) digitized image showing the development of fingers for soil sample number 4, where we can observe the occurrence of 6 fingers with almost constant and thin widths; (c) profile of the fingers that occurred by employing soil sample number 5. Five thin fingers, with geometric characteristics similar to those in (b), may be observed.

Soil sample	Mean pore size x10^{-4} (cm)	Number of observed fingers	Maximum average velocity of fingers (cm/min)	velocity of flow in saturation (cm/min)	average widths of fingers (cm)
2	90.00	2.0 ± 1.0	5.8 ± 0.6	5.0 ± 0.5	6.8 ± 2.5
4	102.00	6.0 ± 1.0	15.6 ± 1.5	16.0 ± 2.5	2.0 ± 0.5
5	130.00	6.0 ± 1.0	18.0 ± 2.0	20.0 ± 3.0	1.5 ± 0.3

Table 3: Some morphological and physical characteristics of fingers obtained from experiments in 2D.

Time (min)	Maximum length of the fastest finger: d_{max} (cm)	Fractal dimension of the wetting front: D_p
2.15	7.40	1.09
4.46	21.00	1.11
6.08	29.00	1.15
7.40	38.60	1.17
9.70	46.70	1.20
12.20	58.00	1.25
13.80	68.00	1.30
14.60	80.00	1.34

Table 4: Finger displacement and its fractal dimension D_p, for soil sample number 2.

Figure 5b shows the profile of the fingers for soil sample number 4. In this, we can observe the generation of six quite thin fingers with approximately constant diameters. Also, we can observe the lack of an induction zone, and the appearance of fingers occurring immediately on the interface between the two layers. In spite of that, we could observe a brief paralyzation of the flow on the interface, for some seconds, before penetrating in the second layer.

The time taken by the fastest finger to reach the bottom of the soil column was approximately 7 minutes, much faster than that in the experiment with sample number 2 (Figures 4 and 5a). This time is comparable to the time the fluid would take to travel the same distance under saturated conditions.

Figure 5c shows the profile of the fingers that occurred when sample number 5 was used. We can observe five well-defined fingers, with similar characteristics to those in Figure 5b, with slightly shorter diameters. We can also observe a small induction zone and a certain sinuosity in the trajectory of the fingers. It can be a consequence of the existence of non-homogeneous regions that appeared, surely, in the process of filling the column with the coarse layer. In the case of samples of coarse-textured soils it is more difficult to visually detect the appearance of non-homogeneous regions during the process of filling, as a consequence of the size of their constituting particles.

In all the experiments, we verified that the displacement velocities of the fingers, after being generated, increased until they reached maximum velocities. Subsequently, their displacement velocities remained constant. Such maximum velocities practically coincided with those of the flow in conditions of saturation, as shown in Table 3.

In the experiments using soil samples 4 and 5, it was observed that the generation of each finger is separate from that of the others. That is, a finger may appear at the instant the other has reached the bottom of the column. However, once generated, they remain fixed, thus defining the preferential paths for the water flow. In this situation, if the initial conditions are changed, for instance,

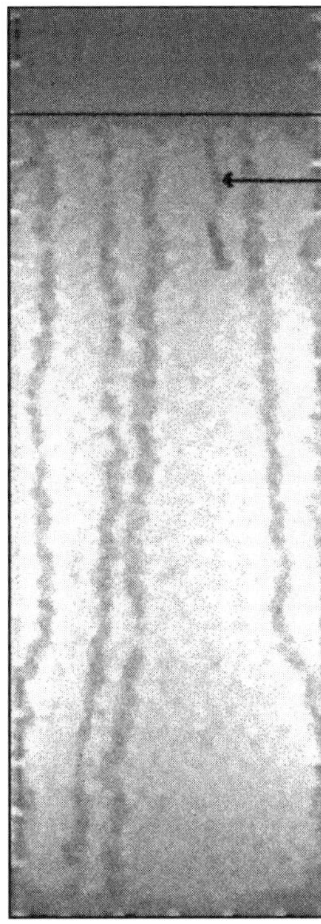

Figure 6: "Freeze" of preferential paths, after being "open" in the initial experiment in Figure 7. After the column rested for 24 hours, without application of water, the experiment shown in Figure 7 was repeated. The fingers initially generated, which had reached the bottom of the column, remained constant, and those that were halfway reached the bottom. It is interesting to observe the fingering that occurred during the process of redistribution of liquid.

by increasing the pressure of the water entering the surface or by making it intermittent (turning the faucet on and off), we can observe that the profile of the fingers, generated initially, remains constant. Hence, the fingers generated in the initial phase of the flow function to by-pass or short-circuit the process of infiltration of water in the soil.

With the purpose of studying the behavior of fingers, in the phases of redistribution of the fluid, which reproduces, for example, the situation after an irrigation system has been turned off, the column of soil shown in Figure 5c, where fingers already existed, stayed for 24 hours in the open. After this time the same experiment was repeated, which resulted in Figure 6. Figure 6, which represents this situation, indicates that the path followed by the liquid was the same as that traced in the initial experiment (Figure 5c), but now one more finger was observed. However, this extra finger was generated after the liquid supplying on the soil surface was stopped. That means that fingers may also occur in processes of liquid redistribution in the soil. Most probably, the new finger observed in Figure 6 was caused by the modification of the boundary conditions during the experiment. It is important to take into consideration that at the instant the water flow on the surface of the top layer was interrupted, the top layer continued furnishing water to the bottom layered, but with a decreasing flow rate. The decreasing flow rate may have produced a new hydrodynamic instability which, in turn, may be responsible for generating the new finger shown in Figure 6.

Table 3 presents the summary of the finger characteristics generated in the laboratory, for each type of sample. In this table, we can also observe the mean pore size (measured through a mercury porosimeter), the number of fingers, their widths and the velocities of the fastest ones, which first reached the bottom of the column. For the aim of comparison, the velocity of the flow in conditions of saturation is also shown in this table.

IV.C Characterization of Fingers via Image Processing Techniques, and Calculation of their Fractal Dimension D

To process the images obtained from the experiments, software especially devised for this purpose was used, such as Khoros (Donohoe, 1992). Each experimental image was transformed, digitized and processed, permitting the observation of finger dynamics through the soil column and the relation of their distribution to the sample granulometry. After they were converted to binary images, the fractal dimension D was calculated for each one. In the calculation of their fractal dimension D, the "box-counting" and the "sand-box" methods (Barnsley, 1988) were used.

In the interpretation of the results (*cf.* Figures 7 to 9), considering the fractal dimension D of each structure in relation to time and soil texture, it is appropriate to draw attention to the fact that D does not respond to a statistically self-similarity structure, but to a statistically self-affine structure (Mandelbrot, 1986a,b). This happens because the displacement of the fingers follows a privileged direction, that of gravity, generating structures with characteristics of "self-affine" (anisotropic) and not of "self-similarity"(isotropic), where D is the total fractal dimension (Meakin, 1989).

Below we present some results pertaining to the establishment of the fingers showing the "bulk" fractal characteristic D_b (where "bulk" is the area occupied by the fingers) and the fractal dimension of the profile, D_p (associated to the displacement front of the fingers), of the distribution (Chang et al, 1994).

Table 4, sample number 2, shows the finger displacement front as a function of its fractal dimension D_p. The dynamics of the front is characterized at several instants of time, as well as its corresponding length in the soil column. The maximum length ($d_{\max}$) is defined as the length reached by the fastest finger

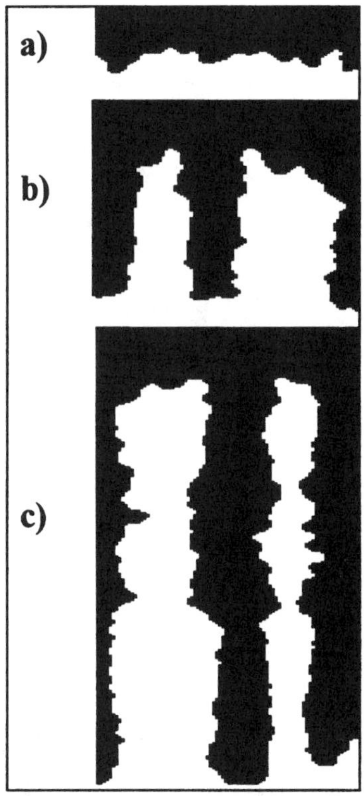

Figure 7: Processed image of the fingers for soil sample number 2, for three different instants of time t, length of the fastest finger $d_{\max.}$ and fractal dimension of the profile D_p. a) Beginning of instability, time $t = 2.15$ min, $d_{\max.} = 7.40$ cm and $D_p \approx 1.09$; b) time $t = 6$ min., $d_{\max.} = 29$ cm and $D_p \approx 1.15$; c) end of the experiment: time $t = 14.6$ min, $d_{\max.} = 80$ cm and $D_p \approx 1.34$. As we can observe from this sequence of images, the structure of the fingers does not present fractal characteristic, for it is compact. Nevertheless, the surface or the profile of the structures presents a "statistical self-similarity", constituting "statistically self-affine" surfaces.

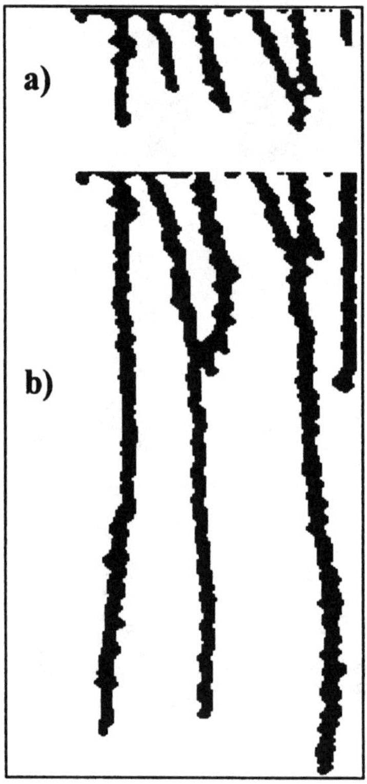

Figure 8: Similarly to Figure 7, processed and digitized image showing two instants of the development of the fingers for soil sample number 4. a) Structure of the fingers for $t = 2.0$ min, $d_{\max.} = 17.3$ cm, $D_b \approx 1.42$ and $D_p \approx 1.30$. b) Structure for the final instant, $t = 6.0$ min., $d_{\max.} = 80$ cm, $D_b \approx 1.48$ and $D_p \approx 1.39$, with D_b representing the fractal dimension of the "bulk".

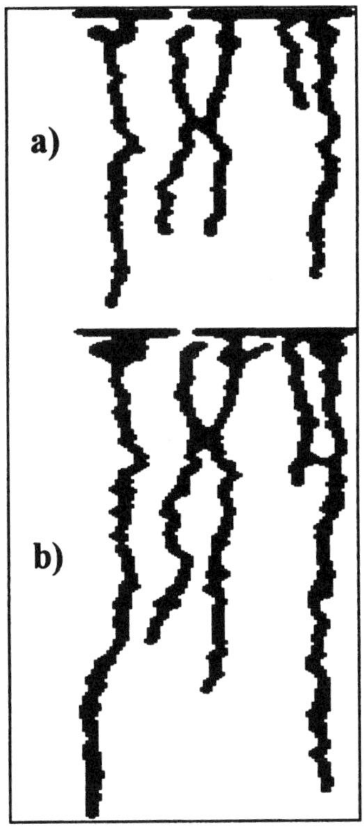

Figure 9: Similarly to Figure 7, two instants of space-time evolution of the finger structure for soil sample number 5. a) $t \approx 3.0$ min., $d_{\mathrm{max.}} = 48.8$ cm, $D_b \approx 1.43$ and $D_p \approx 1.30$. b) Final instant of the experiment: $t = 4.4$ min, $d_{\mathrm{max.}} = 80$ cm, $D_b \approx 1.45$ and $D_p \approx 1.40$.

Soil sample	Fractal dim. of the bulk	Fractal dim. of wetting front	Capillary number x10^{-5}	Bond number	Fraction of wetted area (%)
2	-----	1.34 ± 0.03	1.15	0.0011	53 ± 5
4	1.49 ± 0.03	1.40 ± 0.04	3.60	0.0014	20 ± 3
5	1.45 ± 0.03	1.42 ± 0.04	4.60	0.0023	15 ± 3

Table 5: Finger characterization through fractal dimension for each soil sample where the phenomena occurred at the final instant of the experiment, capillary number, bond number and the mean fraction of wetted area.

at a given time. The morphological structure of these fingers, that is, the area of occupation of these fingers, is quite compact and may not be considered to be of the fractal type. These structures are very similar to the fingers generated in the Hele-Shaw cells (Avnir, 1989; Feder, 1988), when the structures are not fractal.

Table 5 shows the characterization results, through the fractal dimension, of the structures generated by employing soil samples number 4 and 5 for the final instant of the experiment. We can observe that the structures of the fingers, corresponding to samples number 4 and 5, represent fractal structures, with fractal dimension for the "bulk" D_b equal to 1.49 and 1.45, respectively. Also, Table 5 presents the fractal dimensions of the profiles D_p and the fraction of wetted area of the soil column by the fingers. The fraction of wetted area refers to the value of the area occupied by the fingers within the soil column, considered from the interface between the two layers. The results presented in Table 5 related to the capillary and bond number are meaningful regarding to its interpretation. For all samples the capillary number is very small. This is a strong indication that the flow mechanism is dominated by capillary forces instead of the viscous ones. We may also notice the dependence of capillary number with bond number, confirming that the gravitation effect increases as the pore size increases in a porous medium. These results will be very useful to adjust the theoretical model to the experiment as described at the end section of this chapter.

It was documented during the several experiments carried out that the fractal characteristics of the fingers, under the same initial conditions, are related to the soil granulometry or the pore distribution profile (Posadas and Crestana, 1995), whereas for coarse-textured soils, the structure of fingers presents statistically self-affine or fractal characteristics. From these results, the end section will briefly describe the theory of invasion percolation and the modifications introduced in the morphological simulation of the fingers.

V Experimental Measurements of the Fingering Phenomenon by Means of MR and X-ray Tomography and Images

Below we will describe some results obtained from experimental measurements of the fingering phenomenon, using the techniques of X-ray and MR tomography and spin-echo MR images (Crestana *et al.*, 1985; Crestana, 1992; Posadas and Crestana, 1995; Tannús *et al.*, 1995).

V.A Experiments Performed Using a Medical X-ray Tomograph

Now we present experimental results regarding the study of fingering dynamics, obtained with a fourth-generation medical X-ray tomograph employed to scan a cubic column ($15 \times 15 \times 15$ cm^3) of stratified soil (Posadas *et al.*, 1993; Posadas *et al.*, 1996). These studies present some experimental opportunities that may be explored by the use of a medical X-ray tomograph designed for hospital applications (Crestana *et al.*, 1985; Posadas *et al.*, 1993; Posadas and Crestana, 1995).

Figure 10 depicts the instant of time $t = 1.58$ min, of the water displacement through the cubic soil column, considered from the beginning of the infiltration. In this digitized image, not yet processed, we can clearly notice, in light gray, the phenomenon morphology, the wet soil-dry soil interface, the first completely saturated layer of soil (fine texture: sample number 1) and the beginning of flow instability (inferior layer of coarse texture: sample number 2), after a certain induction zone under the interface, represented by the dark gray line. We may also notice a finger displacing rapidly along the wall of the container, as well as the non-planar surface between the two layers, even though all the experimental care had been taken as to having it built as planar. This image corresponds to a vertical slice, 1.5 mm thick, 1.0 cm apart from the center of the column. The reconstruction matrix used was one of 512×512 pixels and a field of view equal to 200 mm. The spatial resolution of these images, considered as volume elements (voxels), was $0.430 \times 0.430 \times 1.5$ mm^3 or 0.28 mm^3, with 0.185 mm^2 pixels.

Figure 11 presents processed images at three instants of time during the flow displacement through a column central slice of 1.5 mm thick. These images show the evolution of finger development from the interface between the two layers, as seen in Figure 10. We can notice the development of a compact and wide finger with an average velocity of $\overline{V} \approx 5.5$ cm/min. We can also notice the presence of the induction zone before the finger beginning to grow. In order to prevent particles coming from the fine layer from penetrating into the second layer of coarser texture, in all the tomography experiments, filter paper was placed on the interface between the two layers during the filling process of the column with soil. In the 2-D experiments we managed to lower the height of this induction zone by placing the filter paper on the interface between the two

Figure 10: Digitized image without processing, obtained with a medical X-ray scanner. This image depicts a vertical section 1.0 cm apart from the center of the cubic column ($15 \times 15 \times 15$ cm^3) for the instant of time $t = 1.58$ min, after the infiltration through the soil surface had begun. We can clearly notice the wet soil-dry soil interface and the beginning of the flow instability, after a certain induction zone below the interface (dark gray line) between both layers. We can also notice a finger displacing rapidly along the wall of the container.

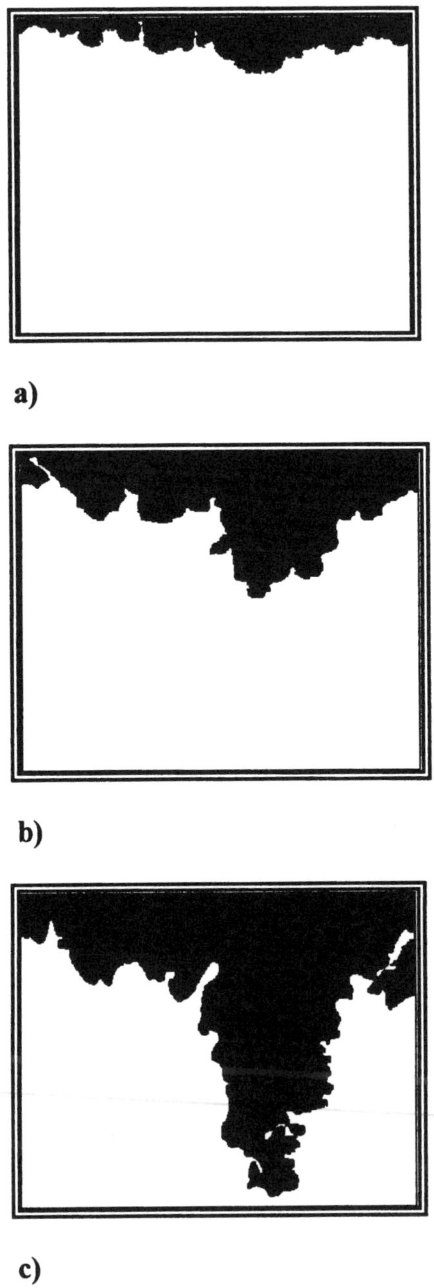

Figure 11: Processed vertical section images obtained through a medical X-ray tomograph, representing the growth of fingers in the central part of the cubic column of soil. It shows a slice, 0.5 mm thick, 15 cm wide and 12 cm high, in three different instants of time: a) $t = 1.25$ min, b) $t = 2.0$ min and c) $t = 2.7$ min, considered from the beginning of the infiltration.

layers. However, there are other physical factors that take part in the generation of this induction zone, especially those of local origin, on the pore level, such as surface tension, capillary pressure and contact angle. There is the need of further experiments in order to learn more about these factors.

The dynamics of the displacement front or infiltration profile through this central slice was also characterized through its fractal dimension D_p. As was observed in Figure 11, the morphological structure of these fingers is rather compact, showing fractal characteristics only on the surface or on the displacement profile of the fluid. This behavior is similar to the case of the planar column with sandy soil of fine texture (sample number 2). The values of the fractal dimension D_p found for the three instants of time, as shown in Figure 11, were a) $D_p \approx 1.015$, for $t = 1.25$ min, b) $D_p \approx 1.04$, for $t = 2.0$ min and c) $D_p \approx 1.07$, for $t = 2.7$ min. This behavior is also similar to that found in the planar column. This points out that the fluid displacement maintains a global geometrical characteristic, of similar behavior, external as well as internal to the soil medium, in this type of texture. The statistical auto-similarity would only be present on the surface or on the infiltration profile. Similar behavior was also found when other internal slices to the soil column were analyzed. Nevertheless, this kind of behavior shall be different for diverse soil texture.

V.B Some Results Obtained with Magnetic Resonance Imaging (MRI)

Experimental results on the fingering phenomenon are presented below, for steady-state conditions (after the moment when the fingers formation was completely paralyzed) obtained with a magnetic resonance scanner, through the same cubic column of sandy soil (with no paramagnetic material present) as employed in the X-ray tomograph experiment shown in Figure 10 (Posadas *et al.*, 1993; Tannús *et al.*, 1995; Posadas and Crestana, 1995).

The MR imaging results are presented in Figures 12 and 13. All images were obtained under the same previous physical conditions as shown in Table 2, and with the same rate of infiltration applied at the soil surface (200 ml/min) that was used during the X-ray imaging experiment. We observed seven 1.8-mm thick slices, their centers being 2 cm apart. Each slice was reconstructed on a 256×256 matrix, with a 25-cm field of view, producing images with voxel volumes of $0.1 \times 0.1 \times 18$ mm^3. The total acquisition time for obtaining the seven slices was about 4.0 minutes (Posadas and Crestana, 1995; Posadas *et al.*, 1996).

Figure 12 represents three acquisitions in a vertical plane of the cubic soil column after the flow reached steady-state conditions, established at the instant one finger (the fastest one) touched the bottom of the box and the water on the top layer was turned off. In these three images we may observe the three-dimensional character of the fingering phenomenon and the great spatial variability associated to it. For example, in Figure 12a the fingers reached the bottom of the container. However, in Figures 12b and 12c, near the central part of the acrylic column, the fingers are still halfway down the column. Also,

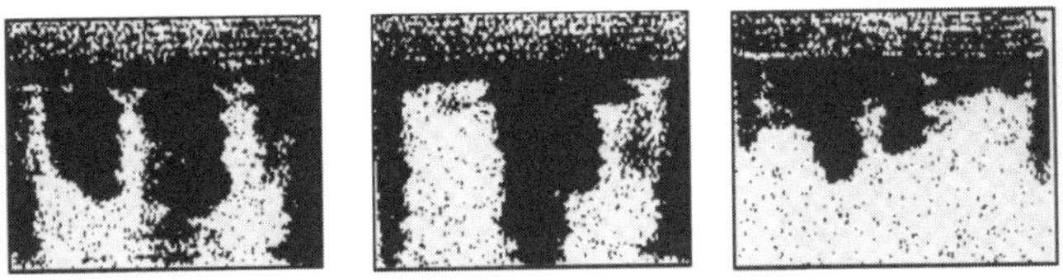

a) b) c)

Figure 12: Images obtained with the *MRI* system, showing three transverse sections of the fingering phenomenon in static conditions. Each section represents a slice, 2 cm thick, 15 cm wide and 15 cm high, of the cubic soil column, $15 \times 15 \times 15$ cm^3. We can clearly notice the spatial variability of the phenomenon in these three images. In a) the fingers are close to the bottom of the box, and in b) and c) sections close to the central part of the cubic column, the fingers are still midway down the column (Posadas *et al.*, 1993; Posadas and Crestana, 1995).

through these images, the geometric dimension of the fingers can be quantified, such as their average diameter, which was approximately 2.5 cm, and their lengths. It can also be observed that the fingers follow rather straight preferential paths, which shows the strong influence of the gravitational force on the system (Posadas and Crestana, 1995).

Figure 13 represents another set of measurements in the horizontal plane under the same conditions as in Figure 12. We can distinctly observe the number of fingers through the different vertical slices. We can also see that the diameters of the fingers that reach the bottom of the cubic column are reasonably constant.

This shows that *MRI* possesses a great potential for the three-dimensional study of the fingering phenomenon, at least under of steady-state flow conditions (Posadas *et al.*, 1993; Posadas and Crestana, 1995).

It is also possible to see different shades of gray, corresponding to a greater (dark gray) or smaller (light gray) quantity of water, as related to the *MR* signal or scale. That is, it gives an idea of the horizontal distribution of water. It would be interesting to fix a horizontal section, and verify the effect of the horizontal diffusion of the water, as a function of time.

The horizontal section images shown in Figure 13 represent the occupation of fingers through each section, following the direction of the infiltration. In repeated experiments, on different packed columns, similar occupation behavior was observed, even though the occupied zones changed in each experiment. This demonstrates the aleatory character of pore occupation, creating an aleatory distribution of fingers for each coronal section. For the sake of speculation, the fractal dimension D was calculated for each plane from (a) to (f) in Figure 13, resulting in values for D for each coronal slice of the finger occupation structure, under static conditions. The first image (Figure 13a), which corresponds to the top layer of fine texture, is a compact structure (completely saturated),

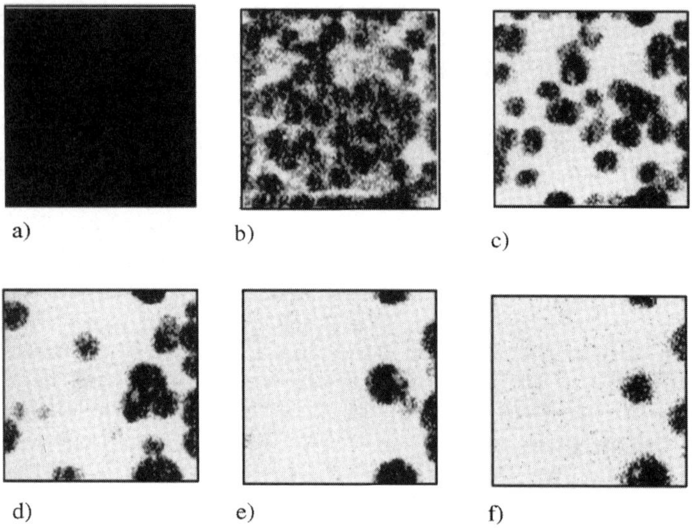

Figure 13: Images obtained by *MRI*. Six transverse sections (slices 1.0 mm apart) of the cubic soil column are presented in the same conditions as in Figure 12. They represent slices of the column following the direction of the infiltration. a) Saturated section near the surface of the column (first layer), and f) section corresponding to the bottom of the soil column. b), c), d) and e) represent intermediate situations, equally 1.0 mm apart. The parts in white represent dry zones and those in gray, wet zones allowing as a consequence, for instance, to enumerate the number of fingers in each transverse section (Posadas *et al.*, 1993; Tannús *et al.*, 1995; Posadas and Crestana, 1995).

and therefore non-fractal. The second image (Figure 13b), which corresponds to the beginning of the infiltration instability (inferior layer of coarse texture: sample number 2), has fractal dimension $D \approx 1.80$. Figures 13c, 13d, 13e and 13f have fractal dimensions 1.70, 1.55, 1.48 and 1.38, respectively. Apparently, these results produce information about the spatial occupation of the column by the fingers along the vertical, demonstrating that there is a relation of fractal character (thus following a law of scale) within the range of scales considered and the assumed steady-state regime. However, further investigation in this respect is still necessary, so as to better characterize these structures, which indeed bear statistical auto-similarity in some scale regions. It would be interesting, for example, to relate the fractal dimension to the infiltration of different chemical substances or even to follow the infiltration dynamics. At the moment this idea is still not very practical, as it assumes the existence of an *MRI* system capable of producing images in real time without being seriously affected by paramagnetic effects, which is practically impossible with contemporary equipment.

V.C Study of the Fingering Displacement by Means of NMR Spin-Echo "Imaging"

By employing the same cubic column considered in the experiments reported in the two preceding sections and assuming that gravity force governs finger propagation, we could also follow the dynamics of fingers through observations of the wetting front advancement. This was done by acquiring the *MR* spin-echo signal as a function of time, using a "phase codification" along the vertical axis in the direction of the gravity force. The *MR* signal increased with time as the infiltration of water ran along the cubic column. One acquisition of the *MR* signal spectrum was accomplished along the vertical axis for each instant of time, in seven vertical slices, each 2 cm thick and with their edges 1.0 mm apart. The spectrum collected in the form of an *MR* signal, for each slice and for each instant of time, was analyzed using the unidimensional Fourier transform (1-D FFT). The time for acquiring each slice was 0.2 s. With this we obtained the relative signal intensities, proportional to the displacement front and the spatial position on the vertical plane, for each measured instant of time during the infiltration. By means of image processing it was possible to treat the signal intensities adequately, allowing us to obtain a spatial composition in the form of images of the wetting front along the cubic column, as presented in Figure 14. Figure 14 is a sequential image obtained from *MR* spin-echo signal reconstruction, corresponding to the spatial distribution of the wetting front at three different instants of time. We can clearly see in these profiles the spatial and temporal character of the fingering phenomenon, which presents great spatial and temporal variability. It is interesting to note that the tips of the fingers, which appeared initially, grew rapidly, whereas in other regions of the space the growth was minimum. The largest protuberances (fingers) also represent spatially the locations in the regions where the fluid displaced more abundantly (Tannús *et al.*, 1995; Posadas and Crestana, 1995).

VI Invasion Percolation

Looking at Table 5 (fourth column), we can observe that all of our experiments are performed under conditions of very small capillary numbers. As a consequence, the invasion percolation model is a good candidate for describing the water flux. Also, modified DLA (Liang, 1986) could be tested as a model for simulating the soil fingering phenomenon, although this was not performed in the present work.

Invasion percolation is a dynamic percolation process (see percolation chapter in this book) introduced by Wilkinson and Willemsen (1983), motivated by the study of the flow of two immiscible fluids in porous media (de Gennes and Guyon, 1978; Chandler et al, 1983). Consider the case in which oil is displaced by water in a porous medium. When the water is injected very slowly, then capillary forces completely dominate the viscous forces (low N_{ca}), and therefore the dynamics of the process is determined on the pore level. Thus, the displacement process of the fluid advances in sequences of discrete jumps, whose

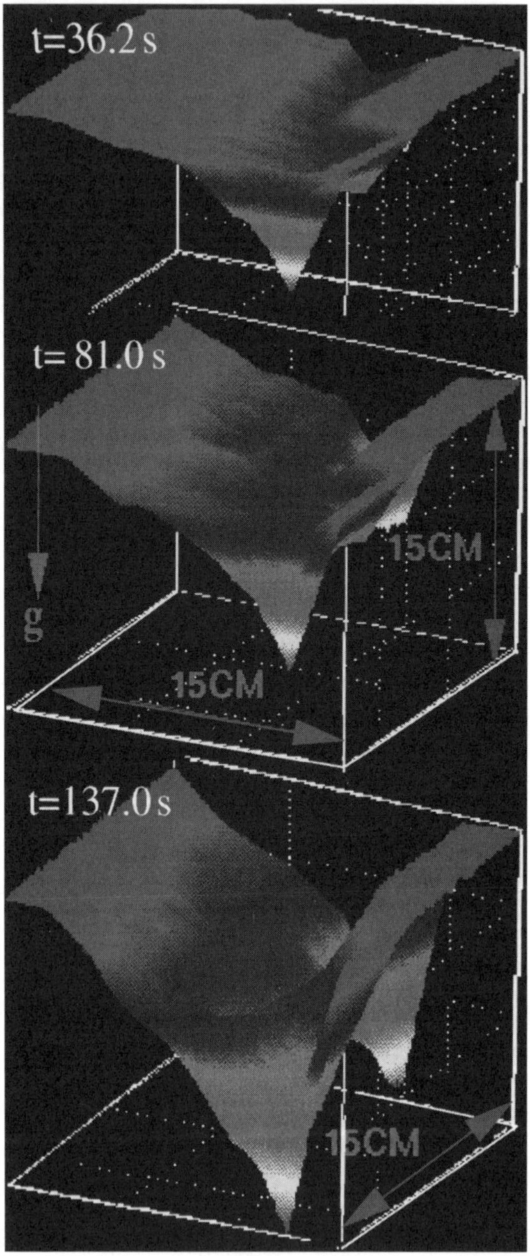

Figure 14: Spin-echo experiment showing spatial distribution of the water front (fingering dynamics) at three different times of the water infiltration into a cubic soil column of $15 \times 15 \times 15$ cm^3.

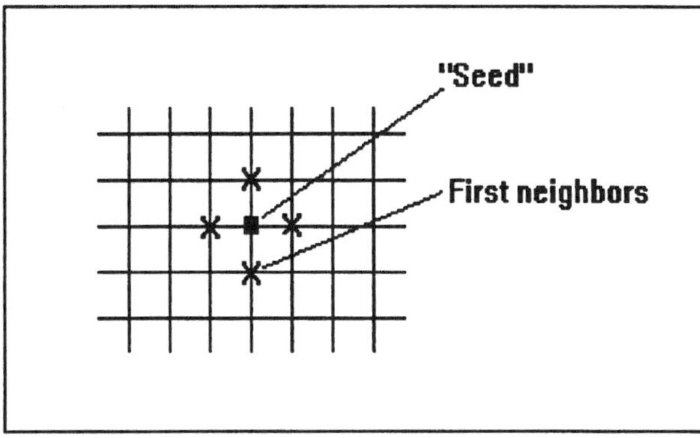

Figure 15: Schematic diagram of a 2-D square lattice showing the "seed" (first invaded site) and the first neighbors (indicated by an "x"), assigned as perimetral sites or growth sites.

locations are chosen by the criterion of least capillary resistance: the smaller pores are filled or invaded, and the fluid displaces rapidly through them. The algorithm of the site invasion percolation, without trapping, in two dimensions, is given by the following rules (Wilkinson and Willemsen, 1983):

(a) Assign random numbers r in the range (0,1) to each site of an $L \times L$ lattice;
(b) Put a seed at the center of the lattice and define a list of **growing sites** composed by its first neighbors (Figure 15);
(c) The invading fluid advances to the growth site (a site that belongs to the perimeter and which is apt to be invaded) that has the lowest random number r;
(d) Update the list of growing sites, and
(e) The invasion process ends when the invading fluid reaches one boundary of the lattice.

VI.A Invasion Percolation Modified for the Case of Fingering Phenomenon in Soils

Invasion percolation by sites, as described previously, was modified to follow, in two dimensions (this model may also be extended to three dimensions), the fingering phenomenon in soils. The modifications were accomplished to take into account the experimental observations carried out in the laboratory. The 2-D experiments were conducted in a flat column (thickness equal to 1.0 cm), permitting to neglect three-dimensional effects (Onody *et al.*, 1995; Posadas and Crestana, 1995).

We first notice that during the experiments the water infiltration through the

Soil sample number	G1	G2	G3	G4
2	- 0.009	0.20	0.35	0.50
4	- 0.011	0.70	0.35	0.20
5	- 0.020	0.75	0.35	0.15

Table 6: Best fitted values of the parameters G for sample number 2 and the corresponding ones calculated for samples 4 and 5.

first layer occurred uniformly, so that the water front equally touched any point of the first line of the second layer at the same instant. This fact allows us to assume that the occupation sites will strictly depend on the pore size distribution at the dry soil-wet soil interface. We incorporates this fact in the body of the theory by having an entire line of seeds (and not just one initial seed at the center of the lattice) at the first line at the top of the lattice, *i.e.*, we developed an edge injection. We should now modify it in order to incorporate the physical quantities responsible for the fingering exhibited in the experiment. This will be performed by four parameters which we have called G_1, G_2, G_3 and G_4, proportional to the bond number, the number of fingers, the surface tension and the fraction of wetted area, respectively.

The results of our experiments indicated that gravity is a dominant force in our system. Therefore, we use a gradient invasion percolation term (Wilkinson, 1984 and Hulin et al, 1988) to describe the dynamics of the fingering phenomenon.

Here, we follow the same scheme used by Frette *et al.* (1992) and Meakin *et al.* (1992). The original random number r_i $(0 < r_i < 1)$ assigned to each site i of the lattice is changed by

$$\overline{r}_i = r_i + G_1 h_i \tag{12}$$

where h_i is the line of the matrix (or the corresponding depth of soil column). The case $G_1 > 0$ corresponds to the stabilized invasion percolation (Birovljev *et al.*, 1991), which does not show any fingers; the destabilizing situation ($G_1 < 0$) was investigated both experimentally (Frette *et al.*, 1992) and via simulations (Meakin *et al.*, 1992), producing only one finger. The parameter G_1 was shown to be proportional to the bond number (Frette *et al.*, 1992). This means that once the G_1 value of any soil is known (for example: sample number 2), then the equivalent G_1 value of a second soil (sample number 4 or 5) can be found and is directly related to the G_1 value previously known and to the respective soil bond numbers (Table 6).

The presence of gravity imposes a privileged direction to the water infiltration. To take this into account, we need to weight the invasion percolation differently, distinguishing the up, right or left direction from the down one. To accomplish this, we introduced parameter G_2: once a site is occupied we assign to the site i beneath (if empty, of course) a new random number $\overline{r}_i = r_i - G_2$.

When $G_2 > 0$, then it favors a downwards flow direction. We have verified that increasing the positive value of G_2 also increases the number of fingers generated by simulation.

An empty growth site in the front can be surrounded by just 1, 2, 3 or 4 occupied sites. In a realistic model, as the number of occupied sites around an empty site increases, the probability of the empty site to the occupied also increases, due to the presence of surface tension. Thus, each time an empty site is found to be surrounded by 2 or more occupied sites, we subtract from its associated random number a factor G_3 ($G_3 > 0$) in order to increase the probability of being invaded. Consequently, in our simulations, G_3 was related to the surface tension and was considered constant for all the tried soil samples.

To bring our model even closer the experiment, we altered the original theoretical assumption that the invasion is done by just one site at each time. Indeed, **many** sites are invaded simultaneously in the real experiment. To perform this task, we introduced parameter G_4 ($0 < G_4 < 1$), which represents the fraction of growing sites to be simultaneously invaded, leading to what we call a **multiple**-invasion percolation model. We have found that this parameter G_4 is proportional to the wetted area of the experiment. So, once we have adjusted the G_4 value for one soil (again, in our case, the soil number 2), we can use the experimental data to determine the corresponding values of other soil samples (for instance number 4 and 5).

VI.B Simulation

We have tried to adhere to the experimental phenomenology. For all simulations we have carried out, we employed the same rectangular lattice compounded of 800 rows and 320 columns, which is proportional to the acrylic column dimensions used in the experiment. Each site in this lattice represents a pore and the assigned random number corresponds to the diameter of the soil pore.

For soil sample number 2 we have adjusted all 4 parameters (G_1, G_2, G_3 and G_4), taking into account the number of fingers and the wetted area. Sailing in the parameters' space, we have learned that G_4 is very easy to locate, G_2 is very sensible and G_1 and G_3 are on an intermediate difficulty level. A lot of effort was necessary to fit them and we have spent several hours of cpu. We only stopped the process of parameters optimization when we arrived at Figure 16a, which reproduces Figure 5a in a convincing way.

Fortunately, for the other soil samples, the task was much easier, due to the **proportionality** between the parameters and the associated experimental data (as discussed in the former section): we could read the parameter values directly from the experimental data. Moreover, G_3, which corresponds to the surface tension, is fixed.

The results of the simulations are presented in Figures 16a, 16b and 16c. In these figures we can clearly notice a good similarity between the simulation fingers and those obtained experimentally (see Figures 5a, 8 and 9, respectively).

Table 6 shows the values of the G parameters. Remember that only the first

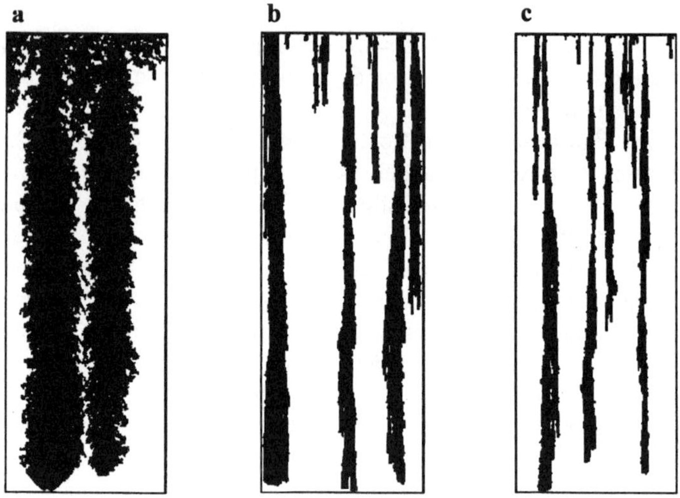

Figure 16: (a) Typical cluster of simulation done through the program of modified invasion percolation. This simulation represents the growth of the fingers for sample number 2, using the parameters $G_1 = -0.009$, $G_2 = 0.20$, $G_3 = 0.35$, $G_4 = 0.50$; for an 800×320 rectangular lattice. We can also notice the development of fingers with similar characteristics to those shown in Figure 5. (b) Typical cluster of the simulation representing the growth of fingers in soil sample number 4. The parameters G's used were: $G_1 = -0.011$, $G_2 = 0.70$, $G_3 = 0.35$ and $G_4 = 0.20$. (c) Typical cluster of the simulation corresponding to soil sample number 5. The G parameters employed in this simulation were: $G_1 = -0.020$, $G_2 = 0.75$, $G_3 = 0.35$ and $G_4 = 0.15$.

line in this table was obtained by fitting (sample number 2). In this table we can observe that the ratio between the parameter G_1 of the soil number 2 and of the soil number 4 is equal to 0.82 and equal to the ratio of the bond number from soil number 2 and soil number 4 (Table 5). This behavior is repeated for the other parameters.

Finally, Table 7 shows the mean values (with the respective standard deviations) of some quantities measured in our simulations employing the G values given in Table 6. Again we observe that it reproduces, in a very reasonable way, the experimental results (Table 3 and 5).

Thus, the modified invasion percolation model was able to describe, in a qualitative and quantitative way, the experimental fingering phenomenon (Posadas, 1994; Onody *et al.*, 1995; Posadas and Crestana, 1995). However, many questions still remain to be answered concerning the geometry, physical mechanisms, and dynamics of gravity-driven fingering phenomena in soils. A further study of these questions will offer opportunities for comparison between experiments and simulations, and therefore, for refinement of the numerical

Soil sample number	Fractal dimension of the bulk	Fractal dimension of the wetting front	Number of observed fingers	Fraction of invaded area (%)
2	----	1.30 ± 0.03	2.0 ± 1.0	54.0 ± 2.0
4	1.50 ± 0.03	1.40 ± 0.03	6.0 ± 1.0	22.0 ± 0.7
5	1.41 ± 0.03	1.43 ± 0.03	5.0 ± 1.0	14.0 ± 0.6

Table 7: Simulation results obtained for samples number 2, 4 and 5 corresponding to the parameters and values shown in Table 5.

model. Another consideration is the possibility of predicting theoretically soil pore distributions, taking into account the G parameters (Onody *et al.*, 1995).

VII Summary

In this chapter we described current scientific research in relation to fingering phenomena in soils, as well as applications of fractal theory, invasion percolation theory and imaging techniques as non-destructive and advanced methods in the study of such phenomena. Those studies showed that the perimeter of the finger displacement fronts represents a behavior of "self-affine" fractal or fractal type, its fractal dimension D_p being dependent on the time and the position along the profile of the soil column. Through the fractal dimension, it was also evident that the "dynamic" surface tension is dependent on D_p and the average pore diameter of the porous system. On the other hand, it was verified, by means of several experiments, that the fractal characteristics of the fingers, under the same initial conditions, are related to the soil granulometry or the pore-size distribution. For fine-textured soils the structure of the fingers is non-fractal, whereas for coarse-textured soils the structure of the fingers presents statistically self-affine or fractal characteristics. In addition to these results, it was also possible to introduce a modified theory of invasion percolation, so as to simulate the fingers structure. The tomography and imaging techniques employed are already available, through nuclear magnetic resonance as well as through X-ray, and presented very innovative and encouraging results for the study of fingering phenomena in soils and porous media in general, allowing us, among other things, to witness, non-destructively, the three-dimensional and aleatory character of the phenomena. Nevertheless, great research challenges remain. It is needed, for instance, to better relate the "effective" surface tension to the fractal dimension of the fingers, the passage from non-fractality to the fractal behavior of the finger structure in relation to the soil granulometry, and the pore profile distribution to the phenomenon occurrence. Concomitantly, the exploration and the development of more appropriate methods to measure their morphological and physical characteristics, in 2 and 3 dimensions, establish an important challenge to be faced in the near future.

Acknowledgments

The authors thank the Funda o de Amparo Pesquisa do Estado de São Paulo - FAPESP, project number 90/3773-7 and process number 90/1770-0, EMBRAPA-CNPDIA-São Carlos-SP-Brazil, project number 801.90.002-28 and 12.0.94.093.00, Conselho Nacional de Desenvolvimento Cient´fico e tecnológico - CNPq, process number 360249/94-9, IFQSC-DFCM-USP, Hospital das Cl´nicas-UNICAMP for financial support and/or allowing the use of equipment and installations. Also, we are indebted to professors R.N. Onody, A. Tannús, H.C. Panepucci (IFQSC-USP) and to R. Lotufo and C. Arruda (UNICAMP) for fruitful discussions and collaboration for obtaining part of the results presented in this chapter.

References

Avnir, A. 1989. The fractal approach to heterogeneous chemistry. John Wiley and Sons Ltd., Great Britain.

Baker, R.S. and D. Hillel. 1990. Laboratory tests of a theory of fingering during infiltration into layered soils. *Soil Sci. Soc. Am. J.* 54:20-30.

Berry, P. and M. Christie. 1991. Computers refine oil production. *Phys. World. Bristol.* 47-51.

Bensimon, D., L.P. Kadanoff, L.L. Shoudan, B. Shraimman and Chaotan. 1986. Viscous flows in two dimensions. *Rev. Mod. Phys.* 58 (4):977-999.

Bear, J. 1972. *Dynamics of Fluids in Porous Media.* American Elsevier, Environmental Science Series. Amsterdam.

Birovljev, A., L. Furuberg, J. Feder, T. Jøssang, K. Maløy and A. Aharony. 1991. Gravity invasion percolation in two dimensions: experiment and simulation. *Phys. Rev. Lett.* 67(5):584-587.

Blumberg, R.L., G. Shlifer and H.E. Stanley. 1980. Monte Carlo tests of universality in a correlated-site percolation problem. *J. Phys. A: Math. Gen.* 13:L147-L152.

Barnsley, M.F., R.L. Devaney, B.B. Mandelbrot., H.-O. Peitgen, D. Saupe and R.F. Voss. 1988. *The Science of Fractal Images.* Springer-Verlag, New York, USA.

Chandler, R., J. Koplik, K. Lerman and J.F. Willemsen. 1982. Capillary displacement and percolation in porous media. *J. Fluid Mech.* 119:249-267.

Chang, W-L.T. 1990. *Fractal Analysis of Wetting Front Instability in Unsaturated Soil.* Unpublished Ph.D. dissertation. University of California, Davis, California.

Chang, W-L, J.W. Biggar and D.R. Nielsen. 1994. Fractal description of wetting front instability in layered soils. *Water Resour. Res.* 30 (1):125-132.

Chen, J.D. and D. Wilkinson. 1985. Pore-scale viscous fingering in porous media. *Phys. Rev. Lett.* 58 (18):1892-1895.

Chen, J.D., M. Dias, S. Patz and L. Schwartz. 1988. Magnetic resonance image of immiscible fluid displacement in porous media. *Phys. Rev. Lett.* 61 (13):1489-1492.

Chouke, R.L., P. Van Meurs and C. Van der Poel. 1959. The instability of slow, immiscible, viscous liquid-liquid displacements in permeable media. *Trans. Am. Inst. Min. Metall. Pet. Eng.* 216:188-194.

Crestana, S., S. Mascarenhas and R.S. Pozzi-Mucelli. 1985. Static and dynamic three-dimensional studies of water in soil using computerized tomographic scanning. *Soil Sci.* 140 (5):326-332.

Crestana, S. 1992. Noninvasive measurements. p. 83-85. In G. Sposito and R.J. Reginato (eds.) *Opportunities in Basic Soil Science Research.* Soil Science Society of America, Inc., Madison, Wisconsin.

Daccord, G., S. Nittmann and H.E. Stanley. 1986. Fractal viscous fingers: experimental results. p. 203-210. In Stanley and Ostrowsky (eds.) *On Growth and Form.* Martinus Nijhoff Publishers. Netherlands.

Daccord, G. 1987. Chemical dissolution of a porous medium by a reactive fluid. *Phys. Rev. Lett.* 58:479-482.

Dias, M.M. and D.J. Wilkinson. 1986. Percolation with trapping. *J. Phys. A.* 19i:3131-3146.

De Gregoria, A.J. and L.W. Schwartz. 1987. Saffman-Taylor finger width at low interfacial tension. *Phys. Rev.* 58:1742-1744.

de Jennes, P.G. and E. Guyon. 1978. Lois générales pour l'injection d'un fluide dans un milieu poreux aléatoire. *J. Mécan.* 17:403-432.

Diment, G.A. and K.K. Watson. 1983. Stability analysis of water movement in unsaturated porous materials: numerical studies. *Water Resour. Res.* 19 (4):1002-1010.

Donohoe, G.W. 1992. Image processing short course with Khoros. University of New Mexico. USA.

Dupont, T., R. Goldstein, L. Kadanoff and Zhou. 1993. Finite-time singularity formation in a Hele-Shaw system. *Physical Review E.* 47 (6):4182-4196.

Engeberts, W.F. and Klinkenberg. 1951. Proceedings of the third world petroleum congress. Sec. 2. (unpublished manuscript).

Feder, J. 1988. Fractals. Plenum Press Publishers, New York.

Frette, V., K.F. Maløy, F. Boger, J. Feder and T. Jøssang. 1990. Diffusion-limited-aggregation-like displacement structures in a three-dimensional porous medium. *Physical Rev. A* 42 (6):3432-3437.

Frette, V., K.F. Maløy, F. Boger, J. Feder and T. Jøssang. 1991. Displacement structures in 2- and 3-dimensional porous media at low viscosity contrast. *Physica Scripta.* T38:95-98.

Frette, V., J. Feder, K.F. Maløy, T. Jøssang and P. Meakin. 1992. Buoyancy-driven fluid migration in porous media. *Phys. Rev. Lett.* 68 (21):3164-3167.

Glass, R.J., T.S. Steenhuis and J.-Y. Parlange. 1988. Wetting front instability as a rapid and far-reaching hydrologic process in the vadose zone. *J. of Cont. Hydrol.* 3:207-226.

Glass, R.J., J.-Y. Parlange and T.S. Steenhuis. 1989a. Wetting front instability: theoretical discussion and dimensional analysis. *Water Resour. Res.* 25 (6):1187-1194.

Glass, R.J., T.S. Steenhuis and J.-Y. Parlange. 1989b. Wetting front instability: experimental determination of relationships between system parameters and two-dimensional unstable flow field behavior in initially dry porous media. *Water Resour. Res.* 25 (6):1195-1207.

Glass, R.J., J.-Y. Parlange and T.S. Steenhuis. 1989. Mechanism for finger persistence in homogeneous unsaturated porous media: theory and verification. *Soil Sci.* 48 (1):60-70.

Glass, R.J. 1990. Wetting front instability in unsaturated porous media: a three-dimensional study in initially dry sand. *Transp. Porous Media.* 5:247-268.

Glass, R.J., J.-Y. Parlange and T.S. Steenhuis. 1991. Immiscible displacement in porous media. Stability analysis of three-dimensional, axisymmetric disturbances with application to gravity-driven wetting instability. *Water Resour. Res.* 27 (8):1447-1956.

Hasen, J. and A. Skjelter. 1988. Fractal pore space and rock permeability implications. *Phys. Rev. B.* 38 (4):2635-2638.

Hill, D.E. and J.-Y. Parlange. 1972. Wetting front instability in layered soils. *Soil Sci. Soc. Am. Proc.* 36 (5):697-702. Hillel, D. and R.S. Baker. 1988. A descriptive theory of fingering during infiltration into layered soils. *Soil Sci.* 146 (1):51-56.

Homsey, G.M. 1987. Viscous fingering in porous media. *Ann. Rev. Fluid. Mech.* 19:271-311.

Hosang, J. 1993. Modeling preferential flow of water in soil - a two-phase approach for field conditions. *Geoderma.* 58:149-163.

Hullin, J.P., E. Clément, C. Baudet., J.F. Gouyet and M. Rosso. 1988. Quantitative analysis of an invading-fluid invasion front under gravity. *Phys. Rev. Lett.* 61 (3):333-336.

Jacquin, Ch. G. and P.M. Adler. 1985. The fractal dimension of a gas-liquid interface in a porous medium. *J. Colloid Interf. Sci.* 107(2):405-417.

Jensen, M.H., A. Libchaber, P. Pelc,. and G. Zocchi. 1987. Effect of gravity on the Saffman-Taylor meniscus: theory and experiment. *Phys. Rev. A.* 35:2221-2227.

Katz, A.J. and A.H. Thompson. 1985. Fractal sandstone pores: implications for conductivity and pore formation. 1985. *Phys. Rev. Lett.* 54 (12):1325-1328.

Lenormand, R. 1986. Pattern growth and fluids displacements through porous media. *Physica A.* 140:114-123. Lenormand, R., E. Toubal and C. Zarcone. 1988. Numerical models and experiments on immiscible displacements in porous media. *J. Fluid Mech.* 189:165-187.

Liang, S. 1986. Random-walk simulations of flow in Hele Shaw cells. *Phys. Rev. A* 33 (4):2663-2674. Mandelbrot, B.B. 1982. *The Fractal Geometry of Nature.* 2nd ed. W.H. Freeman and Company. New York.

Mandelbrot, B.B. 1986a,b. Self-affine fractals sets. p. 3-28. In L. Pietronero and E. Tossati (eds.) *Fractals in Physics.* North-Holland. Amsterdam.

Maløy, K.J., J. Feder and T. Jøssang. 1985a. Viscous fingering fractals in porous media. *Phys. Rev. Lett.* 55:2688-2691.

Maløy, K.J., J. Feder and T. Jøssang. 1985b Radial viscous fingering in a Hele-Shaw cell. Report series. Cooperative phenomena project. Department of physics. University of Oslo. 85-9, 1-15.

Maløy, K.J., J. Feder and T. Jøssang. 1985. Viscous fingering fractals in porous media. *Physical Review Letter* 58 (24):2688-2691.

Meakin, P. 1989. Fractals and disorderly growth. *J. Mater. Educ.* 11:105-167.

Meakin, P., J. Feder, V. Frette and T. Jøssang. 1992. Invasion percolation in a destabilizing gradient. *Phys. Rev. A* 46 (6):3357-3368.

Nittmann, J., G. Daccord and E. Stanley. 1985. Fractal growth of viscous fingers: quantitative characterization of a fluid instability phenomenon. *Nature* 314 (14):141-144.

Onody, R.N. 1994. Invasion percolation in porous media with a biased growth rule. *Phys. Rev. Lett.*

Onody, R.N., A.N.D. Posadas and S. Crestana. 1995. Experimental studies of fingering phenomena in two dimensions and simulation using a modified invasion percolation. *J. Applied Physics*

Paterson, L. 1981. Radial fingering in a Hele-Shaw cell. *J. Fluid. Mech.* 113:513-529.

Philip, J.R. 1975. Stability analysis of infiltration. *Soil Sci. Soc. Am. Proc.* 39:1042-1049.

Posadas, D.A.N. and S. Crestana. 1991. Teoria fractal e técnicas de imagem na caracterização da infiltração instável em um meio poroso não-saturado. *Anais XIX Encontro Sobre Escoamento em Meios porosos.* Vol. I:117-129.

Posadas, D.A.N. and S. Crestana. 1993. Aplicação da teoria fractal na caracterização do fenômeno "fingering" em solos não saturados. *Rev. Bras. Ci. Solo.* 17(1):1-8.

Posadas, D.A.N., S. Crestana, R.N. Onody and L.A.C. Jorge. 1993. Teoria fractal e percolação por invasão validadas na descrição do fluxo preferencial em um meio poroso. *Anais, XX Encontro Sobre Escoamento em Meios Porosos.* Vol. I:141-153.

Posadas, D.A.N., A. Tannús, H.C. Panepucci and S. Crestana. 1993. Estudo tridimensional do fenômeno "fingering" em solos através de imagens de NMR e raios-X. *Anais IV Encontro de Usurios de RMN.* Angra dos Reis. R.J. Brazil. Vol I:120-124.

Posadas, D.A.N. 1994. Estudo do fenômeno "fingering" em um meio poroso através de imagens e teoria da percolação por invasão. Unpublished Ph.D. Dissertation. Instituto de F´sica e Qu´mica de São Carlos. USP. São Carlos. São Paulo. Brazil.

Posadas, D.A.N., A. Tannús, C.H. Panepucci and S. Crestana. 1996. Magnetic resonance imaging as a non-invasive technique for investigating 3-D preferential flow occurring within stratified soil samples. *Computers and Electronics in Agriculture* 14(4):255-267.

Posadas, D.A.N. and S. Crestana. 1995. 2-D and 3-D wetting front instabilities in layered soils investigated through imaging techniques and invasion percolation. Unpublished manuscript.

Raats, P.A.C. 1973. Unstable wetting front in uniform and nonuniform soils. *Soil Sci. Soc. Am. Proc.* 37:681-685.

Ritsema, C.J. and L.W. Dekker. 1993. Preferential flow mechanism in a water repellent sandy soil. *Water Resour. Res.* 29 (7):2183-2193.

Roux, J-N. and P. Willkinson. 1988. Resistance jumps in mercury injection in porous media. *Phys. Rev. A* 37(10):3921-3926.

Saffman, P.G. and S.G. Taylor. 1958. The penetration of a fluid into a porous medium or hele-shaw cell containing a more viscous liquid. *Proc. Roy. Soc. A.* 245:312-329.

Selker, J.S., T.S. Steenhuis and J.-Y. Parlange. 1992. Wetting front instability in homogeneous sandy field under continuous infiltration. *Soil Sci. Soc. Am. J.* 56:1346-1350.

Sposito, G. and R.J. Reginato. 1992. *Opportunities in Basic Soil Science Research.* Soil Science Society of America, Inc. Madison, Wisconsin.

Starr, J.L, H.C. Deroo and C.R. Frink. 1978. Leaching characteristics of a layered field soil. *Soil Sci. Soc. Am. J.* 376-391.

Stauffer, D. 1985. *Introduction to Percolation Theory.* Taylor and Francis. USA.

Tamai, N., T. Asaeda and C.G. Jeevaraj. 1987. Fingering in two-dimensional, homogeneous, unsaturated porous media. *Soil Sci.* 144 (2):107-112.

Tannús, A., D.A.N. Posadas, C.H. Panepucci and S. Crestana. 1995. Dynamic imaging approach for 3-D preferential flow within stratified soil samples, (submitted to Magnetic Resonance Materials in Physics, Biologics and Medicine).

Vicsek, T. 1992. *Fractal Growth Phenomena.* World Scientific Publishing Co., London.

White, I., P.M. Colombera and J.R. Philip. 1977. Experimental studies of wetting front instability induced by gradual change of pressure gradient and by heterogeneous porous media. *Soil Sci. Soc. Am. J.* 41:483-489.

Wilkinson, D. and J.F. Willemsen. 1983. Invasion percolation: a new form of percolation theory. *J. Phys. A: Math. Gen.* 16:3365-3376.

Willkinson, D. 1984. Percolation model of immiscible displacement in the presence of buoyancy forces. *Phys. Rev. A* 30 (1):520-531.

Wilkinson, D. 1986. Percolation in inmiscible displacement. *Phys. Rev. A* 34 (2):1380-1391.

Witten, T.A. and L.M. Sander 1981. Diffusion-limited aggregation, a kinetic critical phenomena. *Phys. Rev. Lett.* 47 (19):1400-1403.

Witten, T.A. and L.M. Sanders. 1983. Diffusion-limited aggregation. *Phys.Rev. B.* 27:5686-5697.

Soil-Water Conductivity of a Fractal Soil

C. Fuentes, M. Vauclin, J.-Y. Parlange and R. Haverkamp

Contents

I Introduction

Most classical models of soil-water conductivity, K, are based on Poiseuille's law. Early and straightforward approaches, like those of Purcell (1949), Gates and Lietz (1950), and Childs and Collis-George (1950), were later improved with often *ad hoc* corrections to fit observations, *e.g.*, Burdine (1953), Fatt and Dykstra (1951), Millington and Quirk (1961), Mualem (1976), and Mualem and Dagan (1978).

Our aim in this chapter is to describe soils using fractal geometry as a unifying concept and re-examine the models above within that context.

II Classical Models

At the pore (microscopic) scale Poiseuille's law and Darcy's law are used for a collection of pores (macroscopic or Darcy's scale).

For a cylinder of radius R, the average velocity V, *i.e.*, the flux through the cylinder divided by πR^2, obeys Poiseuille's law, or

$$V = -(\rho g/8\mu)R^2 \frac{\partial H}{\partial z} \tag{1}$$

where g is the acceleration of gravity, μ the viscosity and ρ the density of the liquid, and H is the total potential. Darcy's law gives the flux q per unit area of soil as

$$q = -K\frac{\partial H}{\partial z} \tag{2}$$

ISBN 1-56670-105-8

where the total potential, H, is usually made up of a gravitational and a matric component and K is the soil-water conductivity.

Tortuosity is normally introduced as an empirical correction for the "tortuous" flow around grains, at the microscopic scale, whereas it is in the z-direction at the macroscopic scale. Then,

$$dz_T = T dz, \qquad T > 1 \tag{3}$$

The value of T depends on the type of soil and the pore size. Constant values like $\pi/2$ (half a circumference of a circle to its diameter), $\sqrt{2}$ (assuming the average flow at 45° to the z-direction) and others have been suggested (Carman, 1956). The same relations applies to the velocities:

$$V_T = TV \tag{4}$$

Poiseuille's equation (1) applies along the tortuous path, then in the z-direction,

$$V = -\frac{\rho g}{8\mu}\frac{R^2}{T^2}\frac{\partial H}{\partial z} \tag{5}$$

If we now integrate V over the water filled pore domain Ω to obtain the flux q, we have that

$$q = -\frac{\rho g}{8\mu}\frac{\partial H}{\partial z}\int_\Omega \frac{R^2}{T^2} da \tag{6}$$

where da is that part of the domain Ω with geometrically defined pathways located between z and $z + dz$, divided by the total soil volume between z and $z + dz$. Note that dz is a differential that must be large enough to be on the Darcy scale, and R and T are functions of the pathway geometry.

It is convenient to define an intrinsic conductivity k, excluding fluid properties, as

$$k = K\mu/(\rho g) \tag{7}$$

i.e.,

$$k = \frac{1}{8}\int_\Omega (R^2/T^2) da \tag{8}$$

The relative conductivity k_r is then

$$k_r = \int_\Omega (R^2/T^2) da \Big/ \int_{\Omega_s} (R^2/T^2) da \tag{9}$$

where Ω_s is the domain Ω at saturation.

Purcell (1949), by analogy with a bunch of capillary tubes, takes $da = d\theta$, where θ is the water content and

$$k(\theta) = \frac{1}{8}\int_0^\theta R^2 d\vartheta \tag{10}$$

its saturated value k_s is obtained when $\theta = \in$, the porosity which we take as the saturated water content θ_s. Purcell (1949) then corrected k_s by an *ad hoc* factor $G(\in)$, which was later introduced in Equation (10) as a function of $G(\theta)$ by Gates and Lietz (1950), or

$$k(\theta) = \frac{1}{8} G(\theta) \int_0^\theta R^2 d\vartheta \tag{11}$$

Childs and Collis-George (1950) made the fundamental observation that any pathway located between z and $z+dz$ is made up of connected pores so that R is a function of the pathway geometry. For simplicity, the pathway is characterized by only two pores: the entrance pore at z radius r, and the exit pore at $z + dz$ radius ρ and R is some function of r and ρ, thus, since Ω is now defined as both the entrance- and exit-pore domains,

$$da = d\theta(r) d\theta(\rho) \tag{12}$$

Equation (8) now becomes, with a $G(\theta)$ correction as in Equation (11)

$$k = \frac{G(\theta)}{8} \int_\Omega \frac{R^2}{T^2} d\theta(r) d\theta(\rho) \tag{13}$$

To push Equation (13) further it is necessary to postulate some relationships between R, T, r and ρ. For instance, Childs and Collis-George take $T = 1, G = 1$ and assume that R is the smaller of r and ρ. Integration by parts then gives (Brutsaert, 1967)

$$k = \frac{1}{4} \int_0^\theta [\theta - \vartheta] r^2 d\vartheta \tag{14}$$

Many other suggestions have been made. For instance, Mualem (1976) considered $G(\theta)$ proportional to $\theta^{\frac{4}{3}}$ or $\theta^{\frac{1}{2}}$ and $R = (r\rho)^{\frac{1}{2}}$, whereas Kunze et al. (1968) take $G \propto \theta$. Purcell's Equation (11) has also been used by Burdine (1953) and Wyllie and Gardner (1958), who take $G \propto \theta^2$ and T dependent on R or constant. Fatt and Dykstra (1951) suggest that $T^2 \propto cR^{-b}$, c and b positive.

Milligton and Quirk (1961) made the fundamental suggestion that Purcell and Childs and Collis-George are limiting cases and that there is a need to interpolate between them. The pathway domain Ω_s is used to define n such that

$$\int_{\Omega_s} da = \in^n \qquad 1 \leq n \leq 2 \tag{15}$$

$n = 1$ being Purcell's result and $n = 2$ Childs and Collis-George's result. Equation (15) is the probabilistic measure to be in the pathway domain and $(1- \in)^{n/2}$ to be in the solid domain ($n/2$ is required since n comes from having two pores necessary to define a pathway). Then,

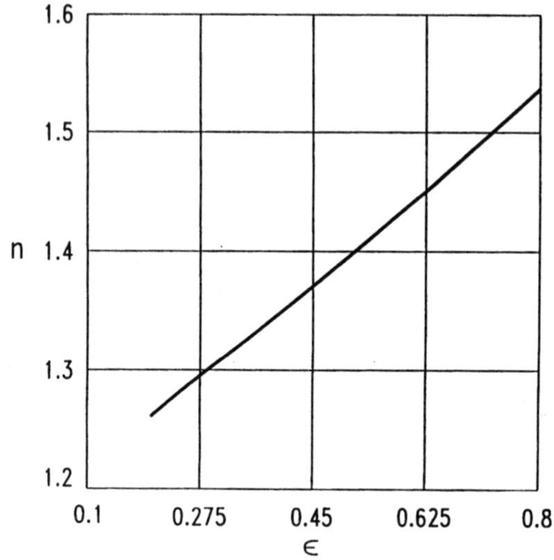

Figure 1: Diagram of the coefficient *n*, defined by Equation (15), as a function of the porosity ∈ .

$$\in^n + (1 - \in)^{n/2} = 1, \qquad 1 \leq n \leq 2 \tag{16}$$

Figure 1 plots n as a function of $\in$. It is clear that for any realistic value of $\in$, n is between 1.3 and 1.5, which is well between 1 and 2.

III Fractal Soils

We assume that the pores, and the complementary grains, form a fractal object. If the pores occupy a volume $\in$, and the gains a volume $1-\in$, by definition of the surface fractal dimension D, the fractal measures of the pathways and grains are $\in^{2D/3}$ and $(1-\in)^{D/3}$, which are the probabilistic quantities used by Millington and Quirk (1961). Again following Childs and Collis-George, the pathways being described by 2 pores, the probability for a pathway is $\in^{2D/3}$. Hence we retrieve Equation (16) and prove it for a fractal soil, giving a fractal interpretation of the coefficient n, *i.e.*,

$$n = 2D/3 \tag{17}$$

By definition of a fractal object we must have $D \leq 3$. Since $1 \leq n \leq 2$, then

$$1.5 \leq D \leq 3 \tag{18}$$

In practice, Figure 1 showed that for reasonable values of $\in$ between 0.3 and 0.6, D must be in a very narrow band,

$$2 \leq D \leq 2.2 \tag{19}$$

Since $D = 2$ means that the fractal would have a dimension low enough to be a surface, this lower limit is quite reasonable. It also means that Purcell's limit, or $D = 1.5$, is never reached in practice.

Equation (12) becomes for a fractal soil

$$da = d\theta^{D/3}(r)d\theta(\rho)^{D/3} \tag{20}$$

which is now consistent with Equation (15) and, of course, reduces to the classical result for $D = 3$. The integrand $(R/T)^2$ in Equation (9) is the effective cross section of the pathway which depends not only on R, the effective radius of the pathway but also on the tortuosity of the path connecting the entrance and exit pores. R/T then is the measure of the path in fractal geometry so that

$$R/T \propto R^{2D/3} \tag{21}$$

The proportionality is not with $R^{D/3}$, because, following the reasoning leading to Equation (15) the path is affected by both entrance and exit pores. We then use Equation (9) to estimate k_r. In the limit, $D = 3$ note that Purcell's result cannot be obtained for $R^2 = \rho r$ but rather its square value. This is as expected as the entrance and exit pores play similar roles in predicting k. Altogether fractal soils should obey

$$k = k_s \int_0^{\rho(\theta)} \int_0^{r(\theta)} R^{4D/3} d\vartheta^{D/3}(r) d\vartheta^{D/3}(\rho) / \int_0^{\rho(\in)} \int_0^{\rho(\in)} R^{4D/3} d\vartheta^{D/3}(r) d\vartheta^{D/3}(\rho) \tag{22}$$

To go further, and as was the case for classical models, we must postulate a dependence between R, ρ and r, *e.g.*:

1. If R is the smaller of r and ρ,

$$k \propto \int_0^{\theta} \{\theta^{D/3} - \vartheta^{D/3}] r^{4D/3}(\vartheta) d\vartheta^{D/3} \tag{23}$$

2. If $R^2 = \rho r$,

$$k \propto \left[\int_0^{\theta} r^{2D/3} d\vartheta^{D/3} \right]^2 \tag{24}$$

3. If R is r or ρ,

$$k \propto \theta^{D/3} \int_0^{\theta} r(\vartheta)^{4D/3} d\vartheta^{D/3} \tag{25}$$

Finally, the dependence between r and ϑ must be given (of course, one could also have given directly the relationship of R with ϑ(r) and $\vartheta(\rho)$). For instance, it is standard to estimate $r(\vartheta)$ by the inverse of the matric potential $\Psi(\vartheta)$.

An especially important case is the Brooks and Corey (1964) result, giving

$$\vartheta(r) \propto r^{\lambda} \tag{26}$$

which is widely used for fractal soils (Tyler and Wheatcraft, 1990; Rieu and Sposito, 1991). The Brooks and Corey (1964) law breaks down near saturation. It is not entirely clear whether this is because a soil cannot be represented as a fractal object near saturation (the similarity law of fractals is limited by the largest pore size, which must be finite) or because Equation (26) is not the most general one for fractal soils. However, there is no doubt that Equation (26) must hold for all fractals when r is small (Falconer, 1990) (in agreement with this it is remarkable that standard water retention laws besides Brooks and Corey, *e.g.*, Brutsaert (1966) and Van Genuchten (1980), are consistent with Equation (26) when ϑ is small).

Thus, either because the Brooks and Corey law holds and certainly at small r, Equation (26) will apply and then (Tyler and Wheatcraft, 1990; Falconer, 1990; Rieu and Sposito, 1991),

$$\lambda = 3 - D \tag{27}$$

Also, quite remarkably, if Equation (26) hold, Equations (23), (24), and (25) all yield the same result,

$$K = K_s(\theta/\in)^{4D/3\lambda+2D/3} \tag{28}$$

with the power varying from 4 to 5, according to Equation (19).

Such a form, with a variety of powers (4 to 5 being fairly representative), has been suggested on semi-empirical ground by Yuster (1951), Irmay (1954), Corey (1954), and Brooks and Corey (1964).

If other relationships are postulated, instead of Equation (26), as applicable to fractals, care must be taken so that the denominator of Equation (22) remains finite. For instance, near saturation, if Van Genuchten's equation (1980) gives, assuming that Equation (27) holds,

$$r \propto (1 - \theta/\in)^{-1/(3-D+C)} \tag{29}$$

then Equations (23) and (24) require for the intregrals to be finite that

$$C > (4D - 9)/3 \tag{30}$$

Van Genuchten (1980) showed that C must be greater than one, which always holds if $D < 2.4$, according to Equation (21), and Figure 1 shows this to be always the case. However, Equation (25) would require the more constraining

$$C > (7D - 9)/3 \tag{31}$$

which requires $C > 2.1$ if D has its largest value around 2.2.

Analytic application of Equation (22) in general and Equations (23) to (25), in particular, is to show the need for a $G(\theta)$ in the classical models, as in Equation (13). For instance, let us compare Equations (14) and (23), which are based on the same assumption that since k is normally a rapidly increasing function of θ, we can make a Taylor expansion of the integrand for ϑ near θ with little error or rewrite Equation (23) as,

$$k \propto r^{\frac{4D-6}{3}} \theta^{\frac{2(D-3)}{3}} \int_0^\theta [\theta - \vartheta] r^2(\vartheta) d\vartheta \tag{32}$$

Hence, according to fractal geometry, a $G(\theta)$ is needed in front of Equation (14). To estimate $G(\theta)$ from Equation (32) it is, of course, required to know $r(\theta)$. Assuming for simplicity a Brooks and Corey law, we deduce:

$$G(\theta) \propto \theta^{2\frac{D-3}{3} + \frac{4D-6}{3(3-D)}} \tag{33}$$

If we had taken $R^2 = \rho r$, Mualem's (1976) assumption, and started with Equation (24), interestingly, we end up with the same $G(\theta)$ as in Equation (33). If again we take $2 < D < 2.2$, the power of θ in Equation (33) varies between 1/3 and 2/3. Of the various suggestions for $G(\theta)$, see discussion below Equation (14), only $G \propto \theta^2$ has a power slightly outside the band suggested by Equation (33) (it would require $D \propto 2.5$, which is too large if Equation (15) holds).

IV Conclusion

Based on fractal geometry we obtained Equation (22), which is similar but simpler than the semi-empirical classical equation requiring some *ad hoc* tortuosity and $G(\theta)$ correction. Its use requires a knowledge of R, and both k_s and the porosity $\in$ must be measured. The fundamental equation of Millington and Quirk was reinterpreted so that the fractal dimension D can be deduced from the measurement of $\in$. It appears that D does not vary very much in practice and might be between 2 and 2.2, at least in most cases.

In the simplest case, when the Brooks and Corey law applies for R, k depends on θ through a power law, and the predicted power seems reasonable. Although this simple case might be the most important for a fractal soil, more complex estimations of R can also be used.

References

Brooks, R.H. and A.T. Corey. 1964. *Hydraulic properties of porous media.* Hydrol. Pap. 3, Colo. State Univ., Fort Collins.

Brutsaert, W. 1966. Probability laws for pore-size distributions. *Soil Sci.* 101:85-92.

Burdine, N.T. 1953. Relative permeability calculation from size distribution data. *Trans. AIME* 198:71-78.

Carman, P.C. 1937. Fluid flow through granular beds. *Trans. Inst. Chem. Eng.* 15:150-166.

Childs, E.C. and N. Collis-George. 1950. The permeability of porous materials. *Proc. Roy. Soc., Ser. A* 201:392-405.

Corey, A.T. 1954. The interrelation between gas and oil relative permeabilities. *Producer's Monthly* XIX, No. 1.

Falconer, K. 1990. *Fractal geometry, mathematical foundations and applications.* John Wiley & Sons, Chichester, England.

Fatt, I. and H. Dykstra. 1951. Relative permeability studies. *Petrol. Trans. Am. Inst. Mining & Met. Eng.* 192:249-256.

Gardner, W.R. 1958. Some steady-state solutions of the unsaturated moisture flow equation with application to evaporation from a water table. *Soil Sci.* 85:228-232.

Gates, J.I. and W.T. Lietz. 1950. Relative permeabilities of California cores by the capillary-pressure method. Drilling and Production Practice. *Amer. Petrol. Inst.* 285-298.

Irmay, S. 1954. On the hydraulic conductivity of unsaturated soils. *EOS Trans. AGU* 35:463-467.

Kunze, R.J., G. Uehara and K. Graham. 1968. Factors important in the calculation of hydraulic conductivity. *Soil Sci. Soc. Amer. Proc.* 32:760-765.

Millington, R.J. and J.P. Quirk. 1961. Permeability of porous solids. *Trans. Faraday Soc.* 57:1200-1206.

Mualem, Y. 1976. A new model for predicting the hydraulic conductivity of unsaturated porous media. *Water Resources Res.* 12(3):513-522.

Mualem, Y. and G. Dagan. 1978. Hydraulic conductivity of soils: Unified approach to the statistical models. *Soil Sci. Soc. Am. J.* 42:392-395.

Purcell, W.R. 1949. Capillary pressures - their measurement using mercury and the calculation of permeability therefrom. *Petr. Trans. Am. Inst. Mining & Met. Eng.* 186:39-48.

Rieu, M. and G. Sposito. 1991. Fractal fragmentation, soil porosity, and soil water properties: 1. Theory. *Soil Sci. Soc. Am. J.* 55:1231-1238.

Tyler, S.W. and S.W. Wheatcraft. 1990. Fractal process in soil water retention. *Wat. Resources Res.* 26:1047-1054.

Van Genuchten, M.Th. 1980. A closed-form equation for predicting the hydraulic conductivity of unsaturated soils. *Soil Sci. Soc. Am. J.* 44:892-898.

Wyllie, M.R.J. and G.H.F. Gardner. 1958. The generalized Koseny-Carman equation. *World Oil* March-April:210-228.

Yuster, S.T., 1951. Theoretical considerations of multiphase flow in idealized capillary systems. *Proc. 3rd World Petrol. Congress* 2:437-445.

A Pseudo-Fractal Model for Hydraulic Property Distributions in Porous Media

F.J. Molz, T.A. Hewett and G.K. Boman

Contents

I Introduction

During the past decade it has become evident that natural porous media are seldom homogeneous or random in a uniform sense (statistically homogeneous). As summarized by Cushman (1990): "Porous media may contain multiple, nested, natural length and time scales or continuously evolving scales. (Property measurements) on one scale may appear to have little relevance to any other scale." This type of natural structure has been called "hierarchical" in nature, and experimental evidence for its existence has been presented by several authors (Osiensky *et al.*, 1984; Fogg, 1986; Molz *et al.*, 1988, 1990; Moltyaner, 1990). One indirect manifestation of such structure would be the well-known phenomenon of scale-dependent dispersivity (Pickens and Grisak, 1981; Molz *et al.*, 1983; Güven *et al.*, 1984; Moltyaner and Killey, 1988a,b; Wheatcraft and Tyler, 1988). A broad overview of hierarchical structure in porous media and various implications is presented in the text edited by Cushman (1990).

At the present time, it is not known with certainty how the concept of "hierarchical structure" in porous media should be represented mathematically, or if a

ISBN 1-56670-105-8

natural (and hopefully general) representation exists (Baveye and Sposito, 1984; Cushman, 1986, 1990a; Plumb and Whitaker, 1990). However, experimental evidence developed by Hewett (1986, 1992) and Molz and Boman (1993) suggests that the hierarchical structure of the three-dimensional porosity function (θ) and hydraulic conductivity function (K) may exhibit the spatial variations of certain self-affine, stochastic fractals called fractional Gaussian noise (fGn) and fractional Brownian motion (fBm). While only a limited amount of field data have been collected, there are other reasons to be interested in the properties of these functions, since they have risen empirically over the past 40 years in several areas of hydrology, geophysics and geology (Korvin, 1992).

A fractal-based function, whether self-similar or self-affine, is a natural candidate for representing a hierarchical structure in either time or space. The present chapter will review the origin of fBm and fGn, and the methodology for using such functions to represent θ and K distributions in saturated porous media. As pointed out originally by Hewett (1986), such methodology serves most naturally as the basis for a conditional-simulation approach for modeling such processes as fluid displacement in petroleum recovery and contaminant transport in porous media. The property to be represented is measured at a finite number of locations, and the chosen function is used to generate an unlimited number of realizations conditioned on the measured data. This process was called "stochastic interpolation" by Molz and Boman (1993). Other fractal-based approaches have been proposed but will not be reviewed explicitly herein (Wheatcraft and Tyler, 1988; Tyler and Wheatcraft, 1990).

It is known that real θ and K distributions are not smooth, except possibly on a very small scale (Hewett, 1986; Molz *et al.*, 1990). However, many contemporary interpolation procedures, such as averaging, Kriging and polynomial interpolation act to smooth the basic data. An advantage of the stochastic realization approach discussed herein is that property distribution irregularity is maintained over as large a range of measurement scales as desired. The conditional simulation procedure becomes progressively more deterministic as the number of property measurements is increased.

II Fractional Brownian Motions and Gaussian Noises

In the hydrologic sciences, applications of fBm and fGn evolved from the work of Hurst (1951), Mandelbrot and Van Ness (1968), and Mandelbrot and Wallis (1968, 1969a,b). Important contributions were made by Voss (1985a,b), and to the authors' knowledge, the ideas were first reviewed, integrated with contemporary stochastic and deterministic hydrologic concepts, and applied to transport processes in porous media by Hewett (1986).

Early applications of fBm and fGn were to streamflow analysis and to the analysis of other types of geophysical time series, although the fBm/fGn terminology was not used at that time (Hurst, 1957; Hurst *et al.*, 1965). Fractal concepts were emphasized later (Mandelbrot and Wallis, 1968, 1969a,b;

Rodriguez-Iturbe *et al.*, 1972). Out of the early work arose a puzzle, since called the Hurst phenomenon or Hurst effect. This puzzle occurs due to the fact that very long range correlations were found in the time-series data analyzed by Hurst, which seemed to contradict the intuitive notion that natural time series have short term memory, so that events separated by several years or decades may be considered independent in a statistical sense. We will return to the spatial analog of the Hurst effect later. Further discussion of the classical Hurst effect may be found in Korvin (1992).

As evident from the previous discussion, the independent variable of interest in early applications was time. In some ways, the definitions of fBm and fGn are most intuitive when introduced within the context of time series. Therefore, the time-based definition will be reviewed briefly and used as a vehicle to introduce basic concepts.

The fractal nature of fBm follows directly from its definition. In its simplest form, $fBm \equiv m_H(t)$ is a single-valued, continuous, function of time, t, having increments $m_H(t_2) - m_H(t_1)$, with a Gaussian distribution and the following statistical properties:

$$E\{m_H(t_2) - m_H(t_1)\} = 0, \text{ and} \tag{1}$$

$$E\{[m_H(t_2) - m_H(t_1)]^2\} = \alpha \left|t_2 - t_1\right|^{2H}, \tag{2}$$

where $E\{X\}$ denotes the mean or expected value of X, α is a constant of proportionality and H is a constant in the range of $0 < H < 1$ (Mandelbrot and Van Ness, 1968; Voss, 1985a,b). The expected values are obtained by averaging over many increments of size $|t_2 - t_1|$, and the assumption is made that the increments are stationary. For true fBm, Equation (2) holds, no matter how small $\Delta t \equiv t_2 - t_1$ becomes. For the special case of $H = 0.5$, Equation (2), the mean of the squared increments becomes

$$E\{[m_H(t_2) - m_H(t_1)]^2\} = \alpha \left|t_2 - t_1\right|, \tag{3}$$

which defines the well-known Brownian motion of classical physics (cBm) if one interprets $m_H(t_2) - m_H(t_1)$ as a spatial increment. Thus, fBm may be viewed as a generalization of the definition of cBm.

In order to develop an appreciation for how the concepts underlying Equations (2) and (3) first arose, it is worth considering cBm in more detail. Brownian motion refers to the erratic movement of small particles, such as smoke, dust, etc., that can be observed under the microscope. The phenomenon was first conceived correctly by Brown (1828). Brownian motion may be observed at different spatial scales simply by changing the magnification of the microscope. Notice, from Equation (3), that cBm is defined in terms of particle displacements (position changes) over fixed time intervals $t_2 - t_1$. Thus to get the particle position as a function of time, one would have to start at a particular particle location and sum the displacements. Under the microscope it was the displacements that Brown studied. They were observed at all magnifications, with smaller displacements observable at higher magnification and occurring

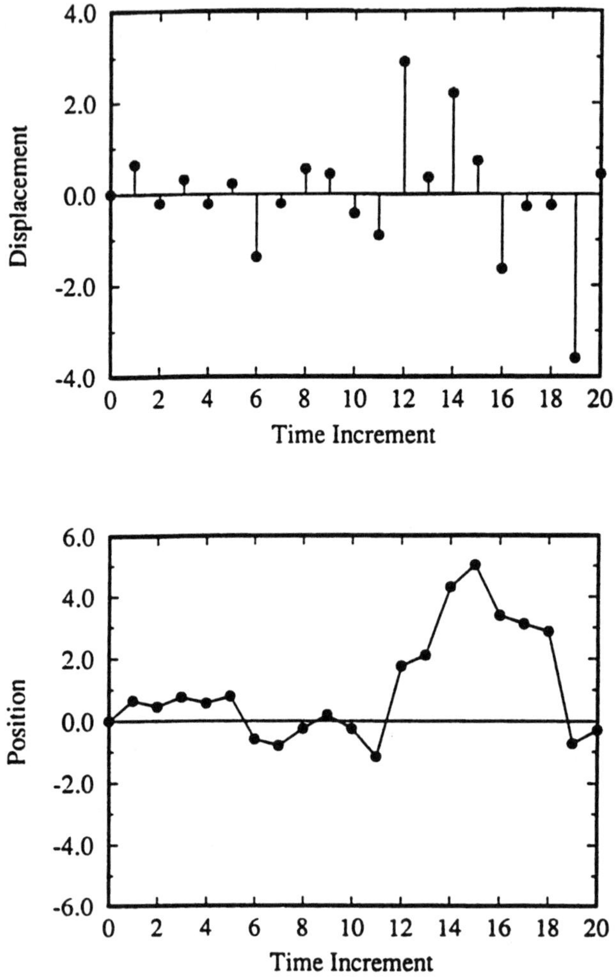

Figure 1: Illustration of the relationship between random displacements (top) and the resulting Brownian motion (bottom).

over a shorter time interval. Each displacement appeared to be uncorrelated with those occurring before and after, and the mean displacement size was proportional to the square root of the selected time increment. This type of behavior is represented by Equation (3).

The relationship between particle displacements and position is illustrated in Figure 1. Twenty displacements were selected from a Gaussian distribution of mean zero and unit variance. (Such displacements are termed Gaussian noise.) These were converted to position by summing and assuming that the particle position was zero at time zero. The lower curve may be viewed as a Brownian

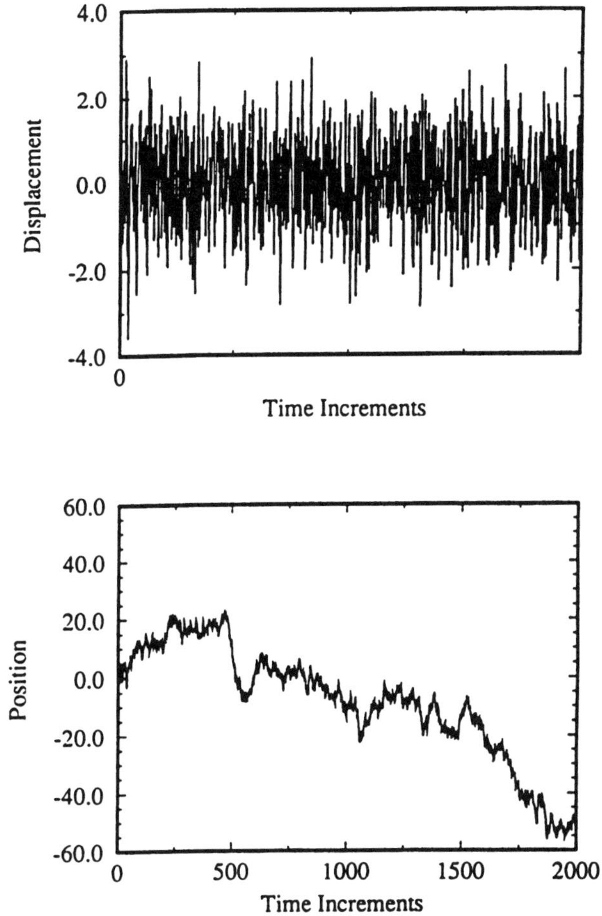

Figure 2: Realistic illustration of cGn (top) and the resulting cBm (bottom).

motion trace at the scale of the time increment Δt. This type of curve is sometimes called a one-dimensional random walk. Note that the straight lines connecting the position points are only an approximation, possibly a poor one, of particle position between measurement times. As illustrated in Figure 2, cBm looks more realistic as a given interval of time is subdivided into more and more time increments. For additional discussion see Feder (1988).

It is Equation (2) that guarantees a fractal nature for fBm, and since (2) holds only on average, the fractals are stochastic (Voss, 1985b). From Equation (2) the following argument may be developed. Let

$$E\{\Delta m_H(t)^2\} \equiv E\{[m_H(t_2) - m_H(t_1)]^2\}, \text{ and}$$

$$E\{[m_H(rt)]^2\} \equiv E\{[m_H(rt_2) - m_H(rt_1)]^2\}, \tag{4}$$

where r is a scaling ratio (multiplier).

Then

$$\begin{aligned} E\{[m_H(r\Delta t)]^2\} &= \alpha\,|rt_2 - rt_1|^{2H} \\ &= r^{2H}\alpha\,|t_2 - t_1|^{2H} \\ &= r^{2H}E\{[m_H(\Delta t)]^2\}. \end{aligned} \tag{5}$$

Equation (5) describes the scaling behavior of fBm. For the case of $H = 0.5$, which yields cBm, the scaling relationship is given by

$$E\{[\Delta m_H(r\Delta t)]^2\} = rE\{[\Delta m_H(\Delta t)]^2\}. \tag{6}$$

This means that the mean-square increments scale in the same ratio, r, as the time increments t. Thus, for cBm, the mean-squared increments are statistically self-similar. Unlike cBm, the type of scaling exhibited by mean-squared fBm increments is not self-similar. Here the general relationship given by Equation (5) holds, wherein the mean-squared increments are multiplied by $r^{2H}(2H \neq 1)$ when Δt is multiplied by r. This type of relationship is called self-affine rather than self-similar.

Strictly speaking, fBm does not have a first derivative. Irregularity is maintained as $\Delta t \to 0$, and a derivative cannot be defined. Thus, in all applications of fBm to physical systems of which the authors are aware, it is necessary to select a high frequency cut-off, which may be accomplished by selecting a set of sufficiently small time increments Δt_s and replacing fBm over each time increment by a straight line. If one insists on using fractal concepts, this requirement is bothersome philosophically, but it will not change the appearance of fBm at the larger time scales of interest. In addition, the resulting irregular curve, which stops being irregular between time increments Δt_s, now has a derivative which is called fractional Gaussian noise or fGn. (It seems appropriate that such modified functions be called something like pseudo-fractals.)

In their pseudo-fractal form, fBm and fGn are very much intertwined with each other, so it is worthwhile to clarify their relationship. Twenty ordinates of a pseudo fBm generated by a procedure to be discussed later, called successive random additions, are given by 0.00, 0.64, 0.45, 0.77, 0.56, 0.79, -0.59, -0.79, -0.25, 0.19, -0.24, -1.15, 1.75, 2.11, 4.31, 5.04, 3.39, 3.12, 2.88, -0.72, and -0.28. These are plotted in Figure 3 for a time increment Δt of one unit, using straight lines to connect the points. From this curve, by taking differences between points, one can calculate the displacements and the average value of the derivative over each Δt. These are also plotted in Figure 3. At the Δt scale chosen, the upper curve represents an fGn trace and the lower curve the corresponding fBm trace. Obviously, we could interpret Figure 1 in this manner, which would yield Gaussian noise upper and Brownian motion lower. Again, for fGn/fBm plots to look realistic, many points must be included as in Figure 4.

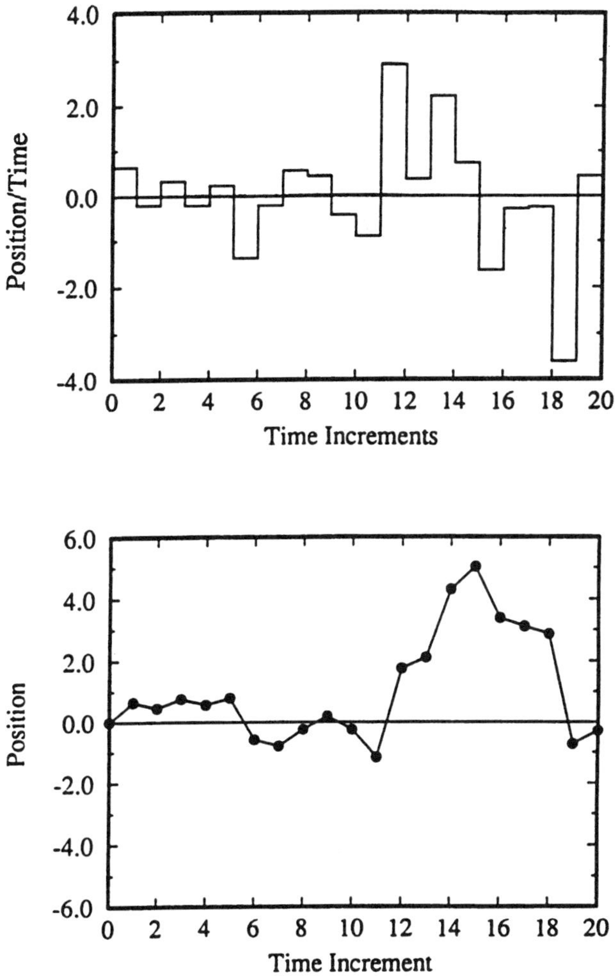

Figure 3: Illustration of the relationship between fBm (bottom) and its increments, fGn (top).

The major feature of fBm as compared to cBm is that sequences of fBm exhibit autocorrelations that can be quantified in terms of the H coefficient (Mandelbrot, 1983, p. 353; Hewett, 1986). If an fBm, $m_H(t)$, is given with $m_H(0) = 0$ and evaluated with a time lag of $\pm\Delta t$ about the arbitrary zero, then it can be shown that the correlation of successive increments is given by

$$\frac{E\{-m_H(-\Delta t)m_H(+\Delta t)\}}{E\{m_H(\Delta t)^2\}} = 2^{2H-1} - 1. \tag{7}$$

For details see Feder (1988). An important property of Equation (7) is that there is no dependence on Δt in the correlation. Thus, values of $m_H(t)$ separated by

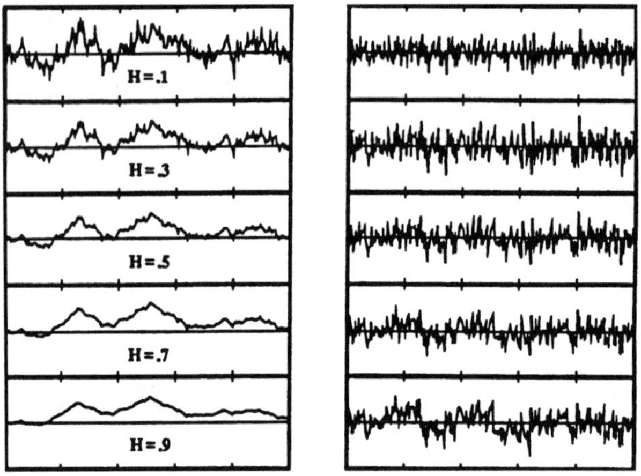

Figure 4: Corresponding traces of fBm and fGn for different Hurst (H) coefficients. Both functions are used herein to represent various aspects of subsurface property distributions.

arbitrarily large Δt will be correlated. It was this property that generated much of the controversy in the area of hydrologic time-series analysis (Rodriguez Iturbe *et al.*, 1972; Mandelbrot, 1983). Apparently, it is also this and related properties that give fBm and the associated increments, fGn, their ability to at least mimic a wide variety of natural phenomena (Mandelbrot, 1983).

Depending on the value of H, Equation (7) defines 3 classes of correlation. For $0 < H < 1/2$ there is negative correlation, for $H = 1/2$ there is no correlation, which is the case of cGn/cBm, and for $1/2 < H < 1$ correlation is positive. It is the positive correlation property that enables fGn/fBm to model phenomena that tend to cluster first on one side of the mean and then on the other (Voss, 1985a). The positive correlation phenomenon also bothered some time series analysts because it meant the function "abhorred" its own mean (Rodrigues Iturbe *et al.*, 1972).

III Detecting the Presence of fBm/fGn in a Data Set

The accepted methodology for calculating H from a data set displaying the properties of fGn or fBm is called rescaled range (R/S) analysis. However, R/S analysis and related topics are still being actively studied and certain aspects remain controversial (Lo, 1991). In hydrology, the initial concept can be found in the work of Hurst (1951), but was fully developed and elaborated by Mandelbrot and Wallis (1969). As developed originally, R/S analysis was applied to the sequence of annual flow volumes into a reservoir (annual volume

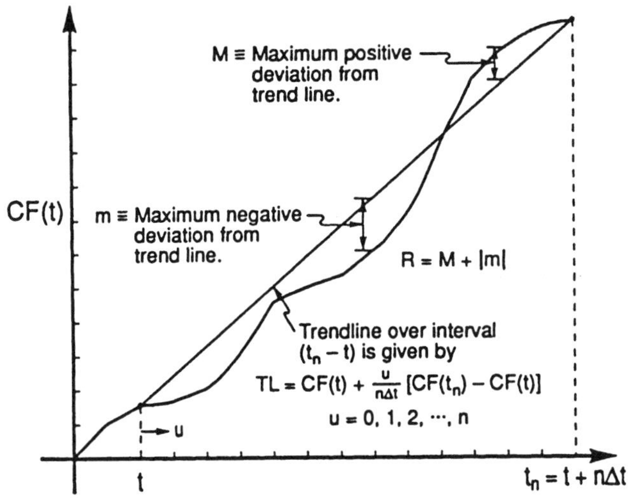

Figure 5: Schematic diagram for illustrating the physical meaning of the range, R, as used originally in storage reservoir analysis. Plotted is cumulative annual inflow to a reservoir, CF, versus time t. For purposes of illustration, it is assumed that at time t the volume of water stored in the reservoir is S and at time $t + n\Delta t$ it is again S.

increments), which is approximated by a fGn process. Let this sequence be given by F_1, F_2, F_3, ... Then the accumulated or "graded" flow over a series of years is given by $GF_i = \sum F_i$, which would approximate a fBm process.

As illustrated in Figure 5, assume we have n years of historical data starting at time t, when the reservoir has storage S, and ending at time t_n, when the storage is once again S. Then, assuming no evaporative losses, the reservoir could be discharged at the constant volumetric rate $d(TL)/dt$ if the storage capacity is such that it could accommodate a deficit of $|m|$ and a surplus of M without going completely dry or overflowing. We could say that such a reservoir required a storage range of $R = M + |m|$. It is intuitive that R would be sensitive to the longer (lower frequency) trends in the F_i data, because one would expect M and m to be reached after a number of wet years or dry years, respectively. One of the discoveries that Hurst (1951) made was to notice that when R for an n-year period, R_n, was divided by the sample standard deviation computed over the same n years, R_n/S_n, averaged over many n-year periods and plotted as a function of n on log-log paper, a straight line often resulted as n became large, with a slope around 0.7 (Mandelbrot, 1983). It was expected that the slope would be 0.5, which would characterize an uncorrelated process. (A general discussion of the definition and interpretation of "range" may be found in Boes (1988).)

Mandelbrot and Van Ness (1968) and Mandelbrot and Wallis (1969b) recognized the generality of the R/S procedure as a method for determining the

H coefficient and suggested fGn/fBm as a model for such behavior, regardless of the physical nature of the independent variable. Mandelbrot (1975) has shown that for an underlying fBm process, the following limit holds (see also Korvin, 1992):

$$\lim_{n \to \infty} (n\Delta z)^{-H} R(z, n\Delta z)/S(z, n\Delta z) = \text{constant}, \tag{8}$$

where z is the fBm or fGn independent variable, Δz is the "distance" between data points, and $n\Delta z$ ($n = 3, 4, ...$) is the lag distance over which R and S are evaluated. If Equation (8) holds, then a plot of $\log(R/S)$ versus $\log(n\Delta z)$, with each R/S value averaged over many lags of length $n\Delta z$, should yield a straight line as n becomes large, with a slope equal to H. If the underlying sequence is fBm, then the function is degraded (differenced or changed to fGn) before applying the R/S procedure. If one is working with fGn, then the R/S procedure is applied directly, which involves grading the fGn. In either case, the plotting procedure enables one to simultaneously test for consistency of the data with an fBm or fGn process (straight line as n becomes large) and to determine the H coefficient (slope of line). It does not prove, however, that fBm/fGn is the only model that could apply.

IV Determination of H Values from θ and K logs

Hewett (1986) and Molz and Boman (1993) applied R/S analysis to θ and K data, respectively. The approach was motivated by observations that vertical variations of geophysical properties, such as surface elevation, sand isopachs, ore grade and some additional variables, are approximated by fGn (Burrough, 1981; Agterberg, 1982; Mandlebrot, 1983; Pentland, 1984; Voss, 1985; Turcotte, 1986). Thus the objective of the R/S analysis was to look for fGn-like behavior in the $\theta(z)$ and $\log[K(z)]$ data, where in our application z is elevation.

Before the R/S values were calculated, both sets of data were normalized (common mean removed and difference divided by the common standard deviation) and the resulting sequence graded (running sum formed), as shown in Figure 6. Computationally, the variance and R/S values were obtained for each lag, $n\Delta z$, using the formulas illustrated in Figure 6 and given below.

$$\begin{aligned} R(z, n\Delta z) \;=\; & \max_{0 \le u \le n\Delta z} [GF(z+u) - \\ & \qquad [GF(z) + \frac{u}{n\Delta z}(GF(z+n\Delta z) - GF(z))]] \\ & - \min_{0 \le u \le n\Delta z} [GF(z+u) - \\ & \qquad [GF(z) + \frac{u}{n\Delta z}(GF(z+n\Delta z) - GF(z))]] \end{aligned} \tag{9}$$

and

$$S(z, n\Delta z) = \left[\frac{\sum F^2(z+u)}{n\Delta z - 1} - \frac{\left(\sum F(z+u)\right)^2}{(n\Delta z)^2 - n\Delta z} \right]^{1/2}. \tag{10}$$

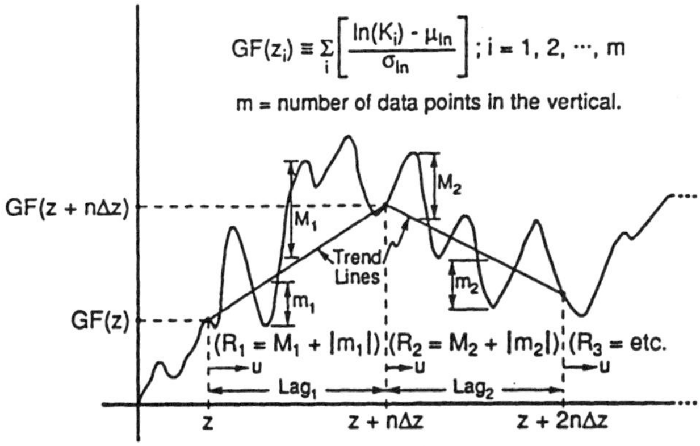

Figure 6: Schematic illustration of how the range, R, of a graded, discrete function, GF, is identified for each lag of size $n\Delta z$.

IV.A R/S Analysis Applied to Porosity Data

Shown in Figure 7 are 2,189 porosity values obtained from a density-based porosity log (assumed grain specific gravity of 2.65) from a late Miocene-early Pleistocene sandstone deposited originally as a submarine fan. In Figure 8 the sequence has been transformed to have a zero mean, unit variance and a sample spacing of 1 (Hewett, 1986).

Shown in Figure 9 is the result of R/S analysis applied to the transformed data in Figure 8. The solid line is a least squares fit for $n\Delta z > 20$, and the dashed reference line has a slope of $1/2$. The plot is consistent with a fGn process having an H coefficient of about 0.86. There is no tendency to approach a slope of 0.5 for large $n\Delta z$, which implies correlations over a vertical sedimentary sequence in excess of 305 m! As pointed out by Hewett (1986), this result suggests an even larger range of horizontal correlations, given the anisotropic nature of sedimentary deposits.

IV.B R/S Analysis Applied to Hydraulic Conductivity Data

Data were obtained using the spinner meter method (Molz *et al.*, 1989) at a field site near Mobile, Alabama. The study formation was a sandy, confined aquifer approximately 21.3 m (70 ft) thick, which rests on a Tertiary - Quaternary contact as illustrated in Figure 10. Plots of the data used in the present analysis may be found in Molz *et al.* (1990), and numerical values are available from the authors.

In the R/S analysis, z was the distance above the confined aquifer base, Δz was 1 ft (0.305 m), and the fully penetrating wells had screen lengths of 70 ft (21.3 m). There were at least 51 K measurements per well at common depths. The process tested for fGn properties was $\log(K(z))$, since there is

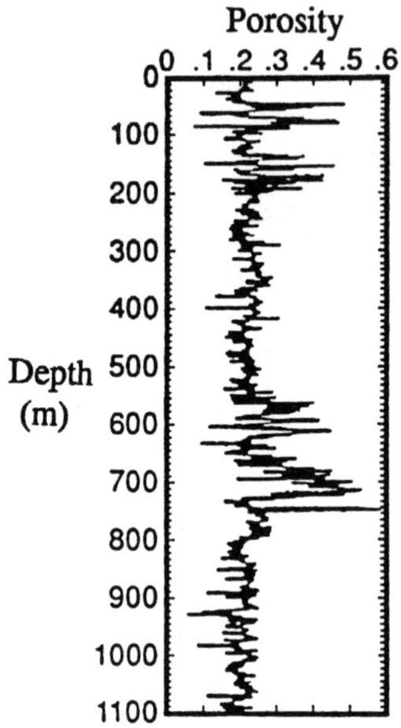

Figure 7: Porosity *versus* depth as determined from a density porosity log.

some evidence that this quantity is Gaussian. Computationally, the various R/S values were obtained for each lag, $n\Delta z$, using Equations (9) and (10) as described previously. A total of 9 wells were used in the analysis, and R/S values were calculated and averaged over the 9 wells for lag lengths varying from 5 ft to 50 ft. (The total number of data points were $9 \times 51 = 459$ points.) The results shown in Figure 11 indicate that the $\log(K)$ data are well represented by a straight line with a slope of 0.82.

IV.C Geostatistical and Spectral Analysis of fBm/fGn

There are important connections between fBm/fGn, geostatistics, and spectral analysis. As illustrated by Hewett (1986) with porosity data, it is possible to use co-variance properties and Fourier transform techniques to refine an estimate for H based on R/S analysis. Here, we will simply review the theory to provide further relevant information and support for topics to be discussed later.

As indicated by Equation (2), fBm is defined in terms of the variance of its increments, with the increments being viewed as fGn. For a typical variable such as porosity, θ, which is assumed to display fGn properties in the vertical z direction, the variogram, $2SV_\theta$, is given by

$$2SV_\theta(h) = E\{[\theta(z+h) - \theta(z)]^2\}, \tag{11}$$

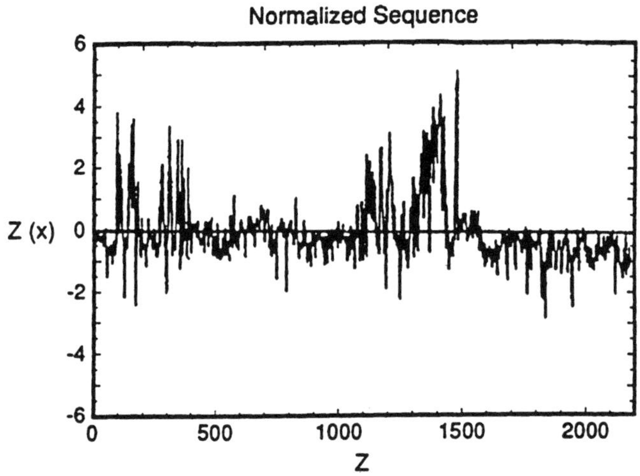

Figure 8: Normalized sequence of values from the porosity log shown in Figure 7.

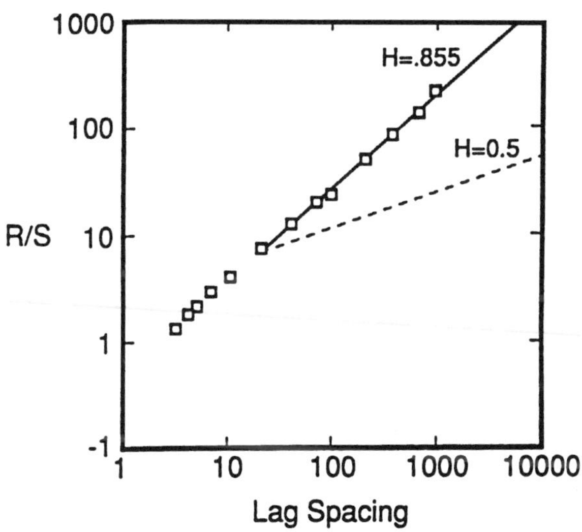

Figure 9: Result of R/S analysis applied to porosity data.

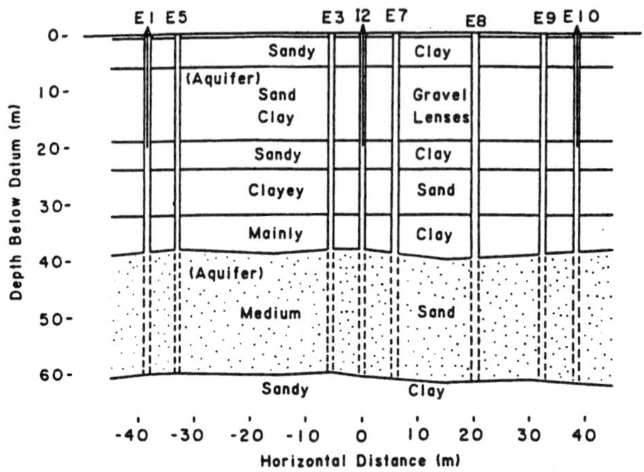

Figure 10: Vertical cross-sectional view through the center of the study area at the Mobile site.

where h is the lag and SV_θ is the semivariogram. This function is related to the autocovariance function, AC_θ , by

$$SV_\theta(h) = \sigma_\theta^2 - AC_\theta(h), \tag{12}$$

where σ_θ^2 is the variance of $\theta(z)$. As discussed by Jenkins and Watts (1968), a realization of a stochastic process does not have a well-defined amplitude spectrum. However, for a statitionary stochastic process, the power spectrum and autocovariance function are Fourier transform pairs, given by

$$\begin{aligned} PS_\theta(f) &= \int_{-\infty}^{+\infty} AC_\theta(h)e^{-2\pi i f h}dh \qquad \text{and} \\ AC_\theta(f) &= \int_{-\infty}^{+\infty} PS_\theta(f)e^{2\pi i f h}df, \end{aligned} \tag{13}$$

where PS_θ denotes the power spectrum, $i = \sqrt{-1}$, f is the frequency and h, the lag. Thus, the definition of fBm/fGn, the variogram, autocovariance function and power spectrum (or spectral density $\equiv PS_\theta/\sigma_\theta^2$) are all intimately connected (Voss, 1988). As discussed by Voss (1988), for H in the range $1/2 \leq H \leq 1$, the power spectra of white noise, fGn, cBm and fBm are all proportional to $1/f^\beta$, with $\beta = 2H + 1$ for fBm and $2H - 1$ for fGn. $H = 1/2$ yields the white noise limit for fGn ($\beta = 0$) and the cBm limit for fBm ($\beta = 2$). $H = 1$ yields $\beta = 1$ noise for fGn and $\beta = 3$ noise for fBm. Thus, for the range of H considered, $0 \leq \beta \leq 1$ for fGn and $2 \leq \beta \leq 3$ for fBm. (The range $1 \leq \beta \leq 2$ corresponds to fBm with $0 < H < 1/2$, the zone of negative correlations.) Thus, both fBm and fGn may be viewed as $1/f^\beta$ noises with $0 < \beta < 3$. Because of these relationships, it appears possible

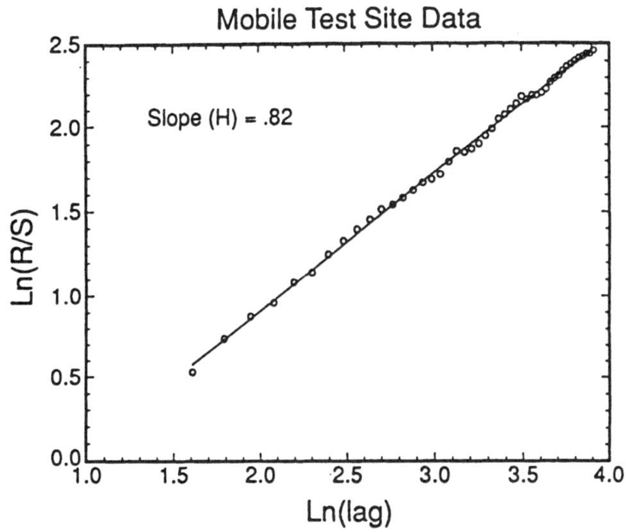

Figure 11: Results of rescaled range (R/S) analysis applied to hydraulic conductivity measurements made at 1-ft (0.3048 m) intervals in 9 wells. The data were collected in a wellfield near Mobile, AL, using the spinner meter method (Molz *et al.*, 1990).

to generate fBm/fGn-like behavior by constructing a function with the proper power spectrum (Voss, 1988).

V Unconditional or Conditional Realizations of θ and K Distributions

For a variety of reasons, most measurements made in the subsurface, such as those of θ and K, are "few and far between". Thus, the need for a realistic and rational approach for interpolating sparse data sets or generating realizations is widespread. Contemporary procedures, such as averaging, Kriging, and various types of polynomial interpolation, act to smooth the basic data and thus do not realistically represent a natural hierarchical structure (Journel and Huijbregts, 1978). This type of smoothing interpolation is used commonly, despite the evidence that real θ, K and other property distributions are not smooth (Hewett, 1986; Hewett and Behrens, 1988; Molz *et al.*, 1990). In recent applications of fGn/fBm to the representation of subsurface property distributions, it has been shown that vertical variations of a particular property are consistent with fGn statistics and then argued that horizontal variations mimic fBm, with the same H coefficient. Hewett (1992) refers to this as the "anisotropic fractal model".

However, recent data discussed by Hewett (1992) indicate that variations in the horizontal direction may also be described by fGn. The discussion was based on R/S analyses of density logs in carbonate rocks taken in horizontal and vertical wells. The vertical data, treated as fGn, resulted in $H = 0.88$, whereas the horizontal data, also treated as fGn, yielded $H = 0.89$. Thus the present situation is somewhat uncertain, although Hewett (1992) argues that over finite spatial scales, fBm and fGn with the same H value can behave in a similar manner. In what follows we will adhere to the anisotropic fractal model while recognizing that future modifications may be necessary. The overall procedure associated with the anisotropic fractal approach consists of the following steps, the first 3 of which have already been discussed.

- i Perform a statistical analysis of the θ or $\log(K)$ measurements to see if the properties of fGn are displayed in the vertical
- ii If step one has a positive result, determine the value of H implied by the data.
- iii Use this H value to define an fBm process for horizontally-distributed or $\log(K)$ values.
- iv Calculate θ or $\log(K)$ values between wells at each elevation of interest by generating semi-random fBm, possibly conditioned by the measured values, and utilizing the H value computed in step two.

V.A Unconditional Realizations in 1 or 2 Dimensions

Possibly the most straightforward procedure for generating an fBm process is that of successive random additions (SRA) as described by Voss (1985b) and used by Hewett (1986) to generate porosity realizations. The procedure itself is a discrete process whereby one starts with a series of θ measurements at a given elevation in 2 or more wells. One then fills in values between the wells so that the variance of increments displays the scaling properties of Equation (5), but with time replaced by horizontal distance. Successive random additions is a generalization of an older process called random midpoint displacement (Voss, 1985b).

A one-dimensional version of SRA may be illustrated by considering the three porosity logs shown in Figure 12. The procedure is applied at a selected series of elevations, one elevation at a time. Each realization starts with 3 data points, and the SRA procedure is summarized in Table 1 for any 2 wells, say wells 1 and 2 in Figure 12. The spacing between these wells is 900 ft (274 m), and the vertical interval is 145 ft (44 m) (Hewett, 1986). fBm generation proceeds independently at each elevation, starting with the measured values and their locations. These values are used to calculate an initial θ variance, σ_0^2, for a particular elevation. For this purpose, all data available at a particular elevation may be utilized. Then for two data points, $\theta_0^{(0)}$ and $\theta_D^{(0)}$, separated by a horizontal distance, D, the recursive SRA procedure fills in values along D, as illustrated in Table 1. Values are calculated recursively at the midpoints

---

(Step 0: σ_0^2 given)

#1 •<——————————D——————————>•#2

$\theta_0^{(0)}$ $\theta_D^{(0)}$

(Step 1: $\sigma_1^2 = \sigma_0^2/2^{2H}$)

x x x

$\theta_0^{(1)}$ $\theta_{D/2}^{(1)}$ $\theta_D^{(1)}$

where $\theta_{D/2}^{(1)} = ((\theta_0^{(0)} + \theta_D^{0})/2 + RN_1$

$\theta_0^{(1)} = \theta_0^{(0)} + RN_2$; $\theta_D^{(1)} = \theta_D^{(0)} + RN_3$, and

$RN_i \in G(0, \sigma_1^2)$; $i = 1,2,3$

(Step 2: $\sigma_2^2 = \sigma_1^2/2^{2H}$)

x x x x x

$\theta_0^{(2)}$ $\theta_{D/4}^{(1)}$ $\theta_{D/2}^{(2)}$ $\theta_{3D/4}^{(1)}$ $\theta_D^{(2)}$

where $\theta_{D/4}^{(1)} = ((\theta_0^{(1)} + \theta_{D/2}^{(1)}/2) + RN_4$

$\theta_{3D/4}^{(1)} = ((\theta_{D/2}^{(1)} + \theta_D^{(1)})/2) + RN_5$

$\theta_0^{(2)} = \theta_0^{(1)} + RN_6$; $\theta_{D/2}^{(2)} = \theta_{D/2}^{(1)} + RN_7$;

$\theta_D^{(2)} = \theta_D^{(1)} + RN_8$; and $RN_i \in G(0, \sigma_2^2)$; $i = 4,5,6,7,8$

(Step 3: $\sigma_3^2 = \sigma_2^2/2^{2H}$)

x x x x x x x x x

$\theta_0^{(3)}$ $\theta_{D/4}^{(2)}$ $\theta_{D/2}^{(3)}$ $\theta_{3D/4}^{(2)}$ $\theta_D^{(3)}$

etc.

Table 1: Illustration of the application of the *SRA* procedure to data related to two arbitrary wells.

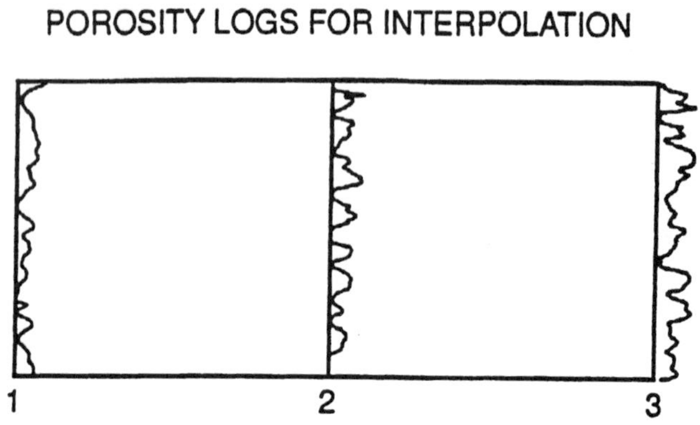

Figure 12: Set of three porosity logs for unconditional stochastic interpolation. The vertical coordinate is elevation.

between the initial or previously calculated values by linear interpolation (other interpolators could be used), followed by the addition of a random number, RN, chosen from a Gaussian distribution of mean zero and variance σ^2. The variance is calculated by taking the variance at the previous step and multiplying it by $(1/2)^{2H}$. This is what builds in the variance structure characteristic of fBm. SRA proceeds by adding a different normal random number at each step to all the interpolated θ values, including the measured values and those calculated at previous steps. Thus, in step 3 of Table 1, RN_i $(i = 9, 10, ..., 17)$ would be added to all nine θ values along the line. (The superscripts represent the number of times a random number has been added to an interpolated quantity.) According to Voss (1985), this procedure generates an approximate fBm process superior to that generated by random midpoint displacement, where at each step a random number is added only to the newly interpolated values. Holding a particular n value along the line constant creates a local violation of the variance structure characteristic of fBm.

Using the SRA procedure, Hewett (1986) generated porosity realizations based on the data in Figure 12. Examples of a few "well logs" through the interpolated field are shown in Figure 13. The data appear quite realistic with larger-scale trends preserved. However, the detailed measurements at wells 1, 2, and 3 are not preserved, which is characteristic of unconditioned realizations.

Molz and Boman (1993) used a similar procedure to generate K realizations. For this application, θ in Table 2 is replaced by $\log(K)$, and variances, denoted by σ_θ^2, correspond to variances of $\log(K)$. Otherwise, the generation procedures are identical.

The interior values of Table 2 were generated using SRA, and the columns shown are equally spaced between two wells (E7 and E8). The left-most column lists the initial values for each layer based on a 9-well average. After an fBm realization was generated, antilogs of the $\log(K)$ values were taken. This resulted in the dimensional K values listed in the table.

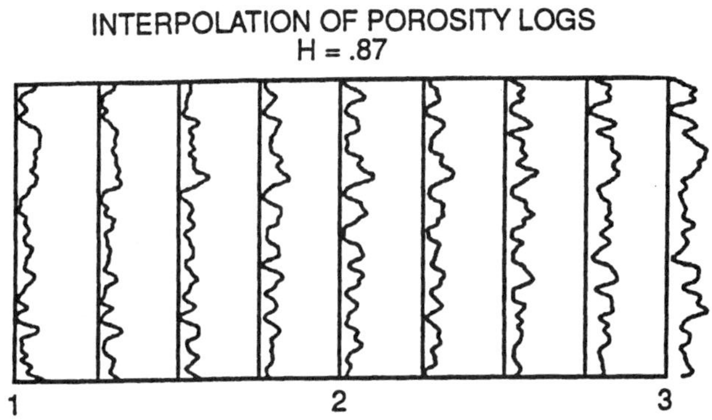

Figure 13: Result of SRA-based interpolation of the porosity logs shown in Figure 12.

$L\sigma_0^2$	E7	(1)	(2)	(3)	(4)	(5)	(6)	(7)	E8
3.59669	11.8	47.5	16.2	61.2	25.1	10.9	8.3	30.0	22.9
2.55382	30.8	92.6	34.7	98.8	43.3	20.0	14.7	40.3	29.8
1.76148	32.5	83.5	38.1	93.4	48.5	26.3	21.0	49.9	40.0
0.17006	24.7	31.8	24.0	30.4	23.9	19.0	17.0	21.4	19.2
0.12221	73.5	84.9	62.2	70.9	53.7	41.2	35.0	39.5	33.6
1.20017	31.4	73.6	41.5	93.6	58.7	38.1	34.1	75.0	67.2
1.10695	4.5	13.4	10.2	29.2	24.5	21.2	25.1	70.3	83.2
0.21054	133.0	173.7	124.8	160.4	120.5	91.9	80.2	101.9	89.0
0.15413	49.9	72.4	62.9	90.1	81.4	74.5	76.5	108.4	111.4
0.15258	52.2	72.2	60.1	82.1	70.9	62.1	61.0	82.5	81.0
0.20447	90.3	130.2	104.1	147.8	123.5	104.8	101.5	142.5	138.1
1.65628	156.5	272.4	88.7	147.8	54.5	21.0	11.8	19.0	10.7
0.68794	47.1	68.9	34.3	48.7	26.2	14.5	10.3	14.3	10.1
0.77672	47.1	71.4	34.4	50.6	26.6	14.3	10.0	14.4	10.1

Table 2: Hydraulic conductivity (K) logs at wells E7 and E8 along with a K realization between the wells based on fBm and successive random additions. The two wells are separated by 42.8 ft (13.05m).

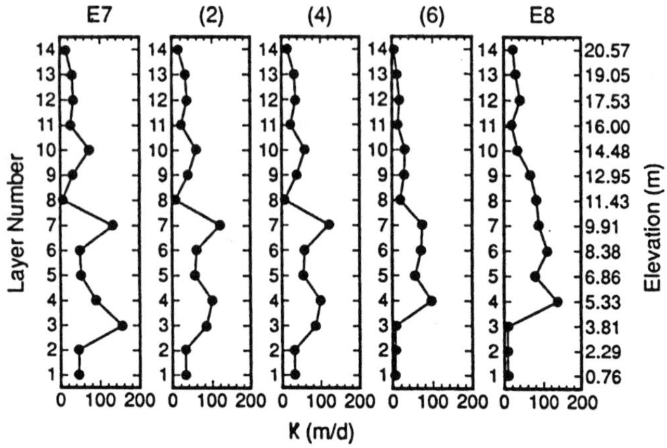

Figure 14: Vertical plots of K realizations derived from SRA applied between wells E7 and E8. The mixture of deterministic and stochastic variations characteristic of fGn is evident.

Shown in Figures 14 and 15 respectively are vertical and horizontal profiles of generated K values. Similar to Hewett's (1986) conclusion concerning porosity, the distributions look realistic. The patterns represent a mixture of randomness and correlation, with the variations, on the average, conforming to the statistics of fGn in the vertical and fBm in the horizontal. As indicated in Figure 15, which shows K variations with horizontal distance for layers 4 and 8, the data are not smoothed. Irregularities are maintained, again displaying the statistics embodied in the original flowmeter data. By continuing the SRA algorithm, K values may be generated at very small horizontal or vertical increments. In order to get a stable realization, the SRA algorithm was continued for 10 steps, each involving a set of bisections between the two wells, to generate the numbers in Table 2. This generated $2^{10} - 1 = 1023$ K values per layer. At that point the variance was so small that the random number additions were insignificant at our scale of interest.

As described by Saupe (1988), the SRA algorithm is easily generalized to two spatial dimensions. One starts with a square grid of points, as illustrated by the circled *1*'s in Figure 16. The individual squares have a side length of D, and the quantity associated with each corner, in our case θ or $\log(K)$, has an initial variance σ_1^2. Analogous to the 1-D case, values at the center of each square, *2* locations, are generated by averaging the values at the four surrounding vertices and adding a Gaussian random number of mean zero and variance $\sigma_2^2 = \sigma_1^2/\sqrt{2}^{2H}$. Recall that in the 1-D case the scaling ratio, r, was 1/2, but for the 2-D case of Figure 16, r is $1/\sqrt{2}$, because the *1* squares have a side length of D while the new squares have a side length of $D/\sqrt{2}$. At each stage, random numbers are added to each point, not just the newly computed points.

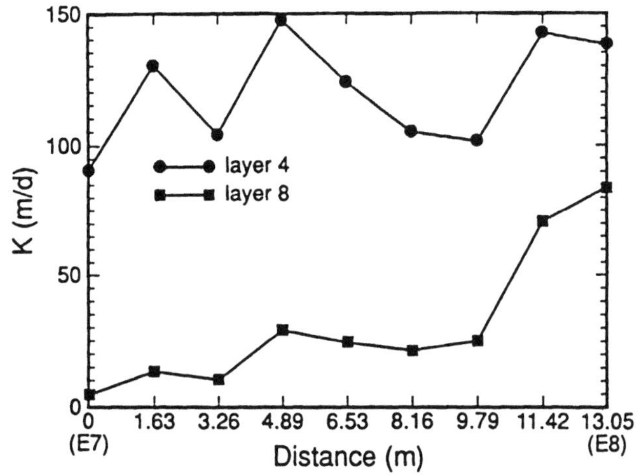

Figure 15: Horizontal plots of fBm K realizations in layers 4 and 8 between wells E7 and E8. This plot shows that the realization scheme does not necessarily smooth the data. Irregularities characteristic of the original measurements are maintained, with the long wavelength variations playing the most important role.

Shown in Figure 17 is a 2-D, unconditioned, horizontal K realization based loosely on flowmeter data collected at the Mobile site at a particular elevation. (We use the term "loosely" because the 9 test-well locations only approximated a square pattern.) Because of the commonly-observed layered structure in granular deposits, there are potential advantages in 2-D fBm generation algorithms (Boman *et al.*, 1995).

V.B Conditional Realizations in 1 or 2 Dimensions

Many of the existing algorithms for generating fBm or fGn (Voss, 1985b; Saupe, 1988) were motivated by the desire to construct synthetic landscapes and other random fractal forgeries. In such applications it is easy, and in fact desirable, to work with regular grids, stay within a particular grid and be unconcerned with having the realization be compatible with any particular data set. However, in hydrologic applications just the opposite is true. Measurements are required in order to determine the H coefficient, and seldom are such measurements made on a regular grid. One would like to take advantage of the measured values when constructing a realization, and values generated outside the zone where the measurements were made are often needed. Voss (1985b; 1988) suggests that the so-called Weierstrass-Mandelbrot random fractal function (WM) is a convenient alternative for such applications. Hewett and Behrens (1990) combined the WM realization approach with geostatistical conditional simulation procedures (Journel and Huijbregts, 1978) to develop a

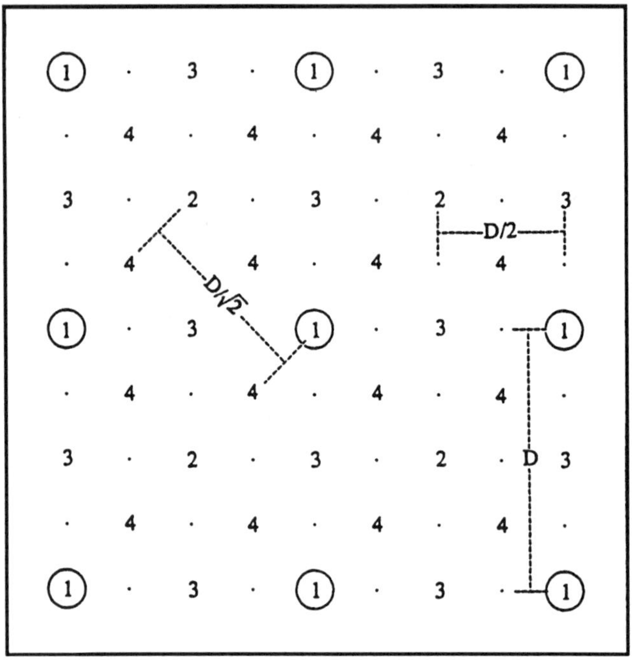

Figure 16: Illustration of the unconditioned 2D SRA procedure. Original data point locations are indicated by circled ones.

1- or 2-D fBm generation procedure that is conditioned on (goes through) the measured data points. This procedure has the flexibility and speed desirable for practical hydrologic applications.

In 1 dimension the WM function may be written as (Voss, 1985b)

$$WM(x) = \sum_{n=-\infty}^{+\infty} A_n r^{nH} \sin(2\pi r^{-n} x + \phi_n), \tag{14}$$

where WM is the Weierstrass-Mandelbrot random fractal function, x denotes the horizontal distance, A_n is a Gaussian random variable with a mean of zero and a constant variance, σ^2, independent of n, r is analogous to the scaling ratio (recall it was 0.5 in 1-D SRA), but it has a more general meaning in Equation (14) and controls the "lacunarity" of the more general class of fractals represented by this equation (Voss, 1985b), H is the Hurst coefficient, or fractal co-dimension, and is related uniquely to the fractal dimension D by $D = 2 - H$, and ϕ_n denotes a random phase angle uniformly distributed on the interval 0-2π. Chu (1993) shows that Equation (14) has the correct variance structure for fBm. It is sometimes desirable to write Equation (14) so that $n \geq 0$, in which case it becomes

$$WM(x) \quad = \quad A_0 \sin(2\pi x + \phi_0)$$

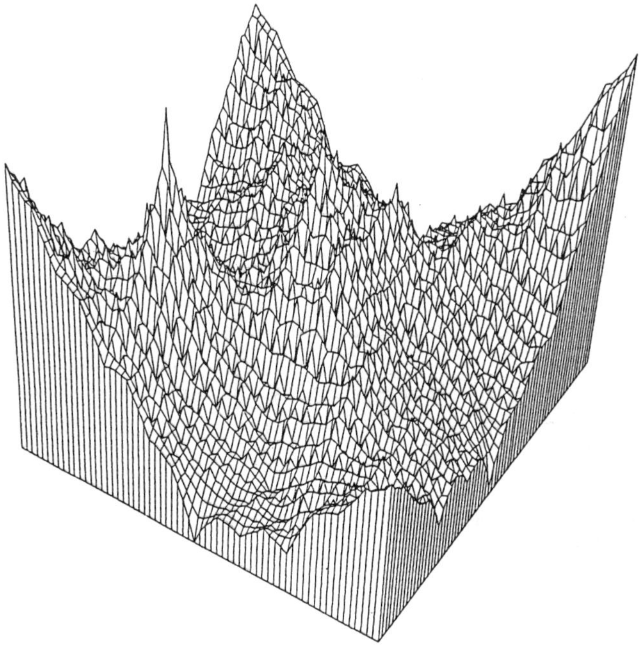

Figure 17: Result of applying 2-D SRA to develop an unconditioned realization of K data.

$$+\sum_{n=1}^{+\infty}[A_{n1}r^{nH}\sin(2\pi r^{-n}x+\phi_{n1})$$
$$+A_{n2}r^{nH}\sin(2\pi r^{n}x+\phi_{n2})]. \quad (15)$$

The WM function represents fBm in the sense that its power spectrum is proportional to $1/f^{\beta}$ with $2 < \beta < 3$ and its mean square increments scale according to Equation (2) or (5). However, the infinite set of functions that satisfies these criteria is very broad. Equation (14) itself contains the unspecified parameter r (the lacunarity) and periodic functions other than sines and cosines could be used in equations analogous to Equation (14). To the authors' knowledge, the physical implications of these observations are yet to be explored fully.

As pointed out by Voss (1985) only a small number of terms from the WM function are needed for any particular realization. Empirical testing has shown that if one wishes to use the WM function over distances ranging from a minimum of δL to a maximum of L, then only terms having frequencies varying between about δL^{-1} and L^{-1} need to be included. Frequencies above the δL^{-1} limit contribute a negligible displacement to $WM(x)$ over scales above δL, whereas those below L^{-1} simply cause a long wavelength displacement of the realization that would ultimately be removed by the conditioning process. Thus, for practical computations, if L^{-1} is the lowest frequency component desired, one can write Equation (15) as

$$WM(x) \approx \sum_{n=0}^{N} A_n r^{nH} \sin(2\pi r^{-n}\frac{x}{L} + \phi_n), \tag{16}$$

again implying a pseudo-fractal nature. To achieve an upper cutoff frequency of δL^{-1}, it is simple algebra to show that

$$N = [\log(\delta L) - \log(L)]/\log(r). \tag{17}$$

Once the number of terms are selected for a particular application, then for each term one must select two random numbers to define the amplitude and phase. When this is completed, $WM(x)$ is defined as a function of x and may be used to calculate a WM value at arbitrary locations, such as the nodes of a numerical model. The use of Equation (16) as the basis for a conditional realization procedure is illustrated in Figure 18. One begins with a number of locations where data (let's say porosity measurements) are obtained. These locations are labeled x_1 through x_6 in Figure 18; there is no need for regular spacing. In a field situation, the locations would be defined by the coordinates of observation wells. The hypothetical data values obtained at x_1 through x_6 are plotted in Figure 18 and a smooth curve fitted to the data points. This could be done in a variety of ways, including Kriging (Hewett and Behrens, 1988), without affecting the outcome materially. Therefore, a computationally simple curve-fitting technique is probably best in a practical sense. The next step is to create an unconditioned porosity realization over the interval $x_1 \leq x \leq x_6$ using Equation (16). This may be done by setting $L = (x_6 - x_1)$ and choosing a variance for the A_n equal to the variance of the data points. This realization is shown schematically as the upper curve in Figure 18. The values falling at the well locations are noted. At this point in the algorithm, the magnitudes of these values will be somewhat arbitrary. One proceeds by fitting a smooth curve through the 6 realized points that fall at the well locations, using the same procedure that was applied to the data points. This curve is shown as a broken line in the upper part of Figure 18. The differences between the WM realization and the smooth dashed line, which are zero at the data locations, are then added to the smooth curve through the data points to produce the conditioned realization shown in Figure 19. The result is a curve that reproduces the data values at locations where measurements were made and embodies the statistical properties of fBm for the H value resulting from R/S analysis of the data.

Because of the layered structure inherent to many granular aquifers, it will often be desirable to generate WM realizations on a layer by layer basis. As discussed by Hewett (1986) and shown by Güven *et al.* (1992) in a particular field application, such a 2-dimensional procedure applied in horizontal planes will be more compatible with the vertical *versus* horizontal anisotropic correlation structure of sedimentary aquifers. One may use the 2-D SRA procedure described previously as the basis for a 2-D conditioned realization, or a 2-D generalization of the WM function may be developed. In what follows, we

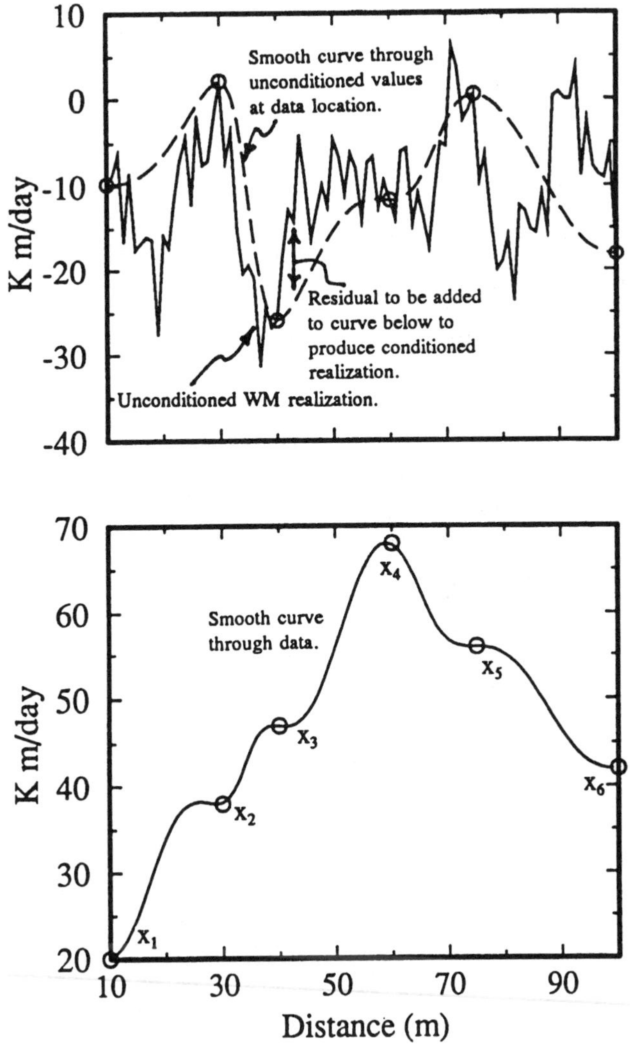

Figure 18: Illustration of the procedure for developing a 1-D conditioned realization based on the WM random fractal function.

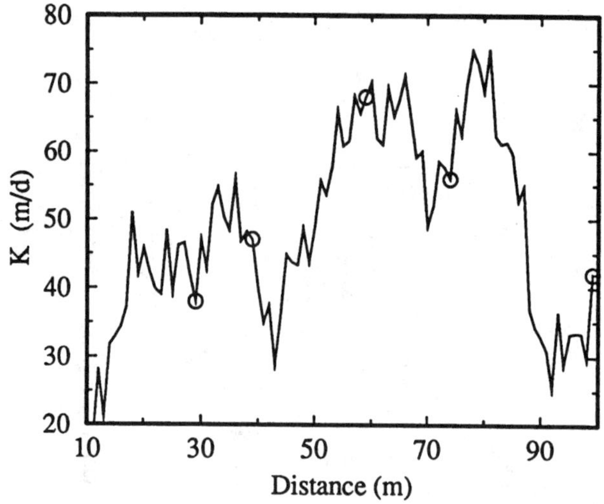

Figure 19: A conditioned stochastic interpolation (realization) superimposed on the conditioning data.

will present a 2-D version of the WM function. Then a detailed example will be developed based on a 2-D SRA algorithm published by Voss (1988).

A two-dimensional generalization of Equation (16) that maintains the concept of upper and lower frequency cut-offs is given by

$$WM(x,y) = \sum_{n=0}^{N}\sum_{m=0}^{N} A_{n,m} r^{(\sqrt{n^2+m^2})H} \sin\left[\frac{2\pi r^{-n} x}{L_x} + \frac{2\pi r^{-n} y}{L_y} + \phi_{n,m}\right], \tag{18}$$

where $A_{n,m}$ and $\phi_{n,m}$ are random numbers similar to A_n and ϕ_n, described previously, L_y^{-1} is the lowest frequency sinusoid in the y direction and L_x^{-1} the corresponding frequency in the x direction. An equation similar to (18) was used by Hewett and Behrens (1988) to generate 2-dimensional conditioned realizations of transmissivity data and by Hewett (1992) to generate pseudo-well logs in a vertical cross-section. The lowest frequency term in (18) is

$$A_{0,0} \sin\left[\frac{2\pi x}{L_x} + \frac{2\pi y}{L_y} + \phi_{0,0}\right]. \tag{19}$$

Chu (1993) generated unconditioned realizations of 2-D fBm using equations similar to (18) and found empirically that the result contained artifacts (banding) running from northwest to southeast in the rectangular realization. This was attributed to the fact that (18) has an isotropic correlation structure and that this was incompatible with using rectangular symmetry to generate any particular $WM(x,y)$ realization. Chu (1993) suggests building up realizations (summing terms) along random radial directions (azimuths) rather than fixed coordinate directions. This also has the advantage of requiring only a single summation.

Starting with Equation (16), one generalizes to 2D by thinking of the frequency r^{-n}/L as a vector along an azimuth defined by the polar angle θ_p and given by

$$\mathbf{f}_p = |\mathbf{f}_p|\, cos(\theta_p)\mathbf{i} + |\mathbf{f}_p|\, sin(\theta_p)\mathbf{j}, \tag{20}$$

where $\mathbf{i}$ and $\mathbf{j}$ are unit vectors in the x and y coordinate directions, respectively. Then for each position vector $\mathbf{u} = x\mathbf{i} + y\mathbf{j}$, Equation (16) generalizes to

$$WM(x, y) \approx \sum_{p=0}^{P} A_p r^{pH} \sin(2\pi \mathbf{u} \cdot \mathbf{f}_p + \phi_p). \tag{21}$$

Written out, Equation (21) becomes

$$WM(x, y) \approx \sum_{p=0}^{P} A_p r^{pH} \sin\left[\frac{2\pi r^{-p} x \cos(\theta_p)}{L} + \frac{2\pi r^{-p} y \sin(\theta_p)}{L} + \phi_p\right], \tag{22}$$

where A_p is a Gaussian random variable with a mean of zero and a constant variance, σ^2, independent of p, θ^p is a random angle uniformly distributed on $0 \leq \theta_p \leq \pi$, L^{-1} is the lowest frequency term desired in the summation, and the upper summation limit $P = (\log(\delta L) - \log(L)]/\log(r)$. One compensates for the single summation in Equation (22) by choosing r closer to 1, thereby increasing P. Equation (22) yields the lowest frequency term

$$A_0 \sin\left[\frac{2\pi x \cos(\theta_0)}{L} + \frac{2\pi y \sin(\theta_0)}{L} + \phi_0\right]. \tag{23}$$

Thus, one can make Equation (22) statistically consistent with field data by choosing the A_p from a distribution with a variance on the L scale equivalent to that of the data. Equation (22) may be conditioned to field data using a 2-D version of the procedure described previously. For further illuminating discussion and results see Chu (1993).

In applications to date based on data from the Mobile site, 2-D K realizations have been generated using the 2-D SRA algorithm discussed previously by Voss (1988). Shown in Figure 20 are hypothetical K values taken at the same elevation in a 160 × 160 m^2 slab of aquifer. The range of K variations is 10 m/d to 125 m/d. Presented in Figure 21 is a conditioned realization based upon the hypothetical K measurements in Figure 20 and a Hurst coefficient of 0.8 associated with the $\log(K)$ process. The results in Figure 21 provide a feel for the nature of the K distributions generated by the procedures discussed herein.

VI Discussion and Conclusions

The definitions and properties of the random fractals fBm and fGn have been reviewed. The one-dimensional forms of these functions were first applied to

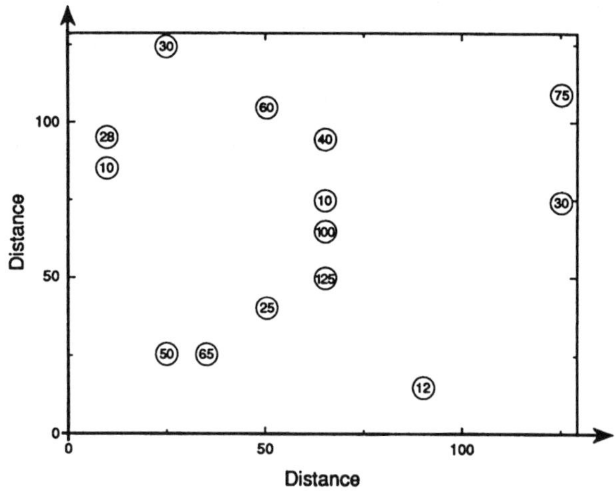

Figure 20: Data values and locations for hydraulic conductivity. All units are meters and days.

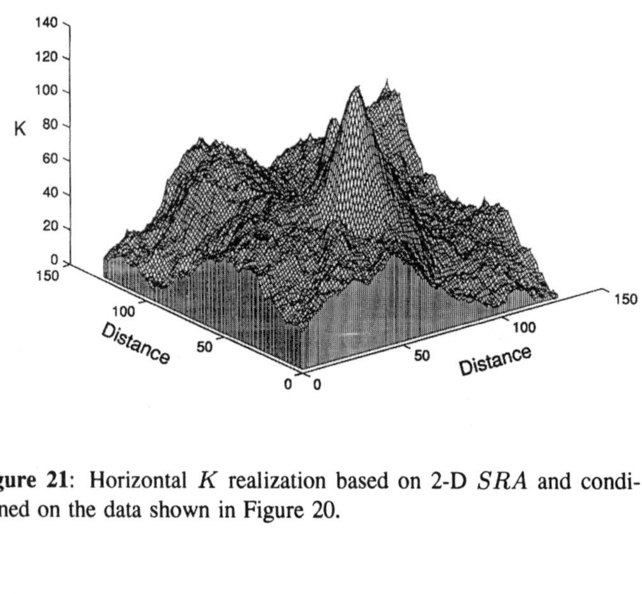

Figure 21: Horizontal K realization based on 2-D SRA and conditioned on the data shown in Figure 20.

the analysis of geophysical time series. Later, 2-D and 3-D versions were used to generate well-known fractal forgeries of landscapes and clouds (Voss, 1988). Herein, we apply 1- and 2-D fBm/fGn to the representation of porosity and hydraulic conductivity distributions in saturated porous media. The physical basis for doing this arises from R/S analyses applied to porosity and hydraulic conductivity measurements, and other geophysical variables (Korvin, 1992). The limited data collected to date suggest that these functions may be natural candidates for representing the "hierarchical structure" that has been postulated to exist in natural porous media.

If further experimental study verifies that fBm/fGn or related functions may be used widely to represent the hydraulic properties of porous media, then this will imply that autocorrelation between properties will extend over very large distances. In a practical sense, it is probably accurate to say that correlation length, like apparent dispersivity, will be scale-dependent. This makes intuitive sense to the present authors and represents the spatial analog of the well-known "Hurst effect" of time series analysis (Korvin, 1992). It is likely that the depositional processes that build up sediments are correlated over large distances (but obviously not infinite), especially in the horizontal direction. It follows from the definition of fBm that it has a power-law variogram given by (Mandelbrot, 1983)

$$2VG(h) = \beta h^{2H}, \tag{24}$$

where β is a constant and h is the lag distance. Obviously, this function does not approach a sill. However, in its pseudo-fractal form with upper and lower cutoffs, the resulting variogram will approach a sill eventually. Strictly speaking, we would now have an irregular function, not a true fractal, representing the property distribution. This also introduces the possibility of non-fractal-based explanations for the function properties, including an explanation for extended but not infinite autocorrelation. All of this must be kept in mind, as pointed out in an excellent discussion by Klemes (1974).

Earth science applications of fBm/fGn are presented within a conditional simulation-stochastic realization framework. The required H parameter is determined from R/S analysis, applied herein to porosity and hydraulic conductivity data. The WM random fractal function was selected as a promising basis for generating 1- or 2-D sequences of fBm or, by differentation, fGn. Each conditioned realization that is generated will, on average, produce the variance structure identified with R/S analysis while reproducing data values at measurement points. As the number of data are increased, stochastic realizations will approach deterministic distributions.

As pointed out by Voss (1985b), fGn- and fBm-related processes are among the most common spatial and temporal variations found in nature, although the physical origin of such processes remains obscure. (For example, attempting to explain long-term memory in thermally-induced, electronic flicker noise is difficult at best (Press, 1978).) Thus, the initial observations that vertical distributions of porosity and $\log(K)$ appear to obey fGn statistics is of theoretical as well as possible practical interest. Our understanding is obviously incomplete, but it is intriguing that these functions seem able to mimic such a wide variety of geophysical temporal and spatial forms. Use of fractal-based simulation procedures in the petroleum engineering field has had positive results (Hewett, 1992).

References

Agterberg, F.P. 1982. Recent developments in geomathematics. *Geo-Process.* 2:1-32.

Baveye, P. and G. Sposito. 1984. The operational significance of the continuum hypothesis in the theory of water movement through soils and aquifers. *Water Resour. Res.* 20:521-530.

Boman, G.K., F.J. Molz and O. Güven. 1995. An evaluation of interpolation methodologies for generating three dimensional property distributions from measured data. *Ground Water* 33(2):247-258.

Brown, R. 1828. On the existence of active molecules in organic and inorganic bodies. *Phil. Mag.* 4:162-173.

Burrough, P.A. 1981. Fractal dimensions of landscapes and other environmental data. *Nature.* 294:240-242.

Chu, J. 1993. *Conditional Fractal Simulation, Sequential Indicator Simulation, and Interactive Variogram Modeling.* Unpublished Ph.D. dissertation, Petroleum Engineering Department, Stanford University.

Cushman, J.H. 1986. On measurement, scale, and scaling. *Water Resour. Res.* 22:129-134.

Cushman, J.H. (ed.) 1990. *Dynamics of Fluids in Hierarchial Porous Media.* Academic Press, New York.

Cushman, J.H. 1990a. Generalized hydrodynamics of microporous media: relationships between the memory function and the scale of observation. p. 485-499. In J.H. Cushman (ed.) *Dynamics of Fluids in Hierarchical Porous Media.* Academic Press, New York.

Feder, J. 1988. *Fractals.* Plenum Press, New York.

Fogg, G.E. 1986. Groundwater flow and sand body interconnectedness in a thick multiple-aquifer system. *Water Resour. Res.* 22:679-694.

Güven, O., F.J. Molz and J.G. Melville. 1984. An analysis of dispersion in a stratified aquifer. *Water Resour. Res.* 20:1337-1354.

Güven, O., F.J. Molz, J.G. Melville, S. El Didy and G.K. Boman. 1992. Three-dimensional modeling of a two-well tracer test. *Ground Water.* 30:945-957.

Hewett, T.A. 1986. Fractal distributions of reservoir heterogeneity and their influence on fluid transport, 61st Annual Society of Petroleum Engineers Technical Conference (New Orleans). Pap. SPE 15386.

Hewett, T.A. 1992. Modelling reservoir heterogeneity with fractals. p. 455-466. In A. Soares (ed.) *Geostatistics Troia* '92, Vol. 1. Kluwer Academic Publishers, Boston.

Hewett, T.A. and Behrens, R.A. 1988. Conditional simulation of reservoir heterogeneity with fractals. 63rd Annual Society of Petroleum Engineers Technical Conference (Houston). Pap. SPE 18326.

Hurst, H.E. 1951. Long term storage capacity of reservoirs. *Transactions of the ASCE*. 116:770-808.

Hurst, H.E. 1957. A suggested statistical model for some time series that occur in nature. *Nature*. 180:494-495.

Hurst, H.E, R.P. Black and Y.M. Simaika. 1965. *Long-Term Storage: An Experimental Study*. Constable, London.

Jenkins, G.M. and D.G. Watts. 1968. *Spectral Analysis and Its Applications*. Holden-Day, Oakland, CA.

Journel, A.G. and Ch. J. Huijbregts. 1978. *Mining Geostatistics*. Academic Press, New York.

Klemes, V. 1974. The Hurst phenomenon: A puzzle? *Water Resour. Res*. 10:675-688.

Lo, A.W. 1991. Long-term memory in stock market prices. Econometrica. 59:1279-1313.

Korvin, G. 1992. *Fractal Models in the Earth Sciences*. Elsevier, New York.

Mandelbrot, B.B. and J.R. Wallis. 1968. Noah, Joseph, and operation hydrology. *Water Resour. Res*. 4:909-918.

Mandelbrot, B.B. and J.R. Wallis. 1969a. Some long run properties of geophysical records. *Water Resour. Res*. 5:321-340.

Mandelbrot, B.B. and J.R. Wallis. 1969b. Robustness of the rescaled range R/S in the measurement of noncyclic long run statistical dependence. *Water Resour. Res*. 5:967-988.

Mandelbrot, B.B. and J.W. Van Ness. 1968. Fractional Brownian motions, fractional noises, and applications. *SIAM Review*. 10:422-437.

Mandelbrot, B.B. 1983. *The Fractal Geometry of Nature*. Freeman, New York.

Moltyaner, G.L. and R.W.D. Killey. 1988a. The Twin Lake tracer tests: longitudinal dispersion. *Water Resour. Res*. 24:1613-1627.

Moltyaner, G.L. and R.W.D. Killey. 1988b. The Twin Lake tracer tests: transverse dispersion. *Water Resour. Res*. 24: 1628-1637.

Moltyaner, G.L. 1990. Field studies of dispersion: radioactive tracer studies at 20, 40 and 260 meters. p. 7-36. In J.H. Cushman (ed.) *Dynamics of Fluids in Hierarchical Porous Media*. Academic Press, New York.

Molz, F.J. and G.K. Boman. 1993. A stochastic interpolation scheme in subsurface hydrology. *Water Resour. Res*. 29:3769-3774.

Molz, F.J., O. Güven, J.G. Melville and C. Cardone. 1990. Hydraulic conductivity measurement at different scales and contaminant transport modeling. p. 37-59. In J.H. Cushman (ed.) *Dynamics of Fluids in Hierarchical Porous Media*. Academic Press, New York.

Molz, F.J., O. Güven and J.G. Melville, 1983. An examination of scale-dependent dispersion coefficients. *Ground Water*. 21:715-725.

Molz, F.J., O. Güven, J.G. Melville, J.S. Nohrstedt and J.K. Overholtzer. 1988. Forced gradient tracer tests and inferred hydraulic conductivity distributions at the Mobile site. *Ground Water*. 26:570-579.

Molz, F.J., R.H. Morin, A.E. Hess, J.G. Melville and O. Güven. 1989. The impeller meter for measuring aquifer permeability variations: evaluation and comparison with other tests. *Water Resour. Res.* 25:1677-1683.

Osiensky, J.L., G.V. Winter and R.E. Williams. 1984. Monitoring and mathematical modeling of contaminated ground-water plumes in fluvial environments. *Ground Water*. 22:298-306.

Pentland, A.P. 1984. Fractal-based description of natural scenes. *IEEE Transactions in Pattern Anal. Machine Intelligence*. 6:661-674.

Pickens, J.F. and G.E. Grisak. 1981. Scale-dependent dispersion in a stratified granular aquifer. *Water Resour. Res.* 17:1191-1211.

Plumb, O.A. and S. Whitaker. 1990. Diffusion, adsorption, and dispersion in heterogeneous porous media: the method of large-scale averaging. p. 149-176. In J.H. Cushman (ed.) *Dynamics of Fluids in Hierarchical Porous Media*. Academic Press, New York.

Press, W.H. 1978. Flicker noises in astronomy and elsewhere. *Comments Astrophys*. 7:103-119.

Rodriguez-Iturbe, I., J.M. Mejia and D.R. Dawdy. 1972. Streamflow simulation 1: a new look at Markovian models, fractional Gaussian noise, and crossing theory. *Water Resour. Res.* 8:921-930.

Saupe, D. 1988. Algorithms for random fractals. p. 71-113. In H. Peitgen and D. Saupe (eds.) *The Science of Fractal Images*. Springer-Verlag, New York.

Turcotte, D.L. 1986. A fractal approach to the relationship between ore grade and tonnage. *Economic Geology*. 81:1528-1532.

Tyler, S.W. and S.W. Wheatcraft. 1990. Fractal processes in soil water retention. *Water Resour. Res.* 26:1047-1054.

Vanmarcke, E. 1983. Random Fields. MIT Press, Cambridge.

Voss, R. 1985a. Random fractals: characterization and measurement. p. 1-11. In R. Pynn and A. Skjeltorp (eds.) *Scaling Phenomena in Disordered Systems*. Plenum Press, New York.

Voss, R. 1985b. Random fractal forgeries. p. 805-835. In R.A. Earnshaw (ed.) *Fundamental Algorithms for Computer Graphics*. NATO ASI Ser. Vol. 17. Springer Verlag, Berlin-Heidelberg.

Voss, R.F. 1988. Fractals in nature: from characterization to simulation. p. 21-69. In H. Peitgen and D. Saupe (eds.) *The Science of Fractal Images*. Springer-Verlag, New York.

Wheatcraft, S.W. and S.W. Taylor. 1988. An explanation for scale-dependent dispersivity in heterogeneous aquifers using concepts of fractal geometry. *Water Resour. Res.* 24:566-578.

Index